P9-BTO-880

Donated by:

Perwez Kalim PhD (KU)
Emeritus Professor of Mechanical Engineering
Wilkes University, Wilkes Barre, PA 18766
December 2023

Introduction to the Design of Fixed-Wing Micro Air Vehicles

Including Three Case Studies

Introduction to the Design of Fixed-Wing Micro Air Vehicles

Including Three Case Studies

Thomas J. Mueller
University of Notre Dame
Notre Dame, Indiana

James C. Kellogg
U.S. Naval Research Laboratory
Washington, D.C.

Peter G. Ifju
University of Florida
Gainesville, Florida

Sergey V. Shkarayev
University of Arizona
Tucson, Arizona

EDUCATION SERIES
Joseph A. Schetz
Series Editor-in-Chief
Virginia Polytechnic Institute and State University
Blacksburg, Virginia

Published by
American Institute of Aeronautics and Astronautics, Inc.
1801 Alexander Bell Drive, Reston, VA 20191-4344

American Institute of Aeronautics and Astronautics, Inc., Reston, Virginia

1 2 3 4 5

Library of Congress Cataloging-in-Publication Data

Introduction to the design of fixed-wing micro air vehicles: Including Three Case Studies / Thomas J. Mueller . . . [et al.].
p. cm. — (AIAA education series)
ISBN-13: 978-1-56347-849-9 (hardcover : alk paper)
ISBN-10: 1-56347-849-8
Includes bibliographical references and index.
1. Micro air vehicles. 2. Reconnaissance aircraft. I. Mueller, T. J.

UG1242.R4.136 2006
623.74′69—dc22

2006050410

Copyright © 2007 by the American Institute of Aeronautics and Astronautics, Inc. All rights reserved. Printed in the United States of America. No part of this publication may be reproduced, distributed, or transmitted, in any form or by any means, or stored in a database or retrieval system, without the prior written permission of the publisher.

Data and information appearing in this book are for informational purposes only. AIAA is not responsible for any injury or damage resulting from use or reliance, nor does AIAA warrant that use or reliance will be free from privately owned rights.

AIA A Education Series

Editor-in-Chief
Joseph A. Schetz
Virginia Polytechnic Institute and State University

Editorial Board

Takahira Aoki
University of Tokyo

Edward W. Ashford

Karen D. Baker
The Brabe Corporation

Robert H. Bishop
University of Texas at Austin

Claudio Bruno
University of Rome

Aaron R. Byerley
U.S. Air Force Academy

Richard Colgren
University of Kansas

Kajal K. Gupta
NASA Dryden Flight Research Center

Rikard B. Heslehurst
Australian Defence Force Academy

David K. Holger
Iowa State University

Rakesh K. Kapania
Virginia Polytechnic Institute and State University

Brian Landrum
University of Alabama, Huntsville

Tim C. Lieuwen
Georgia Institute of Technology

Michael Mohaghegh
The Boeing Company

Conrad F. Newberry
Naval Postgraduate School

Mark A. Price
Queen's University Belfast

James M. Rankin
Ohio University

David K. Schmidt
University of Colorado, Colorado Springs

David M. Van Wie
Johns Hopkins University

Foreword

We are very happy to present *Introduction to the Design of Fixed-Wing Micro Air Vehicles: Including Three Case Studies* by Thomas Mueller, James Kellogg, Peter Ifju, and Sergey Shkarayev. We are confident that this comprehensive and in-depth treatment of this important and rapidly envolving topic will be very well-received by the technical community. The book has seven chapters and about 300 pages.

This groups of authors is extremely well qualified to write this book because of their broad and deep expertise in the area. Their collective command of the material is excellent, and they are able to organize and present it in a very clear manner.

The AIAA Education Series aims to cover a very broad range of topics in the general aerospace field, including basic theory, application, and design. Information about the complete list of titles can be found on the last pages of this volume. The philosophy of the series is to develop textbooks that can be used in a university setting, instructional materials for continuing education and professional development courses, and also books that can serve as the basis for independent study. Suggestions for new topics or authors are always welcome.

Joseph A. Schetz
Editor-in-Chief
AIAA Education Book Series

Table of Contents

Preface

A serious effort to design aircraft that are as small as possible for special, limited-duration missions began in 1996. Vehicles of this type might carry visual, acoustic, chemical, or biological sensors. These aircraft, called micro air vehicles (MAVs), are of interest because electronic detection and surveillance sensor equipment can now be miniaturized so that the entire payload weighs about 18 g or less. Although the long-term goal of this project is to develop aircraft systems that weigh less than 30 g, have about an 8-cm wing span, and can fly for 20 to 30 min at between 30 and 65 km per hour, the original goal was to develop aircraft with a 15-cm wing span that weigh less than 90 g. Fixed-wing MAVs were flying within two years of the original proposal using equipment available from the model airplane community. One of the areas of concern was the aerodynamic efficiency of various fixed-wing concepts because these vehicles are very small and must fly at very low speeds. The corresponding chord Reynolds-number range for a 15-cm vehicle is from 5×10^4 to about 1.5×10^5. Very little information on the performance of various low-aspect-ratio wing planforms existed for this flight regime until MAVs became of interest. The availability of model airplane equipment at relatively low cost accelerated the proliferation of fixed-wing designs, especially by university students. This resulted in a wide variety of configurations and an interest in testing the performance of these vehicles in annual competitions.

This book presents a brief history of unmanned air vehicles that led to micro air vehicles and the student competitions that began in 1997. Elements of aerodynamics for low-aspect-ratio wings, propulsion, and the basic concepts for fixed-wing MAV design and a method for autopilot integration into micro air vehicles are included. Three different wing configurations (i.e., rigid cambered, flexible, and adaptive wings) are presented in the case studies. The purpose of this book is to help the reader design, build, and fly micro air vehicles and in this way add to the development of ever-smaller airplanes.

Although MAVs can be built using inexpensive model airplane components, they should not be flown by inexperienced pilots without the proper outdoor area, flight training, and insurance. They are very difficult to fly and very difficult for bystanders to see. In the United States, the Academy of Model Aeronautics in Muncie, Indiana, can help locate the proper flying field, flight instructors, and insurance.

T. J. Mueller, J. C. Kellogg, P. G. Ifju, and S. Shkarayev
May 2006

1
Overview of Micro-Air-Vehicle Development

Thomas J. Mueller*
University of Notre Dame, Notre Dame, Indiana

1.1 Introduction

1.1.1 Motivation

The RAND Corporation conducted a study for the Defense Advanced Research Projects Agency (DARPA) in December 1992 that considered a wide variety of microdevices for defense applications. This study projected that it would be possible to have flying vehicles with a 1-cm (about 0.4-in.) span and less than 1-g (0.035-oz) payload in 10 years. In 1993 the RAND Corporation performed a feasibility study on very small controlled or autonomous vehicles [1]. A more detailed study followed and was performed at the Lincoln Laboratory in 1995 [2]. This led to a DARPA workshop in the fall of 1995 [2]. Developing 15.24-cm (6-in.) flying vehicles was proposed in the fall of 1995 by R. J. Foch of the U.S. Naval Research Laboratory (NRL) and M. S. Francis (DARPA) (Foch, R. J., "Conversation with T. J. Mueller," U.S. Naval Research Lab., Washington, DC, 11 May 2004). The technological feasibility for these vehicles was a result of advances in several microtechnologies. These technologies included micromechanical systems and microelectronic components. It was envisioned that these very small airplanes were to be carried and operated by one person to perform special limited-duration missions. Vehicles of this type might carry visual, acoustic, chemical, or biological sensors. They were called micro air vehicles (i.e., MAVs) and became of interest because electronic detection and surveillance sensor equipment were miniaturized so that the entire payload weighed 18 g (0.63 oz) or less. The original goal was to develop an airplane with a 15.24-cm (6-in.) maximum dimension that weighed less than 90 g (about 3 oz) ("conversation with T. J. Mueller"). A November 1995 drawing by R. J. Foch of the NRL baseline design micro air vehicle is shown in Fig. 1.1. The Historical Perspective section will present a detailed description of how R. J. Foch set the original goals for the MAV.

The primary missions of interest for fixed-wing MAVs included surveillance, detection, communications, and the placement of unattended sensors. Surveillance missions include video (day and night), infrared images of battlefields (referred to as the "over the hill" problem) and urban areas (referred to as "around the corner"). These real-time images can give the number and location of opposing forces. This

Copyright © 2006 by Thomas J. Mueller. Published by the American Institute of Aeronautics and Astronautics, Inc., with permission.

*Roth-Gibson Professor Emeritus. Fellow AIAA.

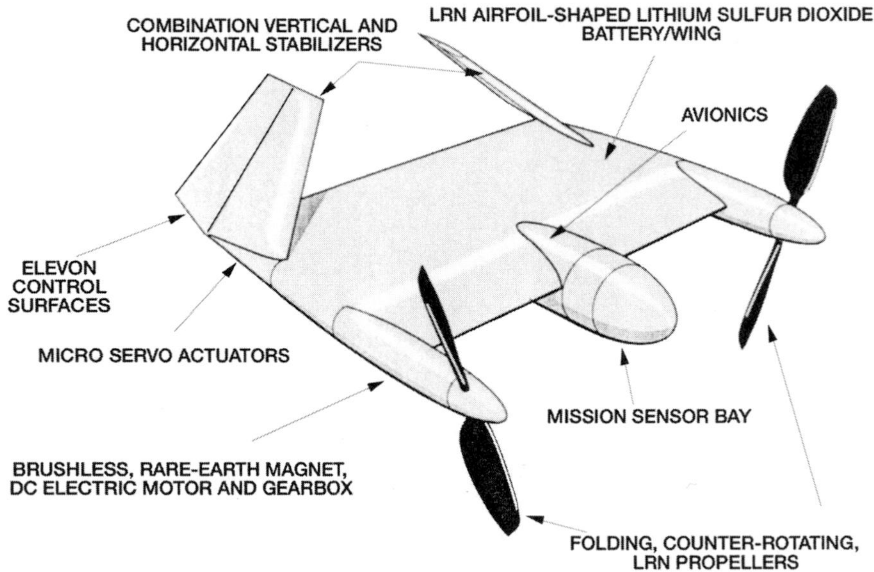

Fig. 1.1 Naval Research Laboratory baseline design micro air vehicle by R. J. Foch (reprinted with permission of U.S. Naval Research Laboratory, Washington, DC).

type of information can also be useful in hostage rescue and counterdrug operations. Because of the availability of very small sensors, detection missions include the sensing of biological agents, chemical compounds, and nuclear materials (i.e., radioactivity). MAVs can also be used to improve communication in urban or other environments where full-time line-of-sight operations are important. Another possible mission is to place acoustic sensors on the outside of a building during a hostage rescue or counterdrug operation [3] and [4].

The requirements for fixed-wing MAVs include a wide range of possible operational environments such as urban, jungle, desert, maritime, mountain, and arctic. Furthermore, MAVs must perform their missions in all weather conditions (i.e., precipitation, wind shear, and gusts). A collision-avoidance system is required because these vehicles fly at relatively low altitudes (i.e., less than 100 m) where buildings, trees, hills, or similar obstructions might be present.

Technical solutions were needed for aerodynamics and control, propulsion and power, navigation, and communications. One area of serious concern was the aerodynamic efficiency of various fixed-wing concepts because these very small vehicles fly at very low speeds, that is, between 24 and 64 km/h (between about 15 and 40 mph). The corresponding chord Reynolds numbers range from about 5×10^4 to approximately 2×10^5. The nondimensional chord Reynolds number is defined as the cruise velocity times the mean wing chord divided by the kinematic viscosity of air. This places MAVs in a regime totally alien to conventional

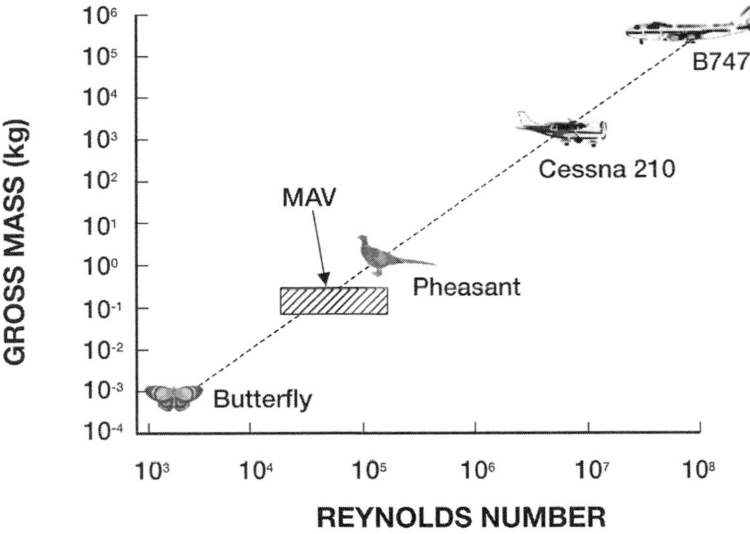

Fig. 1.2 Gross mass vs Reynolds number for micro air vehicles and other flying vehicles.

airplanes. The gross mass of micro air vehicles and other flying objects vs chord Reynolds number is shown in Fig. 1.2 [5] and the payload mass vs wing span for MAVs and small and large unmanned air vehicles (UAVs) is presented in Fig. 1.3 [2]. Until recently, very little information on the performance of various airfoil/wing shapes existed for this low-Reynolds-number flight regime. Also, aerodynamic problems related to vortex lift on wings of aspect ratio two and lower made it difficult to extend existing analytical or empirical techniques to lower Reynolds numbers. It is well known that viscous forces are more significant at low Reynolds numbers. Therefore, the increased drag reduces the lift-to-drag ratios as well as propeller efficiencies.

1.1.2 Objectives

A brief historical perspective is presented in order to give the reader an appreciation for the technical achievements necessary for the development of unmanned air vehicles. These technical developments include automatic stabilization, remote control, and autonomous navigation. Selected highlights of full-sized, small, and micro unmanned-air-vehicles' histories are included. The eventual development of small UAVs and MAVs depended on the development of small internal combustion engines and electric motors, lightweight structural materials, and lightweight electronics and actuators. These critical items were pioneered by the model airplane community and led to the technological feasibility of flying microsystems and specifically the design of fixed-wing micro air vehicles. The university student design and flight competitions that made significant contributions to the development of a variety of MAV configurations are included.

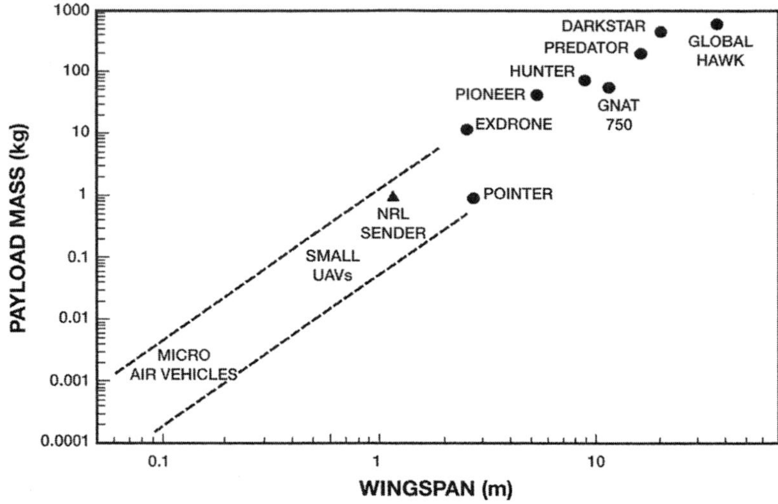

Fig. 1.3 Payload vs wing span for micro air vehicles and small and large unmanned air vehicles (Ref. 2) (reprinted with permission of MIT Lincoln Laboratory, Lexington, Massachusetts).

An extensive experimental study by the University of Notre Dame of the influence of aspect ratio, planform shape, and Reynolds number on the aerodynamic characteristics of thin flat plate and 4% circular camber wings is presented. A review of issues involved in selecting either an internal combustion engine or an electric motor for the propulsion system is examined. The use of the Notre Dame experimental database and the University of Gent numerical methods to design MAVs are briefly discussed. Methodology for system integration of autonomous MAVs that includes an autopilot, a GPS unit, and stability augmentation with infrared sensors is also described.

Although a large number of MAVs have been designed and flown around the world, only three types of vehicles will be described in detail. These three different fixed-wing concepts (rigid cambered, flexible, and adaptive) are included as case studies by recognized experts in this field from the U.S. Naval Research Laboratory, University of Florida, and the University of Arizona.

1.2 Historical Perspective

Throughout history, emperors, kings, scholars, and creative writers dreamed of flying. After the Montgolfier brothers made their first unmanned balloon flight in Paris in 1783, experimental aviation accelerated [6]. By 1799, Sir George Cayley produced the first design for the modern airplane configuration. In 1804 he flew a model glider and was able to fly his full-sized unmanned gliders in 1809 [6]. John Stringfellow, a fellow countryman of Cayley and associate of William Henson, launched a small steam-powered model of Henson's design from an overhead trolley that slid along a wire in a lace factory in 1848. The model did not fly because it lacked stability and could not sustain the momentum of its launch [6].

The following year Cayley tested a boy-carrying glider, and by 1853 he made a controlled flight of a man-carrying glider. Cayley became known as the "Father of Aeronautics" because of these experiments and corresponding publications [6]. Four years later, in 1857, Felix Du Temple began experiments in France with clockwork-powered model airplanes. Alphonse Pénaud, another Frenchman, began selling rubber-band-powered pusher-propeller ready-to-fly model airplanes in 1871 [6]. Pénaud recognized that longitudinal stability was crucial. Most of the activity in aeronautics during the 19th century took place in Europe where there were a large number of pioneers. The serious interest in Europe was demonstrated by the founding of the Société Francaise de Navigation Aérienne in France (1863) and the Aeronautical Society in Britain (the Royal Aeronautical Society) (1866) [6]. Later in the century, in 1879, Victor Tatin flew a tethered monoplane model in France using a compressed air motor. Otto Lilienthal's first successful glider was flown in 1891. He died in a glider accident in 1896 [6]. By the close of the 19th century, there was significant activity in the United States—most notably the work of Albert F. Zahm [7] and [8], Samuel P. Langley [6], Octave Chanute [6], and the Wright brothers [6] and [9].

1.2.1 Unmanned Air Vehicles

Unmanned-air-vehicle (UAV) development was inspired by the evolution of piloted airplanes. Although the history of piloted airplanes has been extremely well documented, until recently there has been little documentation of the history of (UAVs). In 2004, the book entitled *Unmanned Aviation: A Brief History of Unmanned Air Vehicles* by Laurence R. Newcome [10] carefully documented this important area of aviation.

Although this book concentrates mainly on UAV development in the United States, Newcome also covers the early contributions of UAV pioneers in England, France, Germany, Japan, and Russia. According to Newcome, for more than eight decades UAVs have been referred to as robotic airplanes, remotely piloted vehicles (RPVs), aerial torpedoes, and drones. The term RPV was replaced by UAV in the 1990s, and now the U.S. Defense Department dictionary defines a UAV as follows:

> a powered, air vehicle that does not carry a human operator, uses aerodynamic forces to provide vehicle lift, can fly autonomously or be piloted remotely, can be expendable or recoverable, and can carry a lethal or nonlethal payload. Ballistic or semi ballistic vehicles, cruise missiles, and artillery projectiles, however, are not considered unmanned vehicles [10].

Allegedly, Nikola Tesla, a Serbian immigrant and inventor, arrived in New York in 1884 with plans for a remotely controlled unmanned airplane [10]. Fourteen years later he entered a remotely controlled 4-ft-long boat in the Electrical Exposition in Madison Square Garden. Tesla maneuvered his boat to stop or go and turn left or right by blinking its lights using different radio frequencies. Before Tesla showcased his boat in Madison Square Garden, Louis Brennan, an Irish inventor, demonstrated his wire-guided torpedo in 1888. Two decades later, Rene Lorin, a French artillery officer, proposed a jet-powered flying bomb that would be controlled from a manned escort. Because of the possible military applications [10], a growing interest in remote-controlled vehicles emerged in a number of countries.

The success of UAVs required a powered airplane and three additional critical technologies: automatic stabilization, remote control, and autonomous navigation. The age of powered/controlled airplanes began with the Wright Brothers in 1903 and was further accelerated by the approach of World War I several years later. The development of cheaper and more easily produced airplane engines by G. Bradshaw stimulated interest in unmanned aviation in England. Three prototype "pilotless airplanes" that used this engine were designed and built in 1917. Only the design by the Royal Aircraft Factory at Farnborough with radio controls developed by Major A. M. Low reached the flight-test stage. All of these flight attempts ended in crashes [10] ("Letter to Ron Moulton from Peter Collins," Imperial War Museum, Cambridge, England, UK, 14 July 2005). The following year, Elmer Sperry, aided by his son Lawrence, and Peter Hewett, a contemporary inventor, under a contract from the U.S. Navy, developed and demonstrated the first airplane capable of stabilizing and navigating itself without a pilot onboard [10]. This vehicle was called the Curtiss–Sperry aerial torpedo because it used an airframe designed and built by the Curtiss Aeroplane and Motor Company. Five years later, Lawrence Sperry demonstrated a practical radio control system. Charles Kettering obtained a U.S. Army contract in January 1918 to develop an unmanned aerial vehicle that could deliver a 91-kg (200-lb) warhead 80 km (50 miles). It became known as the "Kettering bug" and was half the size of the Curtiss–Sperry aerial torpedo. It was designed to navigate to its target and then fold its wings and dive to the target. Although it was not radio controlled, the "bug" appears to be the first UAV [10]. After World War I, the Air Ministry in Britain continued its interest in unmanned aircraft leading to a target drone (the Larynx) with a radio control system in 1924 [10]. In 1933, the Royal Navy successfully demonstrated a radio-controlled target drone that led to the production of 420 Queen Bee RC target drones, which were derivatives of the DeHavilland Tiger Moth biplane trainer [10] ("Letter to Ron Moulton from Peter Collins," Imperial War Museum, Cambridge, England, UK, 14 July 2005). The conversion of manned airplanes to target drones continued intermittently through the 1920s and 1930s.

Reginald Denny moved to the U.S. from England in 1919. He came from a family of actors and became a Hollywood movie star. Denny's interest in aviation led him to learn to fly and open a hobby shop specializing in model airplanes. In the early 1930s he formed Reginald Denny Industries and expanded his line of model airplanes into radio-controlled target drones. His first target drone, the Radioplane-1 (RP-1), flew in 1935 but crashed during a demonstration for the U.S. Army. It evolved into the RP-2 in 1938 and the RP-3 in 1939. Denny, along with his financial backers Whitney Collins and Harold Powell, formed the Radioplane Company in 1939. The RP-4 successfully evolved into the RP-18 (also known as the OQ-17) and over 15,000 copies of the RP-4 through RP-18 were built and used to train Army and Navy gunners during World War II. The basic Radioplane series evolved into the first reconnaissance UAV in 1955 by adding film cameras [10].

The idea of a glide bomb, first proposed in World War I, returned in the 1930s. After seeing a demonstration of the Queen Bee by the Royal Navy, the U.S. Navy began a program in 1936 at the Naval Research Laboratory to develop a radio-controlled aircraft. This was the beginning of a project that resulted in the radar and television guided assault drone, the TDR-1, which carried a 900-kg (2000-lb) bomb. The TDR-1 was a twin-engine airplane with a cruising speed of 230 km/h

(124 kn). Two-hundred TDR-1s were ordered in March 1942 but were first used successfully in a combat mission on 27 September 1944 at Bougainville in the Pacific. The TDR-1 was controlled by a pilot in a TBM-1 (Avenger Torpedo Bomber) control plane from takeoff to the target [10] and [11]. A more detailed account of the development and eventual use of the TDR-1 can be found in Ref. 11 and by searching the web for "stagone.org."

There also was an effort by the U.S. Army in 1942 to develop glide bombs, called the AZON and RAZON, at Eglin Field, Florida [12]. The Azon was a 430-kg (1,000-lb) bomb with gyro stabilization, a pair of rudders, and a tracking flare that was controlled by the bombardier of a B-24. It was used in Europe in early 1944 with some success. It was also used in late 1944 in Burma to destroy bridges. More information on the development and use of these glide bombs can be found by searching the web for "World War II glide bombs."

Glide-bomb projects began in Germany in 1939 and became the first usable weapons from this concept [12] and [13]. The developers knew that these small unmanned airplanes had to be radio controlled and that they would need a rocket booster to survive enemy fire. The Fritz-X, also know as the FX 1400 or the SD 1400, was designed for armor-plated ships and was fired from 6100 m (20,000 ft) or higher to attain supersonic speeds. It had asymmetrical cruciform stubby wings, and the warhead was a 1400-kg (3100-lb) armor-piercing bomb. The Henschel 293 (Hs 293), the "sister" rocket to the FX 1400, was designed to be fired from 1800 to 3000 m (6,000 to 10,000 ft) for ships without armor plate. The FX 1400 was a glide bomb that was radio controlled from the bombardier of its parent airplane. The remote radio control system was replaced later with a wire control system in order to be more jam proof. The most successful mission for the FX 1400, manufactured by Rheinmetall-Borsig, was the sinking of the Italian battleship Roma with three direct hits in September 1943. The Hs 293 was 3.82 m (12.5 ft) long with a 3.10 m (10 ft) span, had stubby wings, and carried a 500-kg (1100-lb) bomb [13]. These unmanned glide bombs, made by the Henschel Company, were used in combat in 1943. They were dropped from a bomber, steered to the target by the bombardier using a joystick, and traveled up to 900 km/h (560 mph). On 26 November 1943, one Hs 293 dropped from a Heinkel bomber sank the British troop ship Rohna, which had 2193 American, British, and Australian military personnel aboard in addition to an Indian crew of 195 [12] (Moulton, R., E-mail communication with T. J. Mueller, 7 July 2005). The Hs 293 approached the port side of the Rohna approximately 20 to 30 ft above the water line and went through the port side. It finally exploded inside the ship, resulting in a hole in the starboard side. In the end, there were 1149 fatalities. More information about these German glide bombs can also be found by searching the web for "World War II glide bombs." These glide bombs were the first radio-controlled UAVs that could anachronistically be referred to as unmanned combat air vehicles (UCAVs) (Moulton, R., E-mail communication with T. J. Mueller, 7 July 2005).

1.2.2 Radio-Controlled Model Airplanes

Interest in the design and development of small unmanned air vehicles has increased dramatically in the last 25 years. Although the definition of small UAVs is arbitrary, vehicles with wing spans less than 6 m (20 ft) and masses less than

25 kg (55 lb) are usually considered to be in this category [14]. The technology and experience provided by the model airplane community provided the starting point for the design of these vehicles. Model aviation both preceded and evolved alongside of powered flight. Rubber-band-powered and glider models were used by the earliest aviation pioneers to help them understand the aerodynamic forces and ways to control them. Three important technologies were needed to start the development of powered radio control models: 1) small internal combustion engines, 2) appropriate sized radio receivers and transmitters, and 3) actuators to move the airplane control surfaces.

It appears that the first small gas engine was built in 1901 by Charles M. Manly for Langley's powered model airplane [6]. The demand for model and miniature flying toys grew significantly after the Wright brothers' historic first flight, and consequently model enthusiasts began to build their own small gas engines a few years later. Some of these include Ray Arden's four-cycle engine (1907) in the U.S. [15], David Stanger's two-cylinder V-type engine (1908) in England [15], and the Eckert brothers "Baby" engine (1911) in the U.S. [15–17], to name just a few. The Baby had a half-horsepower, a displacement of 43.42 cm^3 (2.65 $in.^3$), and weighed 1.70 kg (3.75 lb). This engine was used in a model airplane competition in 1913. That same year, Max Braune of Leipzig and Josef Zenker of Berlin, Germany, founded a company to manufacture one- and two-cylinder engines for gliders [18]. Also in 1913, David Stanger developed a 60.63 cm^3 (3.7 $in.^3$) Vee-Twin engine, which was used the following year in a model airplane that set an endurance record that was not surpassed for 18 years [19]. By the early 1930s, many model enthusiasts had built their own internal combustion engines. Bill Brown, Bill Atwood, Dan Calkin, and Walter Kratzsch appear to be especially noteworthy [18–23]. Bill Brown made the first successful, limited production, model airplane engine [24] and [25] in 1932, and his first production engine became available in 1934. Thousands of these engines, named the Brown Jr. 60 model B, were made during the 1930s [24]. In this same period, William E. Atwood made a flight of over 20 min in 1932 with a prototype of his "Baby Cyclone" engine. This engine was almost half the size of Brown's engine and was put into production in 1935. The same year the Baby Cyclone engine went into production, Walter Kratzsch founded a company in Goerliz, Germany, and built a series of one- and two-cylinder internal combustion engines for model airplanes. His smallest engine, the F10B, had a displacement of 10 cm^3 (0.61 $in.^3$) (Ref. 21). Yet another contemporary of Bill Brown was Dan Calkin, who was interested in making small engines (Ref. 23). His early ELF engine of 1935 had a 2.26-cm^3 (0.138-$in.^3$) displacement compared to the 9.83-cm^3 (0.60-$in.^3$) displacement of the Brown Jr. and the F10B engines. A more complete history of the development of model airplane engines can be found in Refs. [15–24].

Maxwell Bassett, using a Brown Jr. engine, was the only one to enter a gas-powered free-flight model in the International Wakefield event held in Atlantic City, New Jersey, in 1932 [25]. This event was intended for rubber-band-powered models, but Bassett, nevertheless, managed to take fourth place. The following year, Bassett captured all of the trophies from the rubber-band-powered models in the U.S. Nationals on Long Island, New York. As a result, internal combustion engine models were separately classified as "gas" powered and the first gas-powered free-flight event was held at the Nationals in Akron, Ohio, in 1934 [25]. The stage was set for the invention of radio-controlled (RC) systems for gas-powered models.

Amateur radio hobbyists made important contributions to the development of RC vehicles [25–27]. In the 1930s the only radio frequencies available for RC models were those designated for amateur radio use. To obtain an amateur radio license, the operator had to pass a test in Morse code and radio theory. Thus, the amateur radio (i.e., ham radio) operators were able to apply their knowledge and interest directly to the development of RC models. Enthusiasts were ultimately supplied with the electrical components by the expanding commercial radio industry.

It is difficult to know with absolute certainty when and where in the world the first RC model airplane was successfully flown. It is known, however, that somewhat similar efforts were taking place in both Germany and the United States during the 1930s. Chester Lanzo of Cleveland, Ohio, built a cabin-type gas model in 1934 and experimented with RC using a spark-gap transmitter and a coherer receiver (i.e., a radio-wave detector) [27]. These experiments were unsuccessful. A note in the 10 June 1936 German magazine *FLUGSPORT* [28] describes prizes awarded to Alfred Lippitsch (not to be confused with the airplane designer Alexander Lippisch) and a student named Egon Sykora, both from Dresden, for the successful flight of a RC glider model with rudder control during a competition at the Wasserkuppe, Rhoen, Germany, on 31 May 1936. This appears to be the first reference to a successful RC flight [28] and [29]. A subsequent review article in the magazine *FLUG* und *MODELLTECHNIK* [30] in 1957 and by Matthaus Weidner in his book (1987) confirmed this success [18]. The model, built by Erich Klose and Alfons Menze, shown in Fig. 1.4, had a wing span of 2.5 m (8.2 ft) and weighed 3.5 kg (7.71 lb). Figure 1.5 shows the start of the first flight, Fig. 1.6 shows after it was launched, and Fig. 1.7 was taken after the successful flight. The receiver, batteries, and on/off switch are shown in Fig. 1.8, and Egon Sykora with the ground transmitter is shown in Fig. 1.9. It appears that this successful flight in Germany was unknown in the United States.

C.W. Thompson, Jr., and H. M. Plummer are credited with the first public RC glider demonstration in the United States at the Soaring Society of America's national meet in Elmira, New York, in July 1937 [27]. Walter and William Good of Kalamazoo, Michigan started making rubber-band-powered models when they were 11 years old [31]. Walter's interest in building airplane models continued to increase while William's interest shifted to building radio equipment. Together these twin brothers had the necessary skills and experience to develop a RC model. Their motivation for developing a RC plane came from an article they read about the unmanned target drone airplanes (the aforementioned Queen Bees) built by the DeHavilland Company for the British Navy [31]. Walter and William Good developed a RC rudder control system in the fall of 1936 and mounted it in their free-flight gas model, which they had built in 1935. Their invention was exhibited at the Kalamazoo, Michigan College Science Fair in January 1937 [32]. This is the first reference to a successful gas-powered RC model built by American hobbyists. Figure 1.10 shows Walt and Bill Good with their first RC radio receiver in 1937. Their first radio transmitter used in both the 1937 and 1938 Nationals is shown in Fig. 1.11. In May of 1937 [33], they made numerous flights at the Kalamazoo Airport and garnered national attention. The model was named the "Guff" and in

Fig. 1.4 Radio-controlled glider model built by Erich Klose and Alfons Menze (reprinted with permission of Deutsches Museum, Munich, Germany).

1938 renamed Big Guff when the fuselage was made larger to accommodate a larger radio compartment, shown in Fig. 1.12. It had a 2.44-m (8-ft) wing span, weighed 3.85 kg (8.5 lb), and was powered by a Brown Jr. Gas engine. In July of 1937, Chester Lanzo, with the same plane he had built in 1934, used a new radio receiver with vacuum tubes in a super-regenerative detector circuit, which operated a relay to move the rudder. A total of six RC models by C. Lanzo, P. Sweeney, E. Wasman, W. Good, L. Weiss, and B. Schiffman appeared in the 1937 Academy of Model Aeronautics (AMA) National competition [27]. Chester Lanzo had the lightest model at just less than 2.72 kg (6 lb), including 0.91 kg (2 lb) of radio equipment. His model had a 1.52-m (5-ft) wing span, 35.56-cm (14-in.) wing chord, and a 1.52-m (5-ft) body with only rudder control. It was named the RC-1 and took first place in the first RC National Contest in the United States. Lanzo's model was the only one of the six to fly [27]. Walter and William Good, shown in Fig. 1.13, won the U.S. Nationals in 1938, 1939, 1940, and 1947 with the Big Guff [24]. Interest in RC models was also present in England as indicated in the book by Peter Hunt [34], which was first published in 1942. A photograph of Hunt's RC model, called the Hertzian 11 airborne, is included in Ref. 34, but there is no date given for his first flight or the first flight in England. These events led to the

Fig. 1.5 Start of the first flight of a radio-controlled glider model, 31 May 1936 (reprinted with permission of Deutsches Museum, Munich, Germany).

rapid increase in model development during the decade following World War II [35] and [36]. The first British radio-controlled model airplane event was held at the British Nationals in 1949 [37].

A group of British scientists at the Royal Aircraft Establishment (RAE) and the National Physical Laboratory (NPL) formed the Low Speed Acrodynamics Research Association at Farnborough in 1947 [35]. The first president of this association was Sir Harold Roxbee-Cox, who suggested that they should be concerned with radio controlled dynamically similar models of full-sized airplanes for aerodynamic research. To do this, they needed multichannel, proportional radio control. The system they adopted was that used by John Gardner for his torpedo in 1906 and by Professor. Wagner for several German missiles during WWII. Experiments at RAE in 1948, on the 465-MHz UHF band using super-regenerative receivers in boats, were successful, whereas the same system in a low-wing petrol engine airplane in 1951 was not successful. A 3.1-m (10-ft) wing-span glider known as the TV-8 (i.e., Test Vehicle No. 8) was designed and built in 1952, and the first successful flight took place in the spring of the same year [35]. The TV-8 made approximately 50 flights in 1952 and 1953. After it had been fitted with reliable transistorized servos, it successfully demonstrated the advantages of multichannel

Fig. 1.6 After the launch—it flies (reprinted with permission of Deutsches Museum, Munich, Germany).

proportional control in the United Kingdom. Walter Good produced one of the first successful attempts at proportional control for models using miniaturized components in 1955 [35]. The operating principle of this control system was identical to that used by Professor Kramer for the German wire guided X-4 missile during WWII. Furthermore, in 1950 the first RC model gliders were developed and successfully flown in Russia [38]. Their RC systems had the same major components used in modern models.

Fig. 1.7 After the successful flight (reprinted with permission of Deutsches Museum, Munich, Germany).

The first officially recorded flight of an electric-powered RC model was in June 1957 by H. J. Taplin of Great Britain [39]. Fred Militky of Germany is credited with starting the electric-powered flight movement with his ElectroFlug free-flight design that flew in an eight nation international contest in September 1959 [39] and [40]. The ElectroFlug weighed 127.5 g (4.5 oz), had a wing area of 890 cm^2 (138 in.^{2}), a length of 60.96 cm (24 in.), a Micromax electric motor, and was observed to fly for 22 min. The Graupner Company of Germany produced a larger version of Militky's design, the Silentius, in kit form, in 1960. It was the first commercial electric-powered kit model in the world [40]. The advent of nickel-cadmium (Ni-Cd) rechargeable batteries in the early 1960s further accelerated the interest in electric-powered models. Robert and Roland Boucher, of California, founded AstroFlight, Inc., and combined Ni-Cd batteries and cobalt-type electric motors for model airplanes [39] and [41]. Roland Boucher visited Great Britain in 1973 and joined Peter Russell to make the first demonstrations with powerful electric self-launching models. During the 1970s, 1980s, and 1990s, improvements and innovations were made in model engines and electric motors, propellers, control systems, fabrication methods, and materials along with the

Fig. 1.8 Radio receiver, batteries, and on/off switch (reprinted with permission of Deutsches Museum, Munich, Germany).

introduction of rechargeable NiMH and lithium polymer batteries, digital servos, and piezogyros.

1.2.3 Small and Micro Unmanned Air Vehicles

Although the model airplane community continued the improvement of radio-control equipment, the use of these small airplanes for reconnaissance and surveillance, as well as other useful missions, had to wait for the evolution of miniature video cameras and transmitters and other lightweight electronic sensors. Serious interest in the design of small UAVs in the United States began in the 1970s. In the 1974–1975 time frame, the U.S. Army, Navy, and Air Force began the development of small UAVs. One of the most active groups in the design of small UAVs was the NRL. NRL's Vehicle Research Section was formed in 1975 to conduct research focused on developing technologies and prototype systems that demonstrate the feasibility of small UAVs for Navy electronic warfare missions. Ideally, these small UAVs will operate autonomously, be extremely robust and reliable, and sufficiently inexpensive to be expendable rather than requiring recovery following their missions. In 1975 they designed, built, and flight tested a ship-tethered autogyro platform. By 1978, NRL began the development of their first small UAV for electronic warfare, the long-duration expendable decoy (LODED). As a mission demonstrator the LODED was unsuccessful, but it proved invaluable as a research tool for identifying the limitations in the available technologies for UAVs. Wind-tunnel and flight testing showed that in order to reach the desired levels of performance, reliability, and efficiency, major research investments were still required in the areas of low-Reynolds-number aerodynamics, composite airframe structures,

Fig. 1.9 Egon Sykora with the ground transmitter (reprinted with permission of Deutsches Museum, Munich, Germany).

flight data sensors, mission payload sensors, control and data links, propulsion systems, and onboard electrical power. Since the mid-1970s, NRL has designed and tested over 50 small UAVs and MAVs including the LAURA [42], SENDER [43], the MITE series [44] and [45], and DRAGON EYE [46] vehicles. Figures 1.14 and 1.15 compare the small UAVs, LAURA, SENDER, and DRAGON EYE with five MAVs on the basis of mass vs wing span and chord Reynolds number. Later chapters will address the MITE 2 rigid cambered wing, the UF-flexible wing, and the UA-adaptive wing.

The following description of how the original goals were determined for this new class of vehicles, the MAVs, was written by Richard J. Foch, senior scientist for Expendable Vehicles, U.S. Naval Research Laboratory, Washington, D.C. (Foch, R. J., E-Mail to T. J., Mueller, 20 June 2005). This description is written in the first person to better describe the thought process of Foch and Francis in 1995:

> During 1995, MIT Lincoln Labs developed and demonstrated a moderately high-resolution video imager weighing about one gram. Since payload function plus size, weight and power consumption are the primary design drivers for unmanned air vehicles (UAVs), the Lincoln Labs researchers realized that this imager could enable the development of really tiny, perhaps insect-size,

Fig. 1.10 Walter and William Good with their first radio-control receiver, 1937 (reprinted with permission of Kalamazoo Aviation History Museum).

UAVs. In addition to the one-gram sensor technology being demonstrated, additional technology development efforts would be required to solve other fundamental issues such as: low Reynolds number aerodynamic effects upon performance, stability and control; propulsion and propulsion/airframe integration; energy storage and management; navigation; miniature flight control and navigation sensors; RF data links for guidance, control, and payload data; and any other technologies unique to micro-sized UAV requirements. To illustrate the overall concept of a Micro UAV and provide potential sponsors and users a visualization tool, Lincoln Labs produced a palm-sized display model. This model had a delta winged canard layout with a nose mounted imager and aft-mounted propeller on its cylindrical fuselage. Except for the one gram imager, all other technical details such as type of propulsion system, sensors, antennas, etc., were yet to-be-determined.

In late September 1995, the CBS "60 Minutes" TV show featured Admiral Owens, who was then a member of the Joint Chiefs of Staff, discussing future

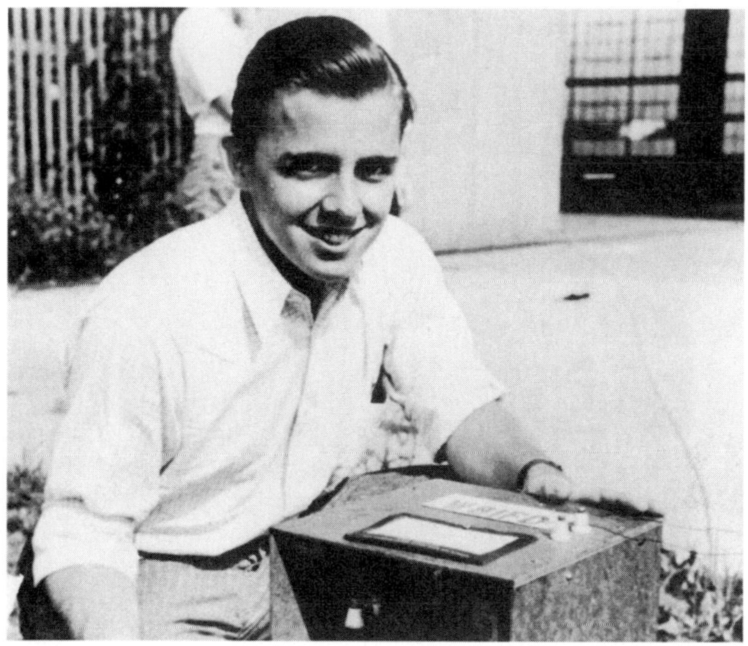

Fig. 1.11 William Good with the 100-W transmitter used in both the 1937 and 1938 Nationals (reprinted with permission of Kalamazoo Aviation History Museum).

technologies that will significantly impact how the US wages war. At one point during the interview, the Admiral picked up the MIT Lincoln Labs Micro UAV display model and said that we (the military) plan to develop "these" so that every soldier could carry one in his backpack and be able to independently perform local reconnaissance missions. I commented to my wife that I'd sure hate to have to make that airplane actually fly, since so many new cutting edge technologies would have to be developed and integrated, just to make it work.

The next day at the Naval Research Laboratory (NRL), I received a telephone call from our Chief Scientist who had just been directed by the Chief of Naval Research—the Admiral in charge of the Office of Naval Research (ONR)—to have NRL develop a plan for a technology-development program for Navy Micro UAVs . Adm. Owens had initiated the development of military Micro UAVs by contacting the Director of the Defense Advanced Research Projects Agency (DARPA) and the Chief of Naval Research, who were aware that my group at NRL were the Navy's most experienced researchers on low Reynolds number aerodynamics and small vehicle systems. We were definitely qualified for this task. I was instructed to develop a plan for a research program to be funded by ONR, to develop the technologies for Micro-UAVs and design, build and fly demonstrations of Navy missions. In addition, DARPA would also be tasked to develop Micro-UAV technologies and flyable systems. I was assigned as the Program Manager for the Navy Micro-UAV program and USAF Col. Mike Francis was detailed to DARPA to be their Micro-UAV Program Manager. A meeting was scheduled for Mike and me, later that week. Prior to this, I had been the chief engineer or program manager for over two dozen

Fig. 1.12 Walter Good with the "Big Guff," the first gas-powered radio-controlled model built by hobbyists, at the Kalamazoo Airport in 1938 (reprinted with permission of Kalamazoo Aviation History Museum).

small UAVs developed by NRL for applied research and Navy electronic warfare missions. Col. Francis had also managed several military programs and had been a key member of the Defense Airborne Reconnaissance Office (DARO) involved in the development of advanced technology UAVs. In addition, we both had done graduate work involving low Reynolds number aerodynamics, and our careers had briefly crossed at the Air Force Institute of Technology at Wright Patterson AFB where Mike was studying the complex aerodynamics of the dragonfly and I was a student summer hire.

Before meeting with Mike, I decided to look into a preliminary concept design for a Micro-UAV to get a feel for such a small system. I wondered what would be the smallest size that a micro air vehicle could be built that was technically feasible, based on a 5 year Government program to "push" technology specifically for micro air vehicles? This would establish the research goals which drive the developmental challenges.

Because of scaling laws, achieving the power required to fly becomes an ever more critical factor when size is reduced. In order to minimize power, an airplane designer desires high aerodynamic efficiency, low wing loading, light weight, low altitude operation and high propeller efficiency. The design challenge is the tradeoffs to equitably balance these parameters and create

Fig. 1.13 Twin brothers William and Walter Good with Big Guff, which won the U.S. Nationals in 1938, 1939, 1940, and 1947 (reprinted with permission of Kalamazoo Aviation History Museum).

an airplane which is the best overall system. These parameters counter each other. For example, a light wing loading requires maximizing the wing area. With a limited wingspan, this means a large chord, low aspect ratio wing. Low aspect ratio wings have poor aero efficiency due to having high induced drag at large lift coefficients. However, for a micro air vehicle low Reynolds number effects significantly limit the aerodynamic and propeller efficiencies. Thus a premium must be placed on low wing loading and the higher induced drag of a low aspect ratio wing can be traded for lower airfoil profile drag from the larger wing chord giving a higher Reynolds number.

I assumed the airframe should be made in one piece to keep the MAV simple. For compactness, thc wingspan and length should be equally as small as possible. The simplest plan form that emerges with the maximum wing area is a square. Since airplanes must balance near the wing quarter chord, a fuselage would be needed that projects ahead of the wing. So, the wing chord must be shorter than the maximum length and the wing planform becomes a rectangle. A practical minimum wing aspect ratio, maximizing wing area but allowing for enough fuselage volume ahead of the wing to balance the airplane is 1.5. So the wing chord could be 2/3 of the wing span.

Based on my model airplane experience, I knew that models with wingspans less than 12 inches (-0.3 m) were approximately an order of magnitude more difficult to adjust and trim to fly than models having wingspans of 24 inches

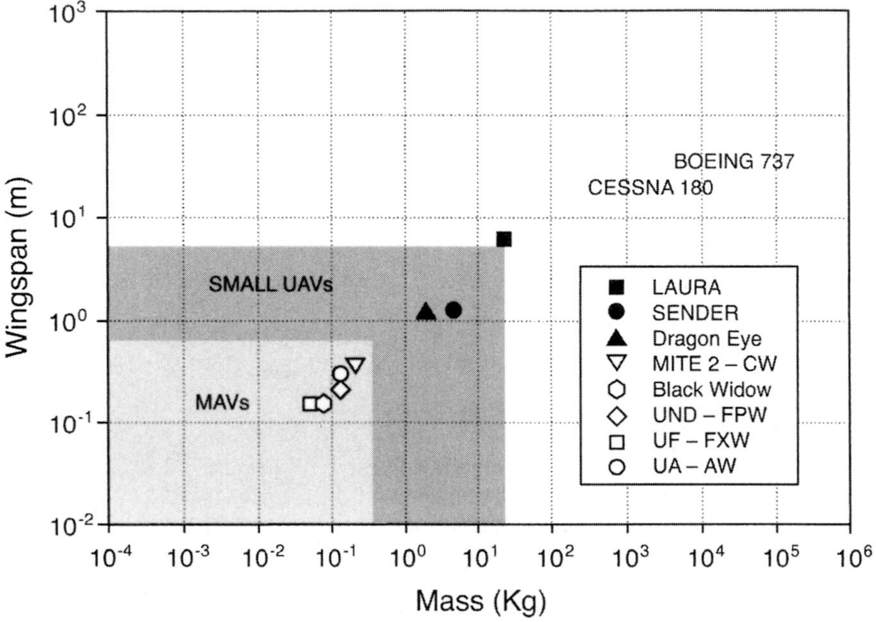

Fig. 1.14 Wing span vs mass for selected small UAVs and MAVs.

(−0.6 m) or greater. There have been many free flight, radio control and control line models built having wing spans near 12 inches. However, by 1995 there had been very few successful powered model airplanes with a 6 inch (−0.15 m) wingspan: presumably they should be at least an order of magnitude harder to fly than twelve inches. I made an additional assumption that, for small model airplanes with a single propeller, it would be difficult to counter torque and slipstream effects for propellers larger than one-half the wing span. Since 3 inch diameter propellers were the smallest that I had seen with reasonably high efficiency, it seemed that a 6 inch wing span, propeller-driven Micro UAV was possible.

Anything much smaller and propellers give way to flapping wings for best propulsive efficiency. Technology was far from ready to tackle solving the problems of tiny wing flappers, so it seemed to me that a practical 6 inch wingspan propeller driven airplane was near the lower size limit. Assuming a 6 inch wingspan, propeller driven airplane is possible, could it carry a useful mission payload and enough stored energy (battery or fuel) for a practical mission? With an aspect ratio of 1.5 the resulting wing chord is 4 inches, so the wing area is 24 sq. in. or 0.167 sq. ft.

In order to determine the airspeed and power requirements, I needed to estimate the airplane's gross weight. Key subsystems were: airframe, mission payload, power plant, fuel, and avionics. Past experience for small electric powered UAVs have established strong relationships between these subsystems that were strongly dependent on available technologies, but relatively independent of size! For example, the gross weight of a liquid fueled, small UAV is about 4 times the mission payload weight. The gross weight of a battery-powered, small UAV is about 5 times the mission payload weight.

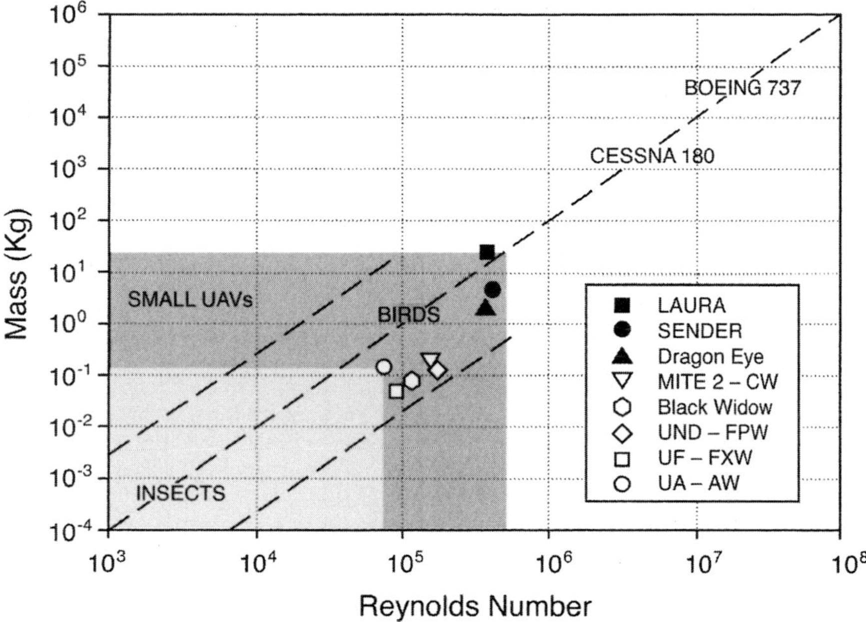

Fig. 1.15 Reynolds number vs mass for selected small UAVs and MAVs.

Since battery powered electric propulsion seemed more sensible to me for a micro airplane than one powered with an engine, I assumed that a practical Micro UAV could be built with a gross weight of 5 times that of the mission payload.

The Lincoln Labs video imager that started the idea of a Micro UAV weighed one gram. Unfortunately, forming a mission payload requires integrating the sensor with a mounting structure, adding a video-bandwidth transmitter and antenna, providing electrical power to the imager and transmitter and all associated wiring, RF shielding and packaging. Based on mid-1990s technology it appeared that it might be possible to achieve a mission payload weight of 15 grams with the one gram imager, not including a battery power source. Therefore, even with a one gram sensor, the mission payload could weigh nearly one-half of an ounce and the gross weight of an electric motor driven Micro UAV carrying it would be about 2.5 ounces.

So, how fast would a 6 inch wingspan airplane weighing 2.5 ounces need to fly? The answer is 18.6 knots (9.56 m/sec) for a maximum lift coefficient of 0.8 and 26.3 knots (13.52 m/sec) for a cruise lift coefficient of 0.4. These lift coefficient values seemed appropriate for this size airplane and the resulting airspeeds would allow flight in winds of up to 10 knots. How much power is required by this airplane? Assuming an L/D of 3.0, a prop efficiency of 50%, and a motor efficiency of 75%; the required battery power would be 8.4 watts. With a battery capable of delivering 90 watt-hr/kg, such as a small lithium sulfur dioxide primary cell undergoing a rapid discharge; 3.3 ounces of battery would be needed for 1 hour of flight. However with a gross weight of 2.5 ounces, one-third or 0.83 ounces were allocated for a battery which for this baseline Micro UAV would provide just over 15 minutes of flight endurance at maximum airspeed.

Now having completed my initial sizing for a challenging, but possible, baseline Micro UAV, I thought about integrating the subsystems, plus stability and control. In order to maximize propeller efficiency, I wanted to use the largest possible propeller. My experience with small free flight models was that torque and slipstream effects were difficult to trim out when the prop was larger than one-half the wing span. However, having now made the assumption that the baseline Navy Micro UAV would be powered by electric motor, why not use two propellers with counter rotation? Since it is always desirable to keep the nose of any UAV available for the mission payload it made the best sense to mount the propellers on the wing leading edge. By placing each at the leading edge of the wing tip, three beneficial features appeared. First, the diameter of each propeller could be increased to the same size of the wingspan, minus the fuselage width. Second, the entire wing would be "blown" with prop wash to help keep the boundary layer attached at its chord Reynolds number of 84,000 at 26 knots. Third, if each propeller rotated in the direction opposite the wing tip vortex, the "Zimmerman Effect" of aerodynamically increasing the aspect ratio beyond that of the wing geometrical value might be realized, with a corresponding reduction of induced drag. Next, I added inward-canted tail surfaces near the trailing edge of the wing tips with elevon controls and a preliminary concept design for an entire Micro UAV emerged. I christened the design the MITE, which stands for Micro Tactical Expendable, with the hope that the Navy's first Micro UAV, "mite" be successful.

Col. Francis and I met later that week. We realized that we were on the forefront of a significant new technology area for the Department of Defense. We agreed that the size goal of fitting into the space of a 6 inch cube like the MITE concept design was sufficiently small to challenge the developers and still have a high probability of success. We discussed the research, development and transition possibilities and decided that DARPA's and the Navy's Micro Unmanned Air Vehicle Programs should be fully complimentary and share their technologies. DARPA would develop micro GPS navigation and tiny imager payloads along with liquid fuel propulsion systems including micro turbines! NRL would develop non-GPS navigation using optic flow and image recognition, plus efficient micro electric motors for electric propulsion and electronic payloads other than imagers.

DARPA kicked-off the Micro UAV Programs by hosting a workshop with university, industry and other government agency participation. Well into the second day of interesting and informative presentations, one of the participants asked us why we called this new class of air vehicles Micro "Unmanned" Air Vehicle since "micro" strongly implies that it was unlikely to be manned. At first we chuckled, then mutually agreed on the name Micro Air Vehicle, or MAV.

The DARPA and Navy MAV programs were highly successful and their technologies were transitioned to industry, enabling the first production Small UAVs such as the USMC "Dragon Eye" and U.S. Army "Raven". Both are used today to provide over the hill reconnaissance for platoons and small units of soldiers serving in Iraq. Currently, DARPA and NRL are conducting follow-on research efforts focused on MAV operation in urban environments, energy harvesting so MAVs can autonomously replenish their power in the field, and multi mode mobility to increase their survivability in hostile environments and enhance their mission effectiveness. Indeed, a new class of tiny, but highly effective military systems were born in 1995.

Fig. 1.16 Naval Research Laboratory MITE 2 with camera payload (reprinted with permission of U.S. Naval Research Laboratory, Washington, DC).

1.2.4 Birth of Micro Air Vehicles

The Naval Research Laboratory (NRL) was funded by the Office of Naval Research for six years starting in 1996 to develop technical solutions needed for the design of practical MAVs. They were a logical choice because of their 25 years of experience in designing, building, and flying small UAVs. Although small UAVs have aspect ratios well above two, MAVs have aspect ratios below two and more often near one. The original NRL MITE vehicle concept (see Fig. 1.1) had a wing span of 15.24 cm (6 in.) and an aspect ratio of 1.25. A number of changes were evaluated to find an optimal configuration. These changes resulted in the MITE 2, which has a wing span of 36 cm (14.5 in.), a chord of 24.5 cm (10 in.), and an aspect ratio of 1.45 [45]. The MITE 2, shown in Fig. 1.16, is one of a series of MAV research vehicles designed to be an affordable, expendable convert sensor platform for close-in short-duration missions. It is an electrically powered vehicle, with dual motors and counter-rotating propellers, that can carry a useful military payload at 32 km/h (20 mph) for 20 min [44] and [45]. The advantages of the two motor design include the following: 1) the dual propellers provide flow over most of the wing for enhanced lift at low speeds, 2) the counter rotating propeller flows oppose the wing tip vortices and reduce the induced drag, 3) the counter rotating propellers cancel the torque effect at high power settings and low speeds for easy hand launching, and 4) the open fuselage nose is ideal for imagers and other payload devices. A case study of the MITE development is presented in Chapter 4. The experience gained during the development of the MITE MAVs was used in the design and fabrication of the Dragon Eye small UAV system [46].

Fig. 1.17 AeroVironment "Black Widow" (reprinted with permission of AeroVironment, Inc.).

AeroVironment, Inc., of Semi Valley, California, is well known for producing successful manned and unmanned unconventional airplanes (see http://www.aerovironment.com [cited 20 September 2005]). They were funded by DARPA in 1996 with a Phase I SBIR contract to study the feasibility of a 15.24-cm (6-in.) MAV. They concluded that a vehicle of this size was feasible and received a Phase II SBIR contract in 1998 that resulted in the Black Widow MAV configuration [47]. The Black Widow, shown in Fig. 1.17, is one of the smallest and most successful MAV systems that can carry a useful payload. This vehicle is electrically powered by one 10-W dc motor with a 4-in. propeller, has an aspect ratio of 1.0, a wing span of 15.24 cm (6 in.), a total mass of ~80 g (2.82 oz), and can carry a color video camera and transmitter. It also has a 3-g (0.10-oz) fully proportional radio control system. A pneumatic launcher and a removable pilot's control unit with a 10.16-cm (4-in.) liquid-crystal display, in a briefcase, were also developed to complete the system. In 1999 the AeroVironment MAV team lead by Matt Keennon received awards from DARPA and *Unmanned Vehicles Magazine* for the Black Widow. The Black Widow set several records for an outdoor flight of a micro air vehicle 10 August 2000 including an endurance of 30 min, a maximum range of 1.8 km (1.11 miles), and a maximum altitude of 234.39 m (769 ft).

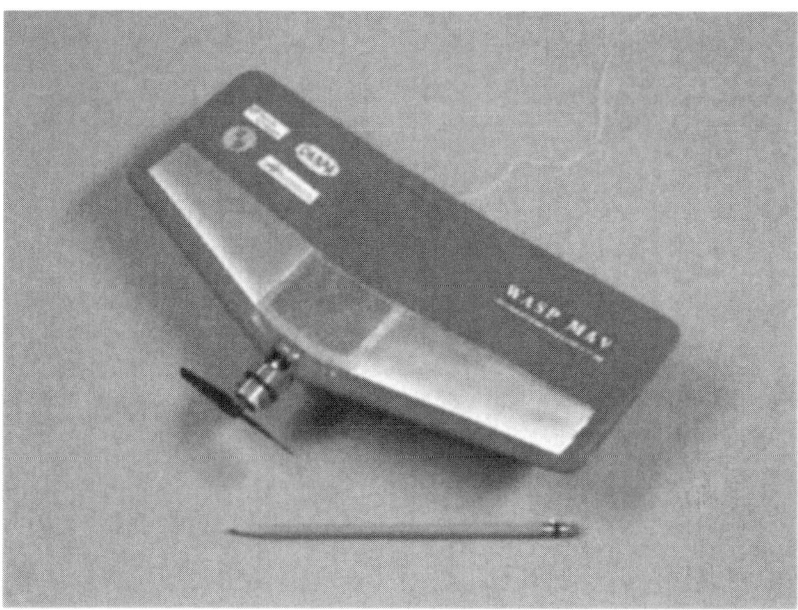

Fig. 1.18 AeroVironment "Wasp" (reprinted with permission of AeroVironment, Inc.).

The success of the Black Widow led to the development of a somewhat larger flying-wing MAV, the Wasp. The Wasp has a root chord of 21.33 cm (8.4 in.), a wing span of 36.57 cm (14.40 in.), and weighs 181.43 g (6.4 oz); it is shown in Fig. 1.18. It is powered by one 10-W dc electric motor, is designed to fly between 40.23 km/h and 48.27 km/h (25–30 mph) at a maximum altitude of 91.44 m (300 ft), and has a color video camera and transmitter. The Wasp is hand launched, has an auto pilot, an endurance of 60 min, and a range of 4 km (2.48 miles) line of sight. This vehicle eliminates the need for the pneumatic launcher of the Black Widow and is easier to fly. A stripped-down RC Wasp set an endurance record of 1 h and 47 min on 19 August 2002 [48–53].

The AeroVironment Hornet MAV [49] and [50] made what appears to be the world's first successful flight of a MAV completely powered by a hydrogen fuel cell on 21 March 2003 according to the company Web site on 19 October 2005 (see Fig. 1.19). The longest flight was for 6 min [50]. The Hornet's radio link, servos, motor, pumps, and other avionic systems were all powered by the fuel cell that also acts as a structural member of the wing. This vehicle is a flying-wing configuration that has a 38.10-cm (15-in.) wing span and a total mass of 170 g (6 oz). The custom fuel cell developed by Lynntech, Inc., of College Station, Texas, produced an energy density higher than similar sized battery-based systems, and the average power during flight was over 10 W. The Hornet is a stable and easy-to-fly vehicle using manually operated ground control of the throttle, rudder, and elevator. An additional radio channel was used to modulate the rate of hydrogen

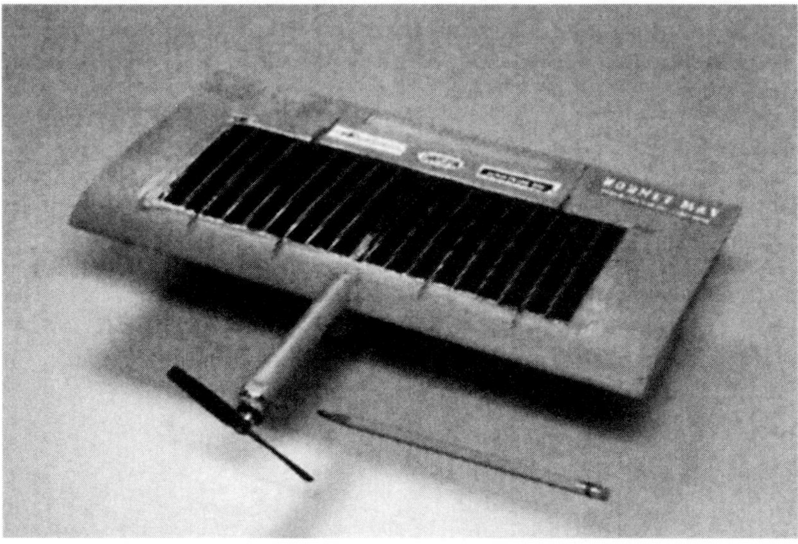

Fig. 1.19 AeroVironment fuel-cell-powered "Hornet" (reprinted with permission of AeroVironment, Inc.).

generated so that the energy released could be varied. The fuel cell and vehicle development were funded by DARPA.

A large number of airplanes have been designed, built, and flown since the beginning of the micro-air-vehicle era in 1996. Government laboratories, industry, and universities in a number of countries have all made contributions to the technologies necessary to field successful MAV systems. The purpose of this brief historical perspective is to highlight selective advancements in aeronautics that led to successful radio-controlled model airplanes by hobbyists, then small unmanned vehicles with useful payloads, and finally micro air vehicles with useful payloads. A chronological list of events is presented in this chapter in Table 1.3. This chronology includes events not discussed in the historical perspective.

1.2.5 Smaller Air Vehicles

In 1996 [2], there was a belief that 7.4-cm (3-in.) micro air vehicles could be developed within a few years and that 1-cm (0.4-in.), 1-g (0.035-oz) vehicles were feasible in the future. There have been fixed-wing MAVs built and flown that have wing spans less than 12.7 cm (5 in.). For example, 12-cm (4.74-in.) MAVs were flown in the endurance competitions in 2002 and 2003, and a 9.43-cm (3.71-in.) model won the endurance competition in 2005. The MAV competitions are described in the next section. None of these fixed-wing vehicles carried payloads.

DARPA organized a workshop on 29 September 2005 to announce a technology development and integration nano-air-vehicle (NAV) program. The goal of this program was to demonstrate the capability to develop an extremely small [less than 5 cm (1.97 in.)] ultralightweight [less than 10 g (0.35 oz)] air vehicle capable of both hover and forward flight carrying a 2-g (0.07-oz) payload. During this workshop, it was decided to relax the size to less than 7.62 cm (3 in.). As already mentioned, a number of 7.62-cm (3-in.) fixed-wing MAVs have been developed and flown. The hover capability requires a rotary-wing, tilt-rotor, or flapping-wing vehicle. However, fixed-wing indoor airplanes of this size have been built and flown although without a payload. John Worth [54] reported that Henry Pasquet has flown a 6.59-cm (2.59-in.) span fixed-wing RC model weighing 2.3 g (0.08 oz). MAV designers can learn a great deal from the indoor modelers and their publications. John Worth started a private newsletter called "Cloud 9 Micro RC" in 1995. In 1999 Air Age contracted to have him produce *Radio Control MicroFlight Magazine*, which started in January 2000 and ended in February 2005. Since this magazine ceased publication, John Worth has started his own online magazine called *Cloud 9 RC Micro World* (see http://www.cloud9rc.com [cited 7 October 2005]). This magazine is a good source of information and equipment available for very small lightweight RC model airplanes.

1.2.6 Micro-Air-Vehicle Competitions

Since the introduction of the micro-air-vehicle concept in the mid-1990s, there have been hundreds of successful configurations built and flown in a large number of countries. A number of companies and government laboratories worldwide have also been active in MAV design and development. An annual competition was held in the United States beginning in 1997 once it became clear that MAVs could be designed and fabricated using equipment available for small model airplanes [55]. In 2000 this competition was named The International Micro Air Vehicle Competition because there was an entry from Korea. In 2002 there were entries from Korea and Germany and an observer from France. The surveillance mission for all of these competitions included the observation of a target behind a barrier located 600 m (1969 ft) from the launch site and transmission of a clear image back to the judges. A design report was also required. The smallest vehicle to accomplish this wins.

The first competition was held in 1997 at the University of Florida. The entries included small radio-controlled airplanes rather than micro air vehicles because the technology available did not permit the miniaturization to the micro level. Steve Morris of the MLBM Company won with a 78.7-cm (31-in.) airplane. The smallest airplane to complete the mission was 66 cm (26 in.) from the University of Florida. However, the design-optimization report accompanying this plane did not meet the judges' requirements. Steve Morris won the competition again in 1998 with a 38.1-cm (15-in.) airplane, and the University of Florida completed the mission with a 40-cm (15.8-in.) plane [55].

The maximum dimension of the vehicles that won the surveillance competition almost decreased every year with the exception of 2003 as shown in Table 1.1. The availability of miniature video cameras and transmitters and an increased understanding of the aerodynamics contributed to the decrease in size of the MAVs. The flexible-wing concept introduced in 1999 by the University of Florida has dominated this competition. The flexible-membrane wing concept will be presented in detail in Chapter 5.

Table 1.1 MAV surveillance competitions: smallest airplane to image target 600 m (1969 ft) away

Year	Host	Winner	cm	in.
1997	University of Florida	MLBM	78.7	31
1998	University of Florida	MLBM	38.1	15
1999	University of Florida	UF	30.5	12
2000	Arizona State University	UF	25.4	10
2001	University of Florida	UF	18.8	7.4
2002	Brigham Young University	UF	14.0	5.5
2003	University of Florida	BYU	18.8	7.4
2004	University of Arizona	UF	13.3	5.25
2005	Konkuk University, Seoul, Korea	KKU	12.7	5.0
2006	Brigham Young University	UF	11.4	4.5

From 1999 to 2001, a payload competition was also held for the smallest vehicle to carry 56.7 g (2 oz) for 2 min. The University of Florida won these competitions in 1999, 2000, and 2001 with vehicles of 36.6, 28.5, and 18.8 cm (14, 11.2, and 7 in.), respectively. This competition was replaced by an endurance competition, with no payload, in 2002 [56–59]. The vehicles scored, from 2002 to 2005, based on flight time per maximum dimension cubed, that is, s/cm^3 (s/in.^{3}). The winners of the endurance competitions are shown in Table 1.2. In 2005 the team from Inha University, Korea, won this competition by flying for 900 s with a 9.43-cm (3.71-in.) vehicle. The rule was changed in 2006 to the smallest packable volume. The University of Florida won in 2006 by flying for 560 s with a vehicle that rolled up into a 61.18-cm^3 (3.73-in^3) tube [60].

An ornithopter competition was added to this annual international event in 2004 [58] and [59], The competition involves building the smallest radio-controlled ornithopter that can fly the most laps around a pylon course in 2 min. The pylons were spaced about 12.19 m (40 ft) apart, and the ornithopters fly either an elliptical course around them or a figure eight through them. The University of Arizona won the 2004, 2005, and 2006 competitions with 28-cm (11-in.), 32.5-g (1.15-oz); 20.3-cm (8-in.), 36-g (1.27-oz); and 15.8-cm (6.2-in.) vehicles, respectively.

The first conference and workshop with the title Micro-UAVs Days was held in Toulouse, France, 26–27 June 2001. This was followed by the second Days MAV 2002 and the first flight competition, the First European MAVs Trophy, in

Table 1.2 MAV endurance competition: flight time per vehicle volume

Year	Winners	s/cm^3	s/in.^{3}
2002	BYU	0.47	7.78
2003	UF	0.30	4.96
2004	UF	0.67	11
2005	InhaU, Korea	1.07	17.5
2006	**UF**	**9.15**	**150**

Toulouse, 18–20 September, 2002. The purpose of the competition was to examine various MAV concepts with tests representative of operational missions such as controlled flight, reconnaissance flight, hovering flight, and autonomous flight. Both indoor and outdoor demonstration flights were made with fixed-wing and rotary-wing vehicles. The reconnaissance flight consisted of flying for 2 min along a triangular course 100 m (328 ft) on each side with a payload of 10 g (0.35 oz) and transmitting a real-time video of a target surrounded by a 2-m (6.5-ft) high fence. The fixed-wing vehicles ranged in size from 20.6 cm (8.1 in.) to 77 cm (30.3 in.). Demonstration flights were made.

The third Days Micro-UAVs 2003 and the Second Trophy Micro-UAVs European were merged and held 1–3 October 2003 in Toulouse, France. This was the first demonstration of autonomous flight. This mission consisted of flying over four GPS points that form a square 300 m (984 ft) on a side. The coordinates were provided in advance for a target located along the flight path. Only one vehicle, a 150-cm (59-in.) minidrone named Twinstar-Paparazzi, from Supaero-ENAC-STNA, Toulouse, France, successfully completed the mission.

The Fourth European Micro UAV Meeting was held 15–17 September 2004 in Toulouse with 240 participants and a total of 40 flights. A conference and the micro-UAV trophy flights were included. Both micro UAVs 20 cm (7.9 in.) and mini UAVs 70 cm (27.56 in.) were flown. The award for the best micro UAV in forward flight was given to the team from the University of Arizona, Tucson, Arizona. Their MAV was named the Blink Junior and was 15 cm (5.90 in.) and weighed 45 g (1.59 oz). The award for the best micro UAV in hovering flight was given to the vehicle named Mosquito from Proxflyer, Norway. This vehicle was 36 cm (14.17 in.) and weighed 110 g (3.88 oz). An award was also given for the best micro UAV in autonomous flight to the team from CENA-ENAC, Toulouse, France, for their 65-cm (25.59-in.), 420-g (14.81-oz) MAV named MicroJet-Paparazzi. This European Micro UAV Meeting now alternates between France and Germany.

The DGA (French Defense Procurement Agency, a part of the French Ministry of Defense) began a three-year funded program named International Universities MINI UAV Competition during the 2002/2003 academic years. This program was open to engineering schools and universities with second-cycle degree programs (or the equivalent for overseas). At the end of the first academic year, 10 university teams were selected on the basis of technical proposals. In the second academic year, an additional eight teams were selected. The purpose of this competition was to display the technical feasibility and performance of mini UAVs. The International MINI UAV Competition was held 12–14 September 2005 at Mourmelon, France. The rules for this competition were different than for the other competitions. The main differences included 1) the maximum size was 70 cm (27.56 in.), 2) bonus points were given for the ability to hover, and 3) the flight used a combat village that required flight out of the line of sight in an urban environment with turbulent aerology and the transmission of data in the urban environment. This was a very difficult flight scenario, and the judges decided that there were no winners; however, three teams were awarded 8000 euros each for special aspects of their projects.

A technical workshop, Micro-Air-Vehicles—Unmet Technological Requirements, was held 22–24 September 2003 in Elmau Castle near Garmisch–Partenkirchen, Germany. The objective was to provide potential users a clear picture of

Table 1.3 Chronological list of events[a]

Year	Event
1783	Montgolfier brothers' first unmanned balloon flight.
1799	Sir George Cayley's first design for the modern airplane configuration.
1809	Sir George Cayley flies full-sized unmanned gliders.
1848	**John Stringfellow's unsuccessful steam-powered model of Henson's design.**
1849	Cayley tests a boy-carrying glider.
1853	Cayley's controlled flight of a man-carrying glider.
1857	**Felix du Temple and Pierre Julien experiment with clockwork and rubber band-powered model airplanes.**
1863	Société Francaise de Navigation Aérienne founded in France.
1866	Royal Aeronautical Society founded in Britain.
1870	**Pénaud develops a rubber-band-powered helicopter.**
1871	**Alphonse Pénaud begins selling rubber-band-powered model airplanes.**
	Wind tunnel invented by Francis Wenham and John Browning.
1872	**Pénaud develops a rubber-band-powered ornithopter.**
1878	**Wright brothers, inspired by father's gift of a Pénaud-type model helicopter, made several copies.**
1879	**Victor Tatin flies a tethered monoplane model using a compressed air motor.**
1880	**Albert F. Zahm begins his experiments with model and full-sized airplanes.**
1884	Nikola Tesla arrives in New York with plans for a remotely controlled unmanned airplane.
1888	Louis Brennan demonstrates a wire-guided torpedo.
1891	Otto Lilienthal makes a number of short glider flights.
1892	**Samuel P. Langley begins tests with steam-powered models.**
1893	Zahm delivers a paper on "Stability of Aeroplanes and Flying Machines" at the Columbian Exposition in Chicago.
1896	**Langley has first successful steam-powered models.**
	Chanute begins successful hang-glider experiments.
	Lilienthal dies in a glider accident.
1898	Tesla demonstrates a remotely controlled boat.
1899	Wilbur Wright builds a biplane kite to test wing warping.
1901	**Langley flies gas-powered model, with engine built by Charles Manly.**
1903	Wright brothers' first successful piloted-powered flight.
1907	**Ray Arden's first four-cycle engine with bore and stroke of 3.17 cm (1.25 in.) and displacement of 25 cm^3 (1.53 $in.^3$).**
1908	**David Stanger develops a two-cylinder V-type engine.**
1911	**Eckert brothers develop their Baby $1/2$-hp engine with a displacement of 43.7 cm^3 (2.67 $in.^3$).**
	CO_2 expansion motors used in model airplane.
1913	**David Stanger develops a 60.7-cm^3 (3.7-$in.^3$) Vee-Twin engine. First gas-model airplane meet in America.**
	Braune and Zenker founded a company to build one- and two-cylinder engines for gliders.
	Gamage, Ltd., model shop in London sells compressed air-powered ready-to-fly model airplanes.
1914	*World War I begins in Europe.*
1917	Royal Aircraft Factory pilotless airplane with radio controls by Major Low reached test flight stage.

Table 1.3 Chronological list of events (Continued)

Year	Event
1918	Elmer Sperry and son and Peter Hewett demonstrate pilotless airplane for the U.S. Navy.
	World War I ends in Europe.
1923	Sperry demonstrated a practical radio control system.
1927	Charles Lindbergh flies nonstop from New York to Paris in $33\frac{1}{2}$ h.
1929	***Model Airplane News* debuts, in United States.**
	Richard Byrd makes the first flight over the South Pole.
1932	**First successful limited production model airplane engine, by Bill Brown.**
	Bell Atwood made a flight in excess of 20 min with a prototype of his Baby Cyclone engine.
	Amelia Earhart becomes the first woman to fly solo across the Atlantic Ocean.
	Maxwell Bassett entered the only gas-powered free-flight model using a Brown Jr. engine in the International Wakefield event in Atlantic City, New Jersey.
1933	**Maxwell Bassett's gas-powered model captured all of the trophies from the rubber-band-powered models at the U.S. National.**
	Royal Navy demonstrated a radio-controlled target drone, the Queen Bee.
1934	**First production model airplane engine, the Brown Jr. 60 Model B (1000 sold).**
	Chester Lanzo's RC experiments were unsuccessful.
	First gas-powered free-flight event at the Nationals in Akron, Ohio.
1935	**Walter Kratzsch founded a company to build one- and two-cylinder combustion engines for model airplanes.**
	First crank shaft rotary induction valve production engine, the Baby Cyclone by Bill and Anderson Atwood (15,000 sold).
	Dan Calkin developed a small engine, the ELF.
1936	**Academy of Model Aeronautics (AMA) founded in the United States.**
	Alfred Lippitsch and Egon Sykora won a prize with the first radio-controlled glider at Rhoen, Germany.
	Walter and William Good develop a RC rudder control system for their free-flight gas model.
1937	**The Good brothers exhibit the first successful American gas-powered RC model.**
	C. W. Thompson, Jr. and H. M. Plummer demonstrate the first public RC glider in the United States in July.
	Lanzo with RC-1 took first place in the first RC National contest in the United States.
1938	**Walt and Bill Good won the RC Nationals in the U.S.A. with "Big Guff." They also won in 1939, 1940, and 1947.**
1939	*World War II begins in Europe.*
	Glide Bomb project begins in Germany.
1940	**Three sizes of four-stroke engines produced before United States was in World War II by Feeney Engine Co., USA.**
1941	**First successful model diesel engine produced, the Swiss Dyno 1.**
	World War II begins in Pacific.

(Continued)

Table 1.3 Chronological list of events (Continued)

Year	Event
1943	German radio-controlled glide bomb made by the Heuschel Co. sinks the troop ship Rohna.
1944	Allied invasion of Europe.
1945	*World War II ends.*
1946	**Methyl alcohol becomes popular fuel for spark ignition engines.**
	Low Speed Research Aerodynamics Association formed at Farnborough.
	Chuck Yeager breaks the sound barrier in the Bell X-1.
1947	**Ray Arden introduces the model glow plug.**
1948	**K & B Infant 0.020, the first mass-produced miniature (1/2 A) model engine.**
	Transistor invented at Bell Labs.
1949	**Synthetic lubricants used in model engines.**
	Charles Stark Draper invents the first inertial navigation system (INS).
1957	**First electric-powered RC model flight.**
	Graupner sells the first electric-powered model airplane kit (the Silentus). First commercially available, proportionally controllable RC system by Quadtroplex.
1959	**Resonant reed multichannel RC systems developed.**
1960	American scientists patent computer chip.
1961	Yuri Gagarin is the first man in space.
1963	President John F. Kennedy is assassinated.
1966	**First production model engine with a muffler, the COX® 0Z 0.049.**
	Monokote introduced.
1969	**Commercial digital proportional RC radio systems emerge.**
	Diaphragm fuel delivery pump.
	Apollo 11 makes first manned landing on the moon.
1973	Decision to develop a satellite navigation system leading to the global positioning system (GPS).
1976	First flight of the CONCORDE supersonic airliner.
1977	First mass-marketed personal computers launched.
	Voyager 1 deep space probe launched.
1979	**In-flight variable-pitch propeller for models.**
1981	The first reusuable spacecraft, the Space Shuttle *Columbia*, was sent into space.
1983	**RC computer flight simulator by Dave Brown.**
	Decision to allow the civilian use of the GPS.
1985	**Conference on low-Reynolds-number airfoil aerodynamics at the University of Notre Dame.**
1986	The first round-the-world flight without refueling by Dick Rutan and Jeana Yeager in the Voyager.
	Royal Aeronautical Society International Conference on Aerodynamics at Low Reynolds Numbers $10^4 < Re < 10^6$, London.
	Aerospace engineering senior design course at Notre Dame begins—design, build, and fly small UAVs.
	RQ-2 Pioneer UAV deployed aboard ship and on shore.

Table 1.3 Chronological list of events (Continued)

Year	Event
1989	**Conference on low-Reynolds-number aerodynamics at the University of Notre Dame.**
	Miniature turbine model engine flies in model airplane in England.
	FMQ-151 Pointer small UAV acquired by U.S. Marines.
	Berlin wall torn down.
1991	**RC receivers converted to narrowband technology.**
1993	**Brushless electric motors for model airplanes become available.**
1994	**Cline proportional control fuel system.**
1995	**Richard Foch and Michael Francis set goals for micro air vehicles.**
	John Worth starts private newsletter "*Cloud 9 micro RC*."
1996	**DARPA and ONR fund MAV research.**
	EPP foam used in RC models.
1997	**First MAV competition at the University of Florida.**
	MQ-1 Predator UAV transitioned to the U.S. Air Force.
1998	**Automatic in-flight mixture control for gas engines.**
	Naval Research Lab MAV test vehicle (MITE) flies.
1999	**First generation of the Black Widow MAV flies for 22 min with a black and white video camera.**
	Piezo-gyros in models.
	Digital servos available.
	3000 Ah NiMH batteries available promising 50% longer flight duration.
2000	**Conference on Fixed, Flapping and Rotary Wing Vehicles at Very Low Reynolds Numbers at the University of Notre Dame.**
	***Radio Control MicroFlight Magazine* founded by John Worth published by Air Age.**
	RQ-4 Global Hawk UAV completed military utility assessment.
2001	**Lithium polymer rechargeable cells become commercially available.**
	First conference and workshop Micro-UAV Days held in France.
2002	First test flights of the X-45 UCAV.
	French Defense Procurement Agency (DGA) begins a three-year funded university program called "International Universities Mini UAV Competition."
2003	Dragon Eye small UAV deployed in Iraq.
	Technology workshop Micro-Air-Vehicles—Unmet Technology Requirements held in Germany.
	Last flight of the CONCORDE supersonic airliner.
2004	Space Ship One, the first private manned spacecraft, exceeds an altitude of 328,000 ft twice within 14 days.
2005	First test flight of the Airbus 380.
	DARPA workshop on Nano Air Vehicles.
	Radio Control Micro Flight Magazine ceases* publication, and John Worth publishes online magazine *Cloud 9 RC Micro World.

[a]Model-airplane-related events are shown in bold.

what technology can provide in the short term and to generate ideas for the design of specific systems. The First U.S.–European MAV Technology Demonstration was held in conjunction with the First U.S.–European Micro-Air Vehicle Workshop 19–22 September 2005 in Garmisch-Partenkirchen, Germany. The objective was to deploy one or more micro air vehicles in an effort to identify a target approximately 300 m (984 ft) from the launch site. The University of Arizona 30.48-cm (12-in.), 185-g (6.5-oz) MAV with a fixed camber wing took first place in the overall best design competition. This MAV was equipped with an autopilot and a video camera. The vehicle loitered autonomously at an altitude of 100 m (328 ft) above ground level over the target for more than 20 min providing video to the ground station, successfully completing the mission. Integration of the autopilot into the MAV will be described in Chapter 3.

It seems clear that similar competitions will continue because of the challenges presented to the mini UAV–MAV community. These competitions have attracted multidisciplinary teams of students from aerospace, electrical, and mechanical engineering as well as from material and computer science. The students have provided an unbiased view of this relatively new field and have contributed new ideas that were transferred to industry and government.

1.3 Conclusion

Interest in unmanned air vehicles was stimulated by World War I. In the beginning full-sized airplanes were used with primitive controls capable of stabilizing and navigating without a pilot onboard. The conversion of manned airplanes to target drones continued during the 1920s and 1930s. A number of radio-controlled, radar-controlled, and television-controlled glide bombs (e.g., the TDR-1, the AZON, the Fritz-X, and the Henschel 293) were used in World War II.

Sir George Cayley and the other pioneers in aviation used model airplanes to help them understand aerodynamic forces and how to control them. The availability of small internal combustion engines and small radio receivers and transmitters plus the invention of control surface actuators in the 1930s led to the era of radio controlled model airplanes. Continuous improvements in RC model equipment, including the introduction of electric motors, plus advances in micromechanical systems and microelectronic components and sensors led to the feasibility of small unmanned air vehicles and then micro air vehicles in the 1990s. Improvements in fixed-wing MAV design and performance have been enhanced by the university student competitions.

Acknowledgments

The author would like to thank the following people and organizations that helped obtain references related to the history of radio-controlled model airplanes: Richard LaGrange of the Academy of Model Aeronautics, Muncie, Indiana; Stewart W. Bailey of the Kalamazoo Aviation History Museum and Linda Brinks of the Kalamazoo Public Library, Kalamazoo, Michigan; Richard Eppler of Stuttgart, Germany, Dietrich Bertermann of Herzebrock, Germany; Claudia Kunze of Hamburg, Germany; Christiane Hennet of the Deutsches Museum, Munich, Germany; Ron G. Moulton of the Royal Aeronautical Society, Bushey, United Kingdom; Ismet Gursul of the University of Bath of the United Kingdom; Robin

East of the University of Southampton, United Kingdom; David Baxter, retired, of the University of Southampton, United Kingdom; Frank H. Anderson of Palm Bay, Florida; Betty Dannels of Buena Vista, Colorado; James Healy of Columbus, Ohio; Holger Menrad of Wolfsburg, Germany; Don Srull of McLean, Virginia; George Cully of Arlington, Virginia; Kevin Rusnak of Dayton Ohio; John Worth of Fairfax, Virginia, and James DeLaurier of the University of Toronto, Ontario, Canada. Information related to the micro-air-vehicle competitions was provided by Jerry Bowman of Brigham Young University, Provo, Utah; Peter G. Ifju of the University of Florida, Gainesville, Florida; Sergey Shkarayev of the University of Arizona, Tucson, Arizona; Robert Michelson of Georgia Institute of Technology, Atlanta, Georgia; Charles Ellington of Cambridge University, Cambridge, United Kingdom; Pierre-Francois Louvigne, Arcueil, France; Jean-Marc Moschetta of the Supaero, Toulouse, France; and Philippe Choy of Palaiseau, France. I would like to thank Richard Foch of the U.S. Naval Research Laboratory, Washington, D.C., for his unique contributions to the development of micro air vehicles from the very beginning. Special thanks go to my daughter Annmarie and my wife Sarah Ann for their comments and suggestions in the early stages of this chapter.

References

[1]Hundley, R. O., and Gritton, E. C., "Future Technology-Driven Revolutions in Military Operations," RAND National Defense Research Institute, RAND Corp., Santa Monica, California, Document No. DB-1100ARPA, 1994.

[2]Davis, W. R., Jr., Kosicki, B. B., Boroson, D. M., and Kostishock, D. F., "Micro Air Vehicles for Optical Surveillance," *The Lincoln Laboratory Journal*, Vol. 9, No. 2, 1996, pp 197–213.

[3]McMichael, J. M., and Francis, M. S., "Micro Air Vehicles—Toward a New Dimension in Flight," *Unmanned Systems*, Vol. 15, No. 3, 1997, pp. 10–15.

[4]Mueller, T. J. (ed.), *Fixed and Flapping Wing Aerodynamics for Micro Air Vehicle Applications*, Progress in Astronautics and Aeronautics, Vol. 195, AIAA, Reston, VA, 2001, p. 2.

[5]Mueller, T. J., "Aerodynamic Measurements at Low Reynolds Numbers for Fixed Wing Micro-Air-Vehicles," RTO EN-9, April 2000, Canada Communications Group, Inc., Quebec, Canada, p. 8-1–8-32.

[6]Wragg, D.W., *Flight Before Flying*, Frederick Fell Publishers, Inc., New York, 1974, 191p.

[7]Zahm, A. F., "Stability of Aeroplanes and Flying Machines," *Proceedings of the International Conference on Air Navigation*, Chicago II, 1893; also *Aeronautical Papers of Albert F. Zahm, Ph.D.*, Vols. 1 and 2, Univ. of Notre Dame, Ind, 1950, 1001p.

[8]Dethloff, H., and Snaples, L., "Who was Albert F. Zahm?," AIAA Paper 2000-1049, Jan. 2000.

[9]Wright, O., *How We Invented the Airplane*, An Illustrated History, edited by F.C. Kelley Dover Publications Inc., Mineola, New York, 1988, 96p.

[10]Newcome, L. R., *Unmanned Aviation: A Brief History of Unmanned Air Vehicles*, AIAA, Reston, VA, 2004, 172p.

[11]Spark, N. T., "Unmanned Precision Weapons Aren't New," *Naval Institute Proceedings*, Vol. 131, No. 2, Sequence Number 1,224, Feb. 2005, pp. 66–71.

[12]Jackson, C., *Allied Secrets: The Sinking of the HMT Rohna*, The Univ. of Oklahoma Press, Norman, OK, 1997, 207p.
[13]Hogg, I. V., *German Secret Weapons of the Second World War*, Greenhill Books, Lionel Leventhal Limited, London, 1999, pp 69–79.
[14]Mueller, T. J., and DeLaurier, J. D., "Aerodynamics of Small Vehicles," *Annual Review of Fluid Mechanics,* Vol. 35, 2003, pp. 89–111.
[15]Anderson, F. H., *An Encyclopedia of the Golden Age of Model Airplanes*, Vol. 1, The Dawn of American Aeromodeling 1907–1935, Frank H. Anderson, Palm Bay, FL 1998, pp. 161–171.
[16]Dannels, T., *American Model Engine Encyclopedia*, Model Museum and the Engine Collectors' Journal, Buena Vista, CO, 2005, p.21.
[17]Anderson, F. H., *Andersons Blue Book*, 4th ed., Frank H. Anderson, Palm Bay, FL, Aug., 2005, p. 6.
[18]Weidner, M., *Flugmodelltechnik: Furher durch d. Abt.*, Deutsches Museum von Meisterwerken der Naturwissenschaft und Technik, Munich, Germany, 1987.
[19]Johnson, J. C., *Flying Model Collectibles and Accessories*, Schiffer Publishing, Ltd., Atglen, PA, 2004, p.11.
[20]Everwyn, G., "Die Deutchen Flugmodellmotore Der Kaiserzeit" ("The German Model Airplane Engines from the Time of the Emperor"), *Modellflug International*, Vol. 4, April 1998 pp. 94–97, and Vol. 5, May, 1998, pp. 68–71,
[21]Everwyn, G., "Die Deutschen Flugmodellmotore der Weimarer Zeit" ("The German Model Airplane Engines from the Time of the First German Republic"), *Modellflug International*, Mai 1999, pp. 62–65 and June 1999, pp. 22–24.
[22]Everwyn, G., *German Motors in the 1920s*, Society of Antique Modellers 35, Yearbook No. 9, Nov. 1996, Peter Michel author, Eric Cooper, London, pp. 72–77.
[23]Brown, J. J., *Dan Calkin and His ELFs*, Model Aviation Books, Santa Ana, CA, 1999, 264p.
[24]Aberle, B., Gierke, D., and Ziroli, N., Sr., "The Century of Radio Control," *Model Airplane News*, Dec. 1999, pp. 28–48.
[25]Good, W., "History of RC Flying," *Model Aviation*, Part 1, March 1986, pp. 56–58 and 123–129.
[26]Phillips, R., *Wireless-Controlled Mechanism for Amateurs*, Cassel and Company, Ltd., London, 1927, pp. 36–46.
[27]Good, W., "History of RC Flying" *Model Aviation*, Part 2, April 1986, pp. 58–63 and 141–148.
[28]*Flugsport*, XXVIII, Jahrgang, No. 12, 10 June, 1936, p. 288.
[29]*Deutsche Luftwacht, Ausgabe Modellflug*, vol. 1, No. 2, 1936, pp. 56–58.
[30]Ledertheil, A., "Vor rund 30 Jahren Begann man in Deutschland . . . Aus der Geschiche der Funkfernsteuerung fur Modelle," *Flug und Modelltechnik* Magazine, vol. 2, No. 32, 1957, pp. 1–3.
[31]"Radio Control Aeromodelling: The Pioneers," Interview with Walter and William Good, video produced and directed by Jay S. Gerber, Kalamazoo, Michigan, 1990.
[32]"500 at Science Demonstrations," *Kalamazoo Gazette*, 13 Jan., 1937.
[33]Hall, R., Jr., "Soaring into History: Twin Brothers Created a New Hobby When They Launched Radio-Controlled Airplane," *Kalamazoo Gazette*, 28 June, 2003.
[34]Hunt, P., *Radio Control for Model Airplanes*, A Harborough Publication, Airplane (Technical) Publications Limited, The Drysdale Press, Ltd., Leicester, England, UK, 1944, p. 61.

[35]Allen, D., "Proportional Beginnings," Part 1, *Radio Control Models and Electronics*, Nexus Publication, currently Highbury Leisure, Berwick House, Kent, England, UK, May 1994, pp. 30–32.
[36]Allen, D., "Proportional Beginnings," Part 2, *Radio Control Models Electronics*, Nexus Publication, Currently Highbury Leisure, Berwick House, Kent, England, UK, June 1994, pp. 49–52.
[37]Laidlaw-Dickson, D. J., "Radio Control," *Aeromodeller Annual*, edited by D. A. Russell, The Aeronautical Press, Ltd., Leicester, England, UK, 1949, pp. 128–137.
[38]Grevski, O. K., *Flying Model Gliders*, DOSAAF, Moscow, 1955, p. 155.
[39]Aberle, B., *Clean and Quiet – The Guide to Electric Powered Flight*, Douglas Charles Press, N. Attleboro, MA, 1995, pp. 1–6.
[40]"Electric Power: German Motor Makes New Model Class Possible," *Aeromodeller,* Dec. 1959, p. 592.
[41]Boucher, R. J., *Electric Motor Handbook*, Astro Flight, Inc., Marina Del Rey, CA, 1994, Chapters 1–10.
[42]Foch, R. J., and Ailinger, K. G., "Low Reynolds Number Long Endurance Airplane Design," AIAA Paper 92-1263, Feb. 1992.
[43]Foch, R. J., "SENDER—A Low Cost Airobotic Platform,"*AUVSI Proceedings*, Orlando, FL, 1996, pp. 863–868.
[44]Kellogg, J., Bovais, C., Dahlburg, J., Foch, R., Gardner. J., Gordon, D., Hartley, R., Kamgar-Parsi, B., McFarlane, H., Pipitone, F., Ramamurti, R., Sciambi, A., Spears, W., Srull, D., and Sullivan, C., "The NRL Mite Air Vehicle," *Proceedings of the 16th, International Conference on Unmanned Air Vehicle Systems,* Univ. Bristol, Bristol, England UK, 2001, pp. 25.1–14.
[45]Kellogg, J., Bovais, C., Foch, R., McFarlane, H., Sullivan, C., Dahlburg, J., Ramamurti, R., Gordon-Spears, D., Hartley, R., Kamgar-Parsi, B., Pipitone, F., Spears, W., Sciambi, A., and Srull, D., "The NRL Micro Tactical Expendable (MITE) Air Vehicle," *The Aeronautical Journal of the Royal Aeronautical Society,* Aug. 2002, pp. 431–441.
[46]Foch, R. J., Dahlburg, J. P., McMains, J. W., Bovais, C. S., Carruthers, S. L., Cole, R., Gardner, J., Kellog, j., Schuetle, L., and Tayman, S., "Dragon Eye, an Airborne Sensor System for Small Unite," *Proceedings of the Unmanned Systems, Florida* [Mira CD-Rom], St. Louis, MO, 2000, pp. 1–13.
[47]Grasmeyer, J. M., and Keennon, M. T., "Development of the Black Widow Micro-Air Vehicle," *Fixed and Flapping Wing Aerodynamics for Micro Air Vehicle Applications*, edited by T. J. Mueller, Vol. 195, Progress in Astronautics and Aeronautics, AIAA, Reston, VA, 2001; also AIAA Paper 2001-0127, 2001.
[48]Lopez, R., "The Revolution Will Not Be Piloted," *Popular Science*, Vol. 262, No. 6, June 2003, p. 61.
[49]Dornheim, M. A., "Small Drones Mature," *Aviation Week and Space Technology*, 15 Sept., 2003, Vol. 159, No. 11, pp. 63, 64.
[50]Dickerson, L., "UAVs Buoyed by Defense," *Aviation Week and Space Technology Source Book,* 17 Jan. 2005, pp. 101–106.
[51]Wilson, J. R., "UAV Worldwide Roundup – 2005," *Aerospace America*, Vol. 43, No. 9, Sept. 2005, pp. 26–34.
[52]Wilson, J. R., "UAV Programs Around the World," *Supplement to Aerospace America*, Vol. 43, No. 9, Sep. 2005.

[53]Aldridge, E. C., Jr., and Stenbit, J. P., "Unmanned Air Vehicles Roadmap 2002–2027," Office of the Secretary of Defense, Washington, DC, Dec. 2002.
[54]Worth, J., "Small Talk," *Flying Models*, July 2005, pp. 28, 29.
[55]Bowman, J. W., "History of the Micro Air Vehicle Competitions," AIAA unpublished Report to student activities committee, Reno, Nevada, January 2003.
[56]"Micro Air Vehicle Design Papers and 6th International MAV Competition," Mechanical Engineering Dep., Brigham Young Univ., Provo, UT, April 2002.
[57]"Micro Air Vehicle Design Papers and 7th International MAV Competition," Dep. of Mechanical and Aerospace Engineering, Univ. of Florida, Gainesville, Florida, April 2003.
[58]"Micro Air Vehicle Design Papers and 8th International MAV Competition," Dep. of Aerospace and Mechanical Engineering, Univ. of Arizona, Tucson, AZ, April 2004.
[59]"Micro Air Vehicle Design Papers and 9th International MAV Competition," Dep. of Aerospace Engineering, Konkuk Univ., Seoul, May 2005.
[60]"Micro Air Vehicle Design Papers and 10th International MAV Competition," Dep. of Mechanical and Aerospace Engineering, Brigham Young Univ., Provo, Utah, May 2006.

2
Elements of Aerodynamics, Propulsion, and Design

Thomas J. Mueller*
University of Notre Dame, Notre Dame, Indiana
Gabriel E. Torres†
Aero Vironment, Inc., Simi Valley, California
and
Donald W. Srull‡
Naval Research Lab, Washington, D.C.

Nomenclature

AR	=	aspect ratio, b^2/S
b	=	wing span
C	=	rated battery capacity
C_D	=	drag coefficient, $D/\frac{1}{2}\rho V\infty^2 S$
C_{D0}	=	drag coefficient at zero lift
C_L	=	lift coefficient, $L/\frac{1}{2}\rho V\infty^2 S$
$C_{L\max}$	=	maximum lift coefficient
$C_{L\alpha}$	=	three-dimensional lift-curve slope, 1/deg
$C_{l\alpha}$	=	two-dimensional lift-curve slope, 1/rad
C_M	=	pitching-moment coefficient about $\bar{c}/4$, $M/\frac{1}{2}\rho V\infty^2 S\bar{c}$
c_{root}	=	wing root chord
c	=	wing chord
c_s	=	specific fuel consumption, weight of fuel consumed per unit power per unit time
$\bar{c}$	=	mean aerodynamic chord, MAC
D	=	drag force
E_e	=	specific energy
E_p	=	specific power
e	=	Oswald span efficiency factor
G	=	gear reduction ratio
H	=	motor heating
h_{CL}	=	nondimensional location of the center of lift, $x_{\text{CL}}/\bar{c}$

Copyright © 2006 by Thomas J. Mueller, Gabriel E. Torres, and Donald W. Srull. Published by the American Institute of Aeronautics and Astronautics, Inc., with permission.

*Roth-Gibson Professor, Fellow AIAA.

†Wasp Project Manager, AeroVironment, Inc., Simi Valley, CA 93063.

‡Consultant to the Naval Research Laboratory, Washington, D.C.

I = current
I_S = motor stall current
I_0 = motor no-load current
K = induced drag factor
K_{Prop} = propeller load constant
K_p = potential component of lift for Polhamus/Lamar approximation
K_v = vortex (nonlinear) component of lift for Polhamus/Lamar approximation or dc motor speed control
L = lift
M = pitching moment
N = motor rotation speed
N_0 = motor no-load rotation speed
n = number
P = shaft power delivered to propeller
P_A = power available from propeller
P_B = power supplied by battery
$P_{\max}$ = maximum power
P_R = power required for level unaccelerated flight
R = resistance
Re = Reynolds number based on wing root chord, $\rho V\infty c_{\mathrm{root}}/\mu$
S = wing area
T = thrust
T_M = motor torque
T_{MS} = motor stall torque
T_{M0} = motor no-load torque
T_R = thrust required
t = time
V = flight speed
V_P = propeller pitch speed
V_S = stall speed
$V\infty$ = freestream velocity
v = voltage
W = weight
W_B = battery weight
W_0 = gross weight
W_1 = weight without fuel
x = distance along chord
x_{CL} = chordwise location of center of lift
$x_{\bar{c}/4}$ = chordwise location of $\bar{c}/4$
x_e = chordwise location for leading-edge and side-edge vortex forces
x_p = chordwise location for potential force
α = angle of attack, deg
$\alpha_{C_{\mathrm{L\,max}}}$ = angle of attack at maximum lift
Δy = leading-edge shape parameter
η = airfoil lift-curve slope efficiency parameter
η_G = gearbox efficiency
η_M = motor efficiency

η_p = propeller efficiency, P_A/P
$\Lambda_{c/2}$ = sweep angle at half-chord
μ = viscosity
ρ = air density
τ = Glauert correction factor

2.1 Introduction

The airfoil section and wing planform of the lifting surface are critically important to the performance of all flying vehicles. Therefore, all small MAVs share the ultimate goal of a stable and controllable vehicle with maximum aerodynamic efficiency. The aerodynamic efficiency is primarily determined by the lift-to-drag ratio of the wing. Most small vehicles are designed for maximum range or endurance at a given cruising speed that is strongly influenced by the lift-to-drag ratio (i.e., aerodynamic efficiency). This chapter discusses the progression from high Reynolds numbers to the low Reynolds numbers encountered in the design of micro air vehicles. Results of the experiments and analysis of low-aspect-ratio (LAR) rigid wings at low Reynolds numbers performed at the University of Notre Dame are presented in some detail because they represent the most extensive study of planform, aspect ratio, and Reynolds number published. Low-aspect-ratio wing theory and physical and numerical experiments are presented as well as their use in the design of MAVs. Because this field is expanding rapidly, not every one of the recent studies could be included. The studies presented were selected to be useful in understanding the unique aerodynamic characteristics of low-aspect-ratio wings at low Reynolds numbers for the design of micro air vehicles.

2.1.1 High-Reynolds-Number Low-Aspect-Ratio Wings

The design of efficient MAVs in 1995 was hindered by the lack of understanding of the aerodynamics associated with LAR wings (aspect ratios less than 2.0) operating at low Reynolds numbers. Classical aerodynamic theory, although accurate for full-scale airplane analysis, was found to be inadequate for this combination of aspect ratio and Reynolds numbers. Also, LAR wings exhibit unique aerodynamic properties such as high values of angle of attack at maximum lift coefficient and nonlinear lift vs angle of attack. The aerodynamics of LAR wings at low Reynolds numbers received little attention before the interest in MAVs developed.

Low-aspect-ratio wings were extensively researched at higher Reynolds numbers in the form of delta wings at subsonic, transonic, and supersonic speeds [1] and [2]. Many of these studies are focused on the influence of vortex formation and breakdown on the unsteady aerodynamics of LAR wings at high angles of attack. Some information was available, however, regarding nondelta LAR wings, with much of the research having been done between the 1930s and the 1960s. Zimmerman [3] and [4], Bartlett and Vidal [5], Wadlin et al. [6], and Neyland and Stolyarov [7] performed experiments of LAR wings at Reynolds numbers greater than 5×10^5. Theoretical and analytical treatises of LAR wing aerodynamics have been performed by Bollay [8], Weinig [9], Lawrence [10], and Ermolenko and Barinov [11], Bera and Suresh [12], Polhamus [13] and [14], and Rajan and Shashidhar [15]. The most complete analysis and review of LAR wings was performed by Hoerner [16] and Hoerner and Borst [17] in the two volume series on lift and drag. Hoerner reviewed many of the theories developed for LAR wings of nondelta planforms. A variety of correlations as well as analytical methods were presented and compared with the available experimental data of the time. Although the information presented by Hoerner and Borst [16] and [17] corresponds to higher Reynolds numbers than those of interest for MAVs, it was shown by Torres [18] that the aerodynamic theory still holds. This theory correctly predicts that a finite wing of a given aspect ratio generates lift from counter-rotating vortical structures near the wing tips. These vortices strengthen as the angle of attack increases. For a low-aspect-ratio wing, the tip vortices might be present over most of the wing area and therefore exert great influence on its aerodynamic characteristics. Wings of aspect ratio below about 1.5 can be considered to have two sources of lift: linear and nonlinear. The linear lift is created by circulation around the wing and is what is typically thought of as lift in higher-aspect-ratio wings. The nonlinear lift is created by the formation of low-pressure cells on the wing's top surface by the tip vortices, such as those observed in delta wings at high angles of attack. This nonlinear effect increases the lift-curve slope as the angle of attack increases, and it is considered to be responsible for the high value of stall angle of attack.

2.1.2 Low-Reynolds-Number High-Aspect-Ratio Wings

No discussion of low-Reynolds-number airfoil aerodynamics would be complete without a reference to the pioneering research that Schmitz performed in the 1930s (published in Germany in 1942 and translated into English in 1967) [19]. He measured the aerodynamic characteristics of several airfoil geometries including a thin flat plate and a thin cambered plate over a Reynolds number range from 2.1×10^4 to 16.8×10^4. His detailed study of the boundary-layer development, including hysteresis, was important for those who became interested in this field decades later. By the 1980s it was well known that the performance of airfoils designed for chord Reynolds numbers greater than 5×10^5 (e.g., McMasters and Henderson [20], Lissaman [21], and Mueller [22]) deteriorates rapidly as the chord Reynolds number decreases below 5×10^5 because of laminar boundary-layer separation. Furthermore, the performance of three-dimensional wings (i.e., finite wings), as measured by $(C_L/C_D)_{\max}$, is less than that for airfoils.

The survey of low-Reynolds-number airfoils by Carmichael [23], though two-and-a-half decades old, is a very useful starting point in the description of the character of the flow over airfoils over the range of Reynolds numbers of interest here. The following discussion of flow regimes from $3 \times 10^4 \leq Re \leq 5 \times 10^5$ is a modified version of Carmichael's original work:

1) The range $3 \times 10^4 \leq Re \leq 7 \times 10^4$ is of great interest to MAV designers as well as model aircraft builders. The choice of an airfoil section is very important in this regime because relatively thick airfoils (i.e., 6% and above) can have significant hysteresis in the lift-and-drag forces caused by laminar separation with transition to turbulent flow [Ref. 22]. Below chord Reynolds numbers of about 5×10^4, the free shear layer after laminar separation normally does not transition to turbulent flow in time to reattach to the airfoil surface. When this separation point reaches the leading edge, the lift decreases abruptly, the drag increases abruptly, and the airfoil is stalled.

2) At Reynolds numbers in the range from 7×10^4 to 2×10^5, extensive laminar flow can be obtained, and therefore airfoil performance improves unless the

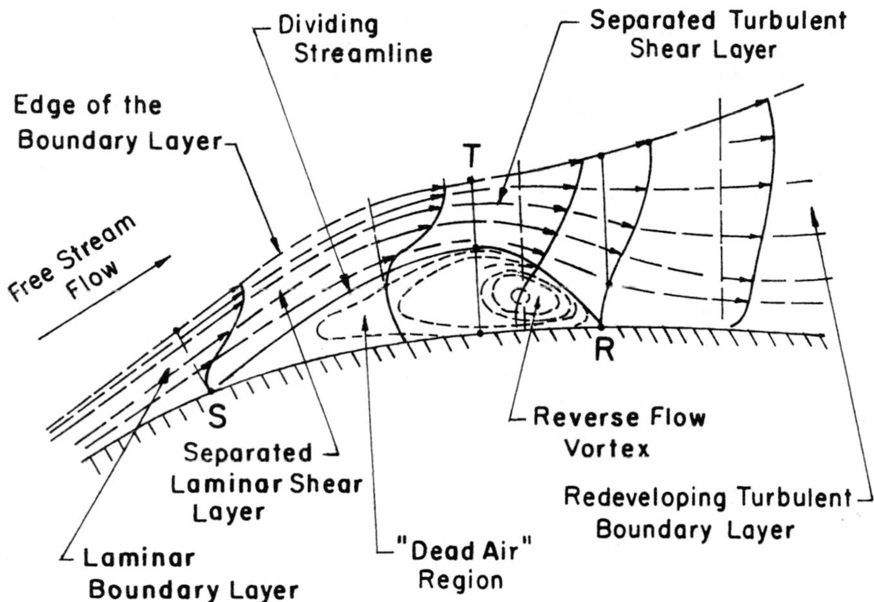

Fig. 2.1 Time-averaged features of a transitional separation bubble [24].

laminar separation bubble still presents a problem for a particular airfoil. Many MAVs and small UAVs fly in this range.

3) For Reynolds numbers above 2×10^5, airfoil performance improves significantly because the parasite drag caused by the separation bubble decreases as the bubble gets shorter. There is a great deal of experience available from large soaring birds, large radio-controlled model airplanes, and human-powered airplanes to support this claim.

The postseparation behavior of the laminar boundary-layer accounts for the deterioration in airfoil performance at low Reynolds numbers. This deterioration is exhibited in an increase in drag and decrease in lift. In this flow regime, the boundary layer on an airfoil often remains laminar downstream of the point of minimum pressure and then separates to form a free shear layer. At Reynolds numbers below 5×10^4, this separated shear layer does not reattach. At Reynolds numbers greater than 5×10^4, transition usually takes place in the separated shear layer. Provided the adverse pressure gradient is not too large, the flow can recover sufficient energy through entrainment after transition to turbulent flow to reattach to the airfoil surface. On a time-averaged basis, a region of recirculating flow is formed, as shown in Fig. 2.1 [24]. A visualization of the laminar separation on the Wortmann FX 63–137 airfoil using titanium tetrachloride ($TiCl_4$) injected at the surface downstream of the separation point [25] is shown in Fig. 2.2. The smooth surface of the airfoil causes a reflected image of the laminar separation and subsequent breakdown into turbulence. Laminar separation occurs at 5% of the chord and is visible. The $TiCl_4$ produces a white smoke-like gas when it comes in contact with the moisture in the air. It was injected through a very small hole downstream of separation (i.e., at 11% of the chord), and the reverse flow carried

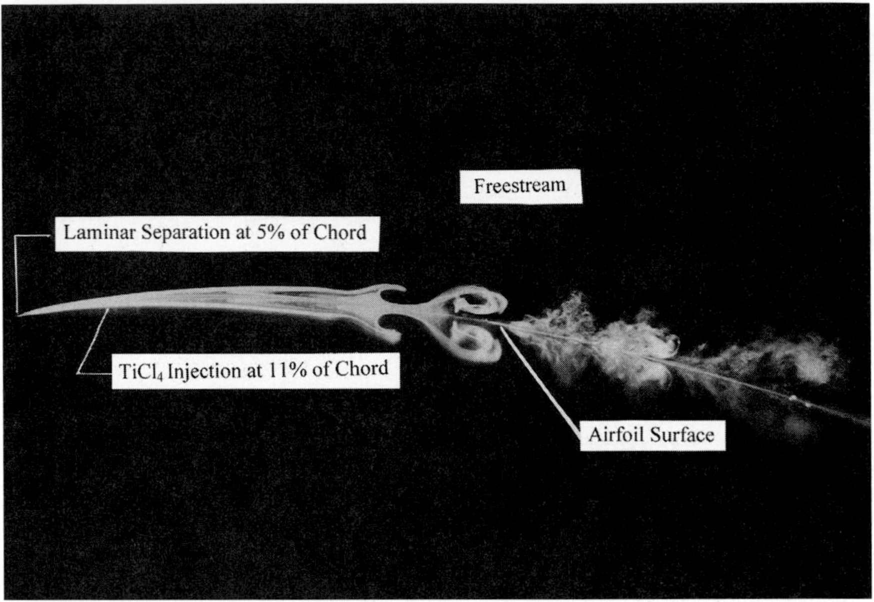

Fig. 2.2 Flow visualization of a transitional separation bubble on the upper surface of a Wortmann FX 63 – 137 airfoil at 13-deg angle of attack at $Re = 10^5$ [25].

the smoke upstream to the separation point. The laminar portion of the bubble is clearly visible. Although the smoke is considered a streakline, if the flow is steady, it coincides with a streamline. The laminar portion of this bubble is steady, and therefore the upper edge of smoke coincides with the dividing streamline. The dividing streamline is defined such that it divides the bubble from the freestream flow. When the free shear layer becomes unstable, at about 21% of the chord, the beginning of the transition region is visible. As this process proceeds, the flow is no longer steady, and the smoke dissipates in the turbulent shear layer that reattaches to the airfoil surface. This is an instantaneous picture; the mean location of reattachment is impossible to locate, and static-pressure, hot-wire, or laser-Doppler-velocimetry (LDV) measurements are needed to determine this location. Because the bubble acts as a boundary-layer trip, the phenomenon is often referred to as a transitional separation bubble. At low Reynolds numbers, the transitional bubble can occupy 15–40% of the airfoil surface and is referred to as a long bubble. The separation bubble often has a dramatic effect on the stalling characteristics (i.e., the drastic decrease in lift and increase in drag) of airfoils. When a short bubble is present, usually at high Reynolds numbers, the lift increases linearly with angle of attack until stall occurs. This is referred to as the bursting of the short bubble. If a long bubble forms on the surface, usually at low Reynolds numbers, stall occurs when it has extended to the trailing edge. The behavior of the separation bubble is also a factor in the occurrence of hysteresis for some airfoils [22]. The flow is unsteady downstream of the maximum vertical displacement of the bubble. In contrast, flow visualization and hot-wire studies that demonstrate a relatively steady flow upstream of the maximum vertical displacement labeled T in

Fig. 2.1 have been conducted (i.e., where transition to turbulent flow is assumed to take place) [26] and [27]. Hence, accurate prediction of its existence and extent, relative to airfoil performance, is necessary in the design of efficient low speed airfoils.

The University of Notre Dame began experiments on low-Reynolds-number airfoil aerodynamics in 1977 under the sponsorship of NASA Langley Research Center. This mostly experimental program continued until 1989 and focused on the laminar separation, transition, and reattachment near the leading edge of airfoils. Measurement techniques were examined at low Reynolds numbers for their influence on drag determination. The effects of freestream disturbances on lift-and-drag performance including hysteresis were also studied. This program was greatly expanded from 1981 to 1989 with the support of the U.S. Naval Research Laboratory and the Office of Naval Research. The effects of different types of surface roughness as well as leading- and trailing-edge flaps, interference from the wake of an upstream airfoil and finite wing, and unsteady approach flow on the performance were studied. A number of airfoils were used, and most of the experiments were performed in the chord-Reynolds-number range from about 7×10^4 to about 3×10^5. Smoke tube, smoke wire, and surface oil-flow-visualization techniques, surface-pressure, hot-wire, LDV, and lift, drag, and pitching moment measurements were used to study these airfoil situations. The First International Conference on low-Reynolds-number airfoil aerodynamics was held at the University of Notre Dame in 1985 [28]. There were papers from Germany, France, and The Netherlands, although the majority of the 23 papers were presented by U.S. participants. An international conference on the aerodynamics at low Reynolds numbers sponsored by The Royal Aeronautical Society and the U.S. Office of Naval Research, London Branch was held in London in 1986. This conference attracted a large number of European and U.S. participants, and the proceedings, which included 44 papers, were published in three volumes [29]. Another international conference was held at the University of Notre Dame in 1989 to assess the state of the art in this field by both U.S. and European scientists and engineers. There were 34 papers and participants from England, Scotland, Belgium, Germany, Italy, and Canada as well as the United States. The proceedings of this conference were published by Springer-Verlag [30].

M. S. Selig and his colleagues began their airfoil design and test program in the late 1980s [31]. Since then, they have tested a large number of airfoils, including many of Selig's own design, in the Reynolds number range from 6×10^4 to 5×10^5 [31–37]. These results, as well as other published data, have been very useful as a starting point for the design of many small UAVs. However, most of the second-generation small UAVs use airfoil sections designed specifically for their application. Although the experimental data obtained for two-dimensional airfoils and large-aspect-ratio wings at low Reynolds numbers were not useful for the design of micro air vehicles, the basic understanding of boundary-layer characteristics and measurement techniques developed provided valuable experience when LAR wings became of interest.

2.1.3 Low-Reynolds-Number Low-Aspect-Ratio Wings

The aerodynamic characteristics of lifting surfaces are critically important in the design of any flying vehicle. Because of the lack of these type of data for

MAV design, each design team must obtain these data by physical or numerical experiments. Although measuring or calculating the lift for a given airfoil cross section and planform is usually not difficult, the determination of the drag is much more difficult because it is usually an order of magnitude smaller than the lift. A limited amount of lift, drag, and pitching-moment data has been published for a few wing cross sections and planforms. There is a need for data that deal with the lateral motions of LAR wings (i.e., rolling, yawing, and sideslipping as well as combinations of these motions) because of their importance in stability and control. The effects of wing geometry such as dihedral and twist on lateral motions and the use of a vertical control surface to stabilize the vehicle, also need to be studied. Many designers of rigid wing MAVs use the Eppler Airfoil Design and Analysis code [38–40] or the Drela XFOIL code [41] and [42] to obtain the wing cross section, then select a planform and experiment to arrive at the final three-dimensional wing shape. Information on these codes can be obtained from an Internet search for "EPPLER code" and/or "XFOIL code." Because most fixed-wing MAVs are essentially flying wings, reflex airfoils such as the Eppler 184 [43], the Selig S5010 [35] and [44], or S5020 [44] have been selected as a starting point by some designers. The Selig S1052 and S1054 are reflex airfoils that might also be of interest [45].

2.2 Rigid Wings

The first-generation MAVs were rigid-wing vehicles, for example, the U.S. Naval Research Laboratory (USNRL) MITE series [46], and the AeroVironment series [47] that evolved into the Black Widow. The questions about the aerodynamic characteristics of a 15.4-cm (6-in.) vehicle that would fly at chord Reynolds numbers around 10^5 were answered as the designs progressed. The MITE vehicles were designed using computational fluid dynamics (CFD) to determine the aerodynamic characteristics of their proposed designs while the AeroVironment vehicles used wind-tunnel experiments.

2.2.1 Thin Flat-Plate Wings

Mueller [48] and Pelletier and Mueller [49] studied the aerodynamics of low-aspect-ratio 2% thick rectangular flat and 4% circular cambered airfoils and semi-span wings at Reynolds numbers between 6×10^4 and 2×10^5. The experiments were performed in a low-speed, low-turbulence (i.e., about 0.05%), indraft wind tunnel using a strain-gauge platform force balance especially designed for the range of forces encountered on these small models at low Reynolds numbers. These wind-tunnel studies examined the influence of two endplates for two-dimensional tests and one endplate for semispan tests by measuring the lift, drag, and quarter-chord pitching moment. All models tested were hung vertically from the force balance and held at the quarter-chord point. The sting, where it was exposed to the flow, was covered by a streamline covering. The gaps between the endplate(s) were approximately 0.8 mm (0.03 in.) [48] and [49]. The effects of freestream turbulence level and trailing-edge shape were also studied. As expected, the 4% cambered airfoil produced better performance than the rectangular flat plate, and the presence of endplates has a significant influence on the results because of the interaction of the boundary layer growing on the endplates and the flow around the airfoil. The freestream turbulence intensity in the tunnel between 0.25 and

1.3% had only a small effect, and the trailing-edge geometry (elliptical or sharp) also had a small effect on performance. The studies by Torres [18] and Torres and Mueller [50] and [51] included a large number of wing planforms and aspect ratios for thin flat-plate wings with elliptically shaped edges all around the model. A preliminary study of the effects of leading-edge shape was conducted by Torres [18] for a rectangular wing with $AR = 2.0$ and Reynolds numbers of 10^5 and 2×10^5. This wing had one of three leading-edge shapes: 5 to 1 elliptical, and a 10- or 20-deg wedge. The 10- and 20-deg wedge models were mounted with the wedge facing either up or down. Therefore a total of five leading-edge shapes were studied. Although the normal-force coefficient vs angle of attack showed only a small variation with leading-edge shape up to stall, the pitching-moment coefficient vs angle of attack showed large variations with leading-edge shape. The results of these experiments clearly indicated that MAV performance was sensitive to leading-edge shape and that more studies were necessary especially at aspect ratios less that two.

Flat-plate wings were chosen as a base in order to study the effects of planform shape, Reynolds number, and aspect ratio without the influence of camber. The four planforms studied were rectangular, Zimmerman, inverse Zimmerman, and elliptical. The Zimmerman planforms are characterized by two half-ellipses joined at either the quarter-chord line (Zimmerman) or the three-quarter line (inverse Zimmerman). The seven values of the aspect ratio studied were 0.50, 0.75, 1.0, 1.25, 1.50, 1.75, and 2.0 as shown in Fig. 2.3. Three primary values of Reynolds number were selected: 7×10^4, 10^5, and 1.4×10^5. Data were obtained for the key performance parameters, such as the lift-curve slope, drag characteristics, comparison with nonlinear lift theory, maximum lift coefficient, angle of attack at maximum lift coefficient, and location of the center of lift. Some tests were also performed at 2×10^5. The angle of attack of the wing models was varied from $\alpha = -10$ deg to either $\alpha = 40$ deg (for wings of $AR \geq 1.50$) or $\alpha = 50$ deg (for wings of $AR \leq 1.25$). The range of α was adjusted whenever possible in order to measure α at $C_{L\,max}$. The wings were then brought back to $\alpha = 0$ deg to determine whether hysteresis was present. No hysteresis was observed in any of the measurements.

The effects of propeller-induced flow on the aerodynamic characteristics, for a tractor propulsion system, were not included in these experiments. Studies of this type demonstrate that propeller-induced flow can have very significant effects on the aerodynamics for tractor installations [52].

Lift and drag are nondimensionalized by the area of the wing and the measured dynamic pressure at each angle of attack. Pitching moment was determined from sting balance measurements and is reported at the quarter-chord location of the mean aerodynamic chord of each wing. It is nondimensionalized by the wing area, the dynamic pressure, and the mean aerodynamic chord (MAC) of the wing. The mean aerodynamic chord represents an average chord, which, when multiplied by the product of the average pitching moment coefficient, the dynamic pressure, and the wing area, gives the moment for the entire wing [18] and [53].

For the later analysis, the moment coefficient was transposed to the leading edge by adding the contributions of the lift and drag coefficients. Unless otherwise noted, moment measurements should be taken to be at the quarter-chord of the MAC.

Force and moment coefficients presented in this work have all been corrected for wind-tunnel blockage (solid blockage, wake blockage, and streamline curvature) according to the techniques presented by Pankhurst and Holder [54] and Barlow,

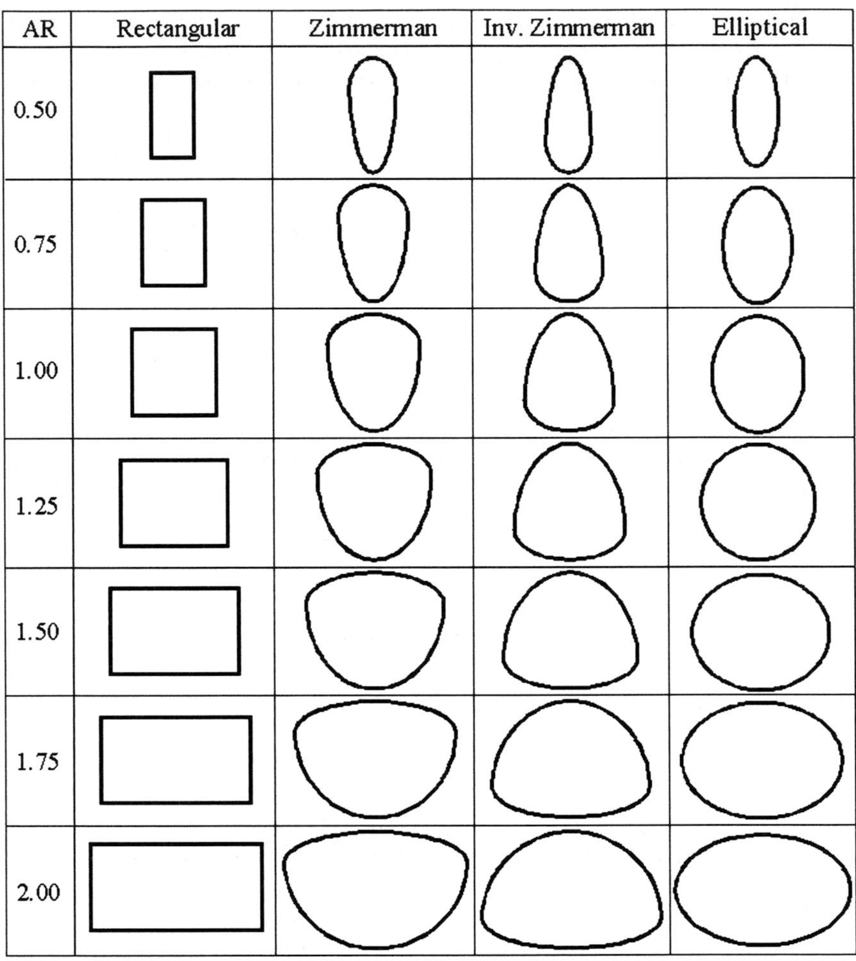

Fig. 2.3 Wing models used in experiments.

et al. [55]. The magnitude of the blockage effects varies depending on the size of the wing model, the wing's aspect ratio, and the angle of attack. The largest blockage corrections were found to correspond generally to wings of aspect ratio 2.00 at angles of attack near stall. For these extreme cases, the blockage correction factors for lift, drag, and pitching-moment coefficients are approximately 10% (meaning that the magnitude of the corrected value is 10% smaller than the magnitude of the uncorrected value). At lower angles of attack, the effects are correspondingly smaller (for example, 6% at 10-deg angle of attack for the rectangular wing of $AR =$ 2.00). The Kline–McClintock technique [56] for error propagation was used to evaluate all uncertainties in the aerodynamic coefficients. Percentage uncertainties are in the order of 5% for C_L, C_D, and C_M. The uncertainty in the angle of attack was determined to be of the order of ± 0.5 to 0.7 deg for most test cases.

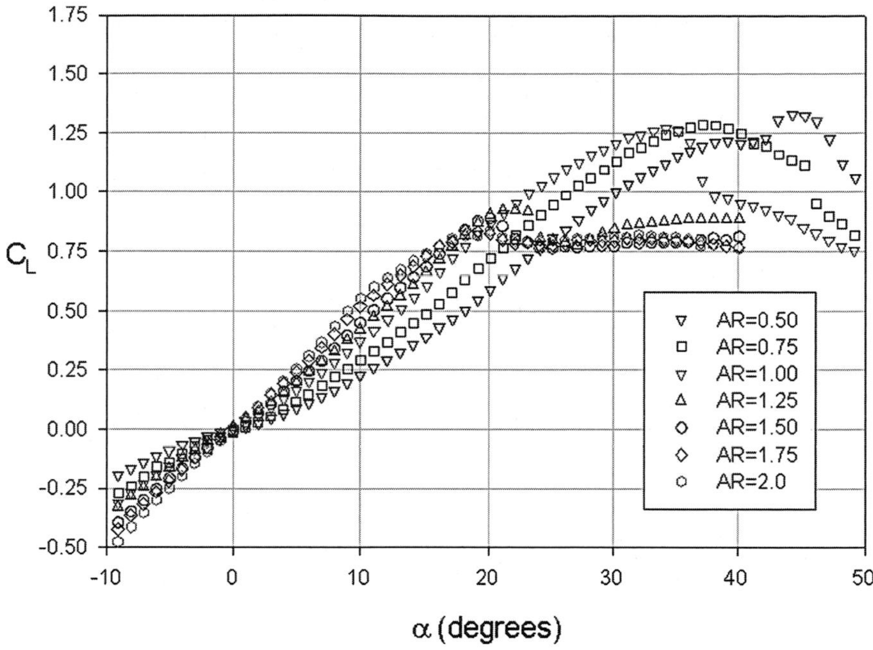

Fig. 2.4 C_L **vs** α **for rectangular planforms at** $Re = 10^5$**.**

2.2.2 General Observations

A complete set of these wind-tunnel data is available in Ref. 18. The most significant aerodynamic characteristics of the thin flat-plate LAR wings just described are as follows:

1) Wings of aspect ratio below 1.25 have highly nonlinear lift curves that are characterized by high values of $\alpha_{C_{L\max}}$ and nonlinear lift-curve slopes as shown in Fig. 2.4 for rectangular planform wings.

2) Above an aspect ratio of 1.25, most planforms exhibit lift curves that are more linear. As predicted by theory, the higher the aspect ratio is, the more linear the relationship between lift and angle of attack.

3) For aspect ratios below 1.00, a performance advantage is seen for the rectangular, inverse Zimmerman and elliptical planforms over the Zimmerman planform at angles of attack greater than 15 deg. An example of this is shown in the C_L vs C_D plot in Fig. 2.5 for an aspect ratio of 0.75.

4) Above aspect ratios of 1.5, the difference between planforms becomes less clear. For higher aspect ratios, the elliptical planform is seen to be more efficient (higher lift and lower drag) than the other planforms. This observation is an indication that as the aspect ratio increases, classical aerodynamic theories become effective, especially in terms of the higher efficiency of elliptical planforms.

5) Aspect ratio is by far the most important parameter affecting the aerodynamic characteristics of LAR wings at low Reynolds number. Wing planform is the next most important factor, followed by Reynolds number.

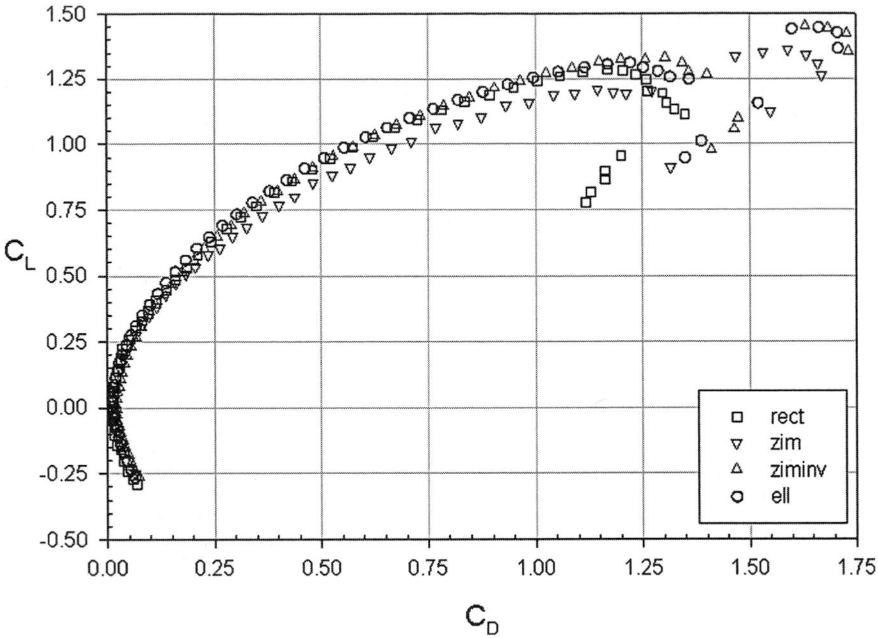

Fig. 2.5 C_L vs C_D for all planforms at $AR = 0.75$ and at $Re = 10^5$.

6) Hysteresis was not found to be present for any of the measurements made in this study. It is expected that the low thickness-to-chord ratio of the wings considered is responsible for this lack of hysteresis.

2.2.3 Lift-Curve Slope

One way of comparing the performance of different planforms is to compare their lift-curve slopes. However, the inherently nonlinear characteristics of low-aspect-ratio wings make defining a lift-curve slope dificult. Rather than assuming a linear relationship, a value for $C_{L\alpha}$ was calculated by first fitting (in a least-square sense) a second-degree polynomial to the data of lift coefficient vs angle of attack for α below $\alpha_{C_{L\,\max}}$. The first derivative of this polynomial with respect to angle of attack evaluated at $\alpha = 0$ deg is used to define $C_{L\alpha}$.

The values of $C_{L\alpha}$ were compared with a number of theoretical predictions of the lift-curve slope. The first one is the classic equation for $C_{L\alpha}$ originating from Prandtl's lifting-line theory:

$$C_{L\alpha} = \left(\frac{1}{57.3}\right) \frac{C_{l\alpha}}{1 + (C_{l\alpha}/\pi AR)(1+\tau)} \left(\frac{1}{\text{deg}}\right) \tag{2.1}$$

where the Glauert correction factor τ is a correction factor used to relate the lift-curve slope for an arbitrary lift distribution to the elliptical distribution [57]. The value of $C_{l\alpha}$ was taken to be $C_{l\alpha} = 5.3743$ rad based on the average of experimentally determined two-dimensional slopes of flat-plate infnite wings with

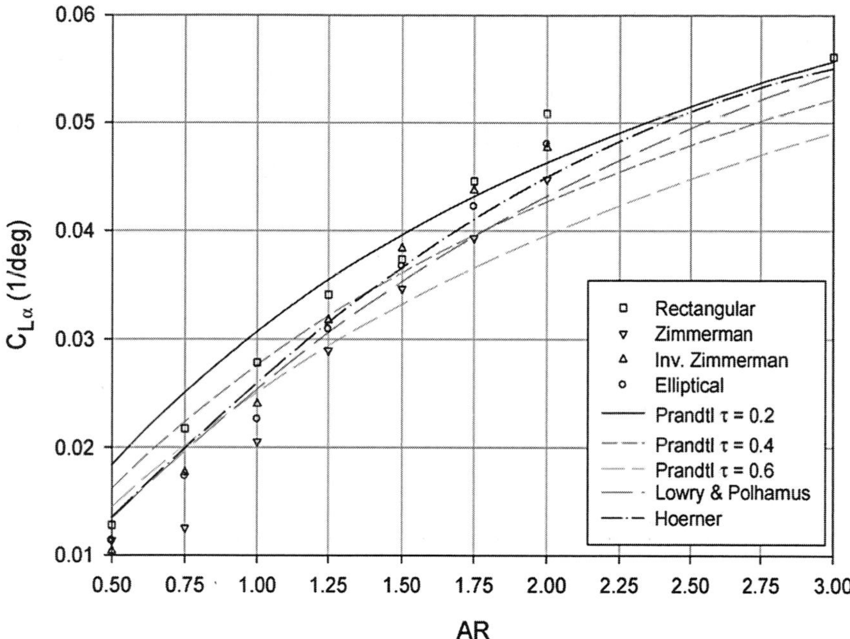

Fig. 2.6 Average lift-curve slope, where $C_{L\alpha}$ vs AR, at $Re = 10^5$.

the same thickness-to-chord ratio and leading-edge shape as the wings used in this work [49].

The second equation used is that discussed by Lowry and Polhamus [58], which is proposed to be more accurate for small aspect ratios (less than 2):

$$C_{L\alpha} = \left(\frac{1}{57.3}\right) \frac{2\pi AR}{2 + \sqrt{(AR^2/\eta^2)(1 + \tan^2 \Lambda_{c/2}) + 4}} \left(\frac{1}{\text{deg}}\right) \tag{2.2}$$

where $\eta = C_{l\alpha}/2\pi$ and $\Lambda_{c/2}$ is the sweep angle at midchord.

Finally, Hoerner and Borst [17] suggest that for thin rectangular plates of low AR (less than 2.5) the lift-curve slope is given by

$$C_{L\alpha} = [36.5/AR + 2AR]^{-1} \tag{2.3}$$

Values of $C_{L\alpha}$ at a Reynolds number of 10^5 are plotted vs AR in Fig. 2.6.

Figure 2.6 also includes data from Pelletier and Mueller [49] for an aspect ratio 3.0 rectangular wing model. The approximations of Eqs. (2.1)–(2.3) are also included. [For the curve of Eq. (2.2), a sweep angle of zero deg was used.]

From Fig. 2.6, it can be concluded that no single theoretical equation can accurately model the lift-curve slope of LAR wings at low Reynolds number. The closest agreement is seen to correspond to Hoerner's low-aspect ratio empirical relationship. Equation (1) with $\tau = 0.2$ or 0.4 also gives relatively good agreement. However, the relationship between lift-curve slope and aspect ratio is almost linear in nature and differs in character from the theoretical approximations. One possible explanation for this difference is the way in which lift-curve slope is defined. As described earlier, $C_{L\alpha}$ is defined as the first derivative of a quadratic equation that is a best fit to the data of lift coefficient vs angle of attack. If a straight line were fitted to the data instead, the value of the slope would be larger, especially for low aspect ratios, and would yield a better match between theory and experiment. This discrepancy underscores the fact that it is of limited value to apply an inherently linear theoretical model to the nonlinear aerodynamic characteristics of low-aspect-ratio wings. Other methods should be considered for a more complete analysis as shown in the next section.

2.2.4 Nonlinear Equations for Lift, Drag, and Pitching Moment

The goal of this section is to provide empirically determined constants from the data of the preceding section that when applied to the nonlinear equations will generate approximate curves of lift, drag, and pitching moment as a function of angle of attack, aspect ratio, and wing planform. These approximate equations will be very helpful for the design of MAVs with flat-plate wings.

One of these methods is the leading-edge-suction analogy adapted by Polhamus [13] and [14] for use in LAR wings (mostly of delta planforms). His method assumed that for delta wings in which the flow is mostly attached (before stall) the total force on the wing associated with the pressure required to stabilize the separated vortex cores is equivalent to the leading-edge suction force required to keep the flow around the sharp leading edges of the wing attached. Polhamus assumed that the leading-edge suction force acts in a direction normal to the plane of the wing and thus gives rise to additional lift associated with the vortices. Using this assumption, he concluded that the total force on a delta wing at moderately high angles of attack before stall is given by an addition of potential lift (calculated from any linear lifting-surface theory) and a vortex lift associated with the leading-edge suction force. He suggested the following equation for lift coefficient:

$$C_L = K_p \sin\alpha\ \cos^2\alpha + K_v \cos\alpha\ \sin^2\alpha \tag{2.4}$$

where K_p is a factor that depends on the aspect ratio, sweep angle, and leading-edge shape of the wing. K_v is a mostly constant factor equal to approximately π. Comparison of this equation with experimental data for delta wings gives remarkably accurate predictions for lift but relies heavily on the empirical factors K_p and K_v.

In 1976, Lamar [59] and [60] extended Polhamus' method to LAR wings of nondelta planforms by taking into account the influence of the side edges. Lamar considered the existence of a force analogous to the leading-edge suction force that acts at the side edges as the vortex cores roll up onto the upper side of the wing. Following the reasoning of Polhamus, Lamar deduced that the side-edge suction force also acts in the direction normal to the plane of the wing and therefore contributes to vortex lift. A modified lifting-surface theory was used by the authors to estimate the relative contributions to the lift force of the leading and side edges, and these coefficients are then used, in a form analogous to Polhamus' equation, to represent the vortex lift of the wing.

The Polhamus equation (2.4) is used for approximating the lift coefficient even though Lamar's analysis shows that K_v varies slightly as a function of aspect ratio and planform. It was found in the present analysis that very good agreement with experimental data could be accomplished by setting K_v constant. This simplification reduces the number of parameters required to recreate the LAR wing aerodynamic data. Therefore, K_v is set equal to π.

A nonlinear equation for the drag coefficient as a function of angle of attack was proposed by Lamar:

$$C_D = C_{D0} + K_p \sin^2\alpha \cos\alpha + K_v \sin^3\alpha \tag{2.5}$$

It was found, however, that agreement with the experimental data was not very good. A much better agreement was found to exist by using the classical equation from Prandtl for induced drag:

$$C_D = C_{D0} + K[C_L]^2 \tag{2.6}$$

where K is the induced drag factor, a function of aspect ratio, planform, and leading-edge geometry.

Lamar's nonlinear equation for the pitching moment about the leading edge is

$$(C_M)_{\text{LE}} = x_p K_p \sin\alpha\, \cos\alpha + x_e K_v \sin^2\alpha, \tag{2.7}$$

where x_p is the location along the chordwise direction at which the potential (linear) lift is assumed to be acting, normalized by the mean aerodynamic chord. Similarly, x_e is the location at which the nonlinear lift is assumed to act, again normalized by the chord. Note that in the Polhamus–Lamar analogy the nonlinear lift is considered to be generated by the leading and side edges, explaining the use of the e subscript for edge. In the present analysis, x_p is taken to correspond to the 25% chord location with respect to the MAC, and it is therefore set to $x_p = 0.25$ and x_e remains as the equation parameter.

The experimental data for the LAR wings presented in this work were used to calculate values for the parameters in Eqs. (2.4), (2.6) and (2.7) that best match the data. This was accomplished by iteratively finding the parameters that minimized the sum of the squares of the errors between the experimental data and the result obtained through the nonlinear equations. This procedure was applied for each of the four planforms and for each of the seven aspect ratios studied in this work. The nonlinear equation approximations are only applicable up to the angles of attack shown in Fig. 2.7. These angles are essentially the stall angles. At greater values of α, the highly nonlinear effects associated with prestall and stall conditions cannot be accurately modeled by the equations.

Figure 2.8 shows a typical comparison of the experimental data and Eqs. (2.4), (2.6) and (2.7) for the rectangular planform of aspect ratio 1.00. Curves for other aspect ratios and planforms are similar in terms of the ability of the equations to approximate the data. Figures 2.9 and 2.10 show comparisons of only the lift curve for all planforms and for aspect ratios between 0.75 and 2.00. Again, agreement between experiment and theory is seen to be good for all aspect ratios.

Figures 2.11–2.13 show the optimum values of K_p, K, and x_e for each of the planforms and aspect ratios. Figure 2.11 reveals that the K_p values for the Zimmerman planform for aspect ratios below 1.25 are noticeably lower than those

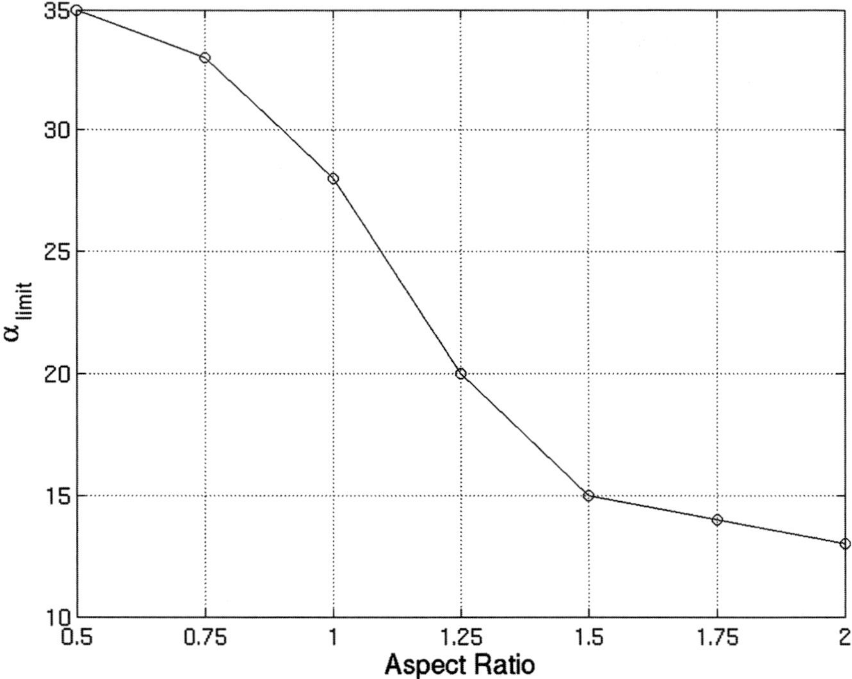

Fig. 2.7 Highest angle of attack for applicability of nonlinear equations.

of the rectangular or inverse Zimmerman planform. This once again points to the conclusion that the Zimmerman planform has lower lift performance for low *AR*. Above an *AR* of 1.5, the difference between the planforms becomes less drastic, but the Zimmerman planform still has a lower value of K_p.

The plot for induced drag parameter *K* (Fig. 2.12) shows that as the aspect ratio increases the induced drag factor decreases, as would be expected from linear theory. Differences between planforms are not as pronounced as in the K_p case; nevertheless, for lower aspect ratio, the Zimmerman planform's *K* value is generally higher than the other planforms. In terms of C_{D0}, the uncertainty in the data for the very low values of C_{D0} indicates no clear trend of this parameter with respect to aspect ratio or planform. The average value of C_{D0} for all wings was found to be approximately 0.015.

Finally, the plot of pitching-moment coefficient parameter x_e can be used as an indicator of the "nonlinearity" of a given planform. The parameter x_e corresponds to the location along the chordwise direction at which the nonlinear portion of the lift can be assumed to be acting. Wings with stronger nonlinear lift contributions will have greater values of x_e as seen in Fig. 2.13. This same trend will also be reinforced by the analysis in the section on Center of Lift.

2.2.5 Maximum Lift Coefficient

An important characteristic of LAR wings is their high value of $C_{\mathrm{L\,max}}$ and $\alpha_{C_{L\,\max}}$. Figures 2.14 and 2.15 plot the values of $C_{L\,\max}$ and $\alpha_{C_{L\,\max}}$ for each of the

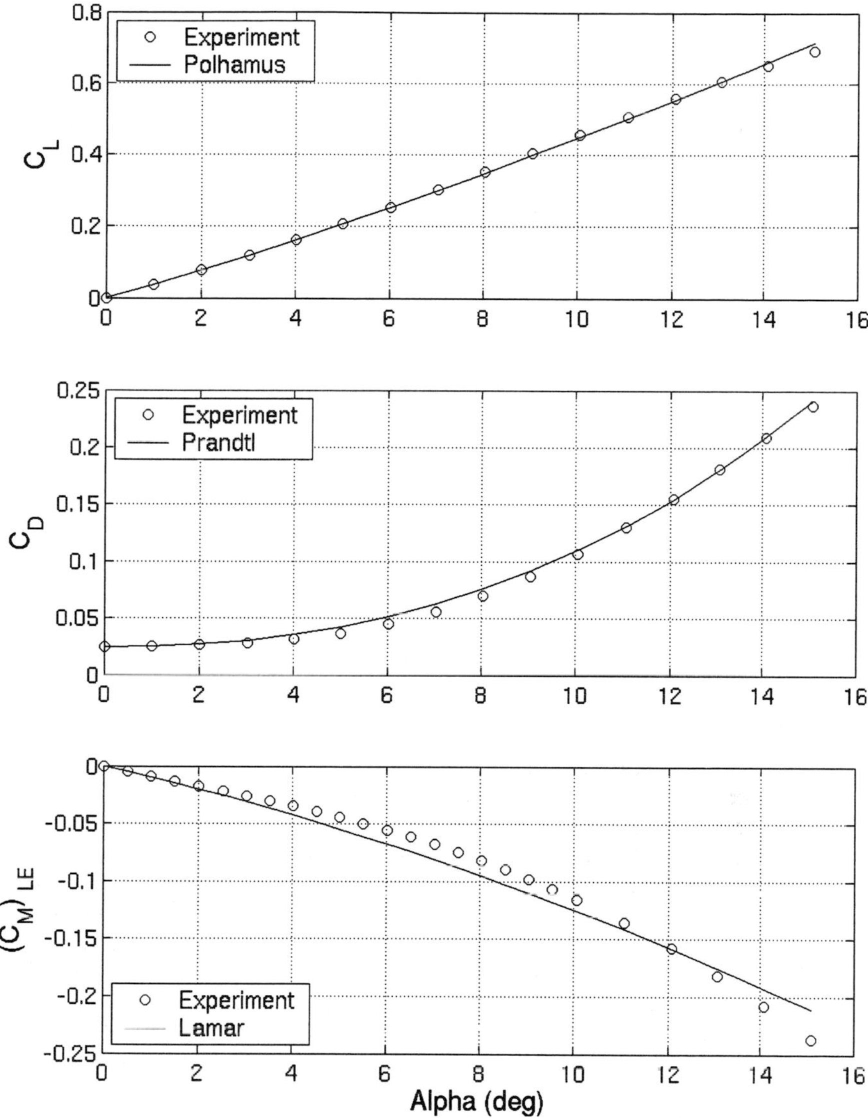

Fig. 2.8 Comparison with nonlinear equations, rectangular planform, $AR = 1.00$ at $Re = 10^5$.

four planforms as a function of aspect ratio. It can be seen from these graphs that aspect ratio and wing planform have significant influence in the maximum lift characteristics of LAR wings.

It is evident that a transition zone exists for $C_{L\max}$ and $\alpha_{C_{L\max}}$ in the aspect-ratio range between 1.25 and 1.50. The reason for the transition is proposed to be as follows: for wings of aspect ratio below 1.25, as the angle of attack approaches

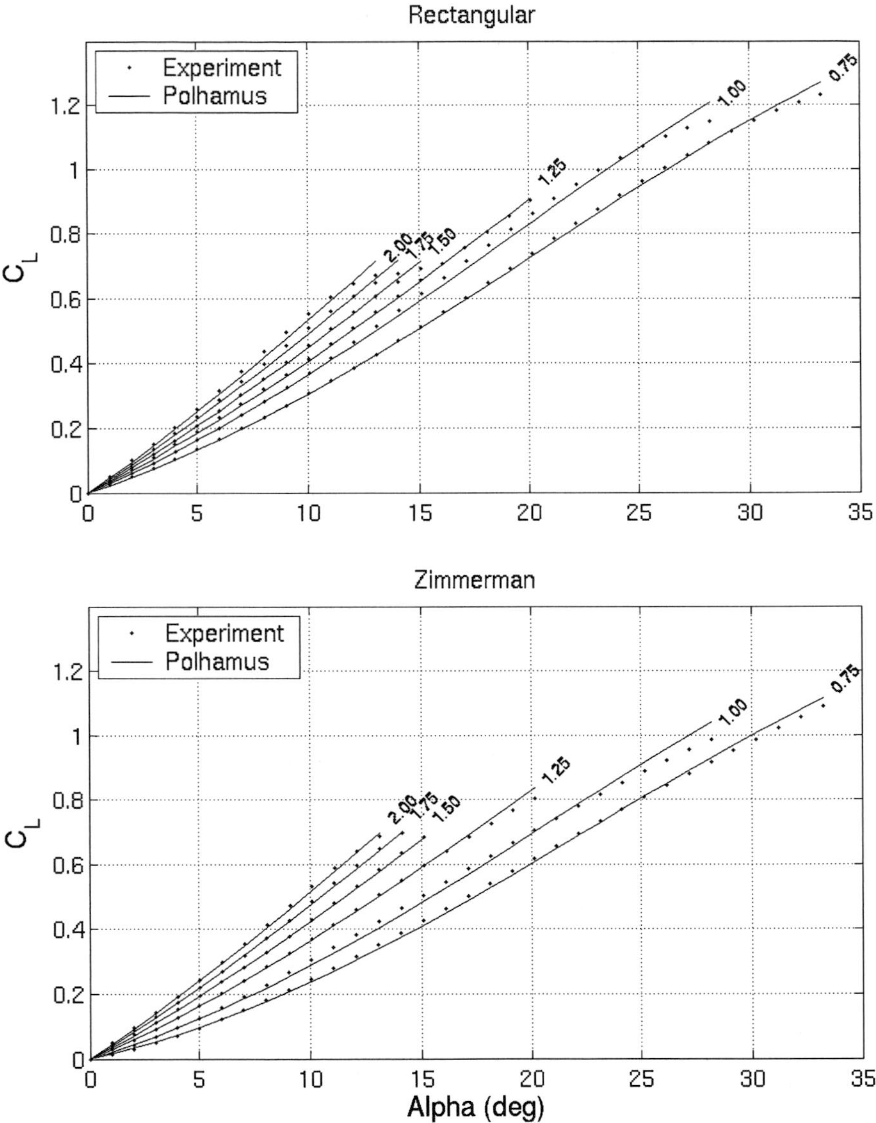

Fig. 2.9 Lift comparison with nonlinear equations, rectangular and Zimmerman planforms, at $Re = 10^5$.

$\alpha_{C_{L\max}}$, the flow induced by the wing-tip vortices is able to energize the flow on the upper surface of the wing and delay the onset of separation. As the aspect ratio increases, however, the strength of the wing-tip vortex structures on the upper surface of the wing decreases. Separation for these wings therefore occurs at lower angles of attack (and correspondingly lower lift coefficients).

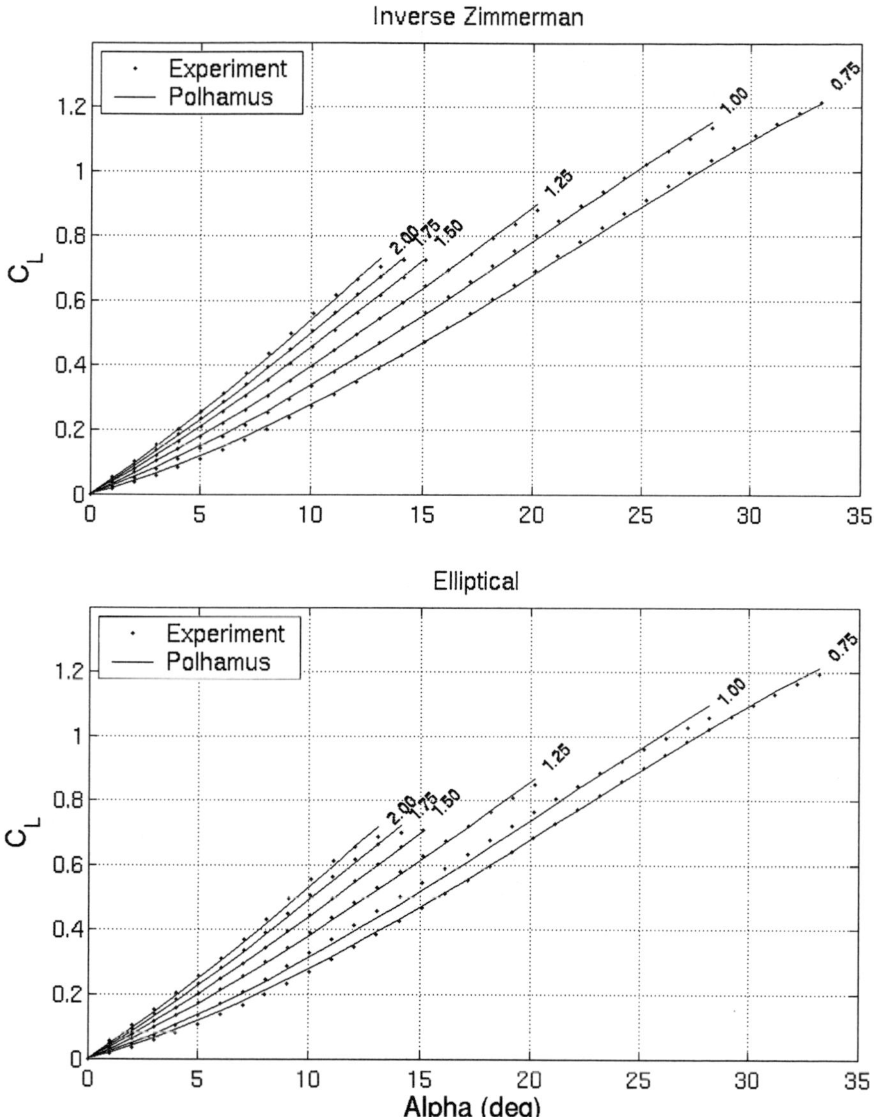

Fig. 2.10 Lift comparison with nonlinear equations, inverse Zimmerman and elliptical planforms, at $Re = 10^5$.

Also, the aspect ratio at which the transition in the $\alpha_{C_{L\max}}$ curve occurs is dependent on wing planform. The elliptical and Zimmerman wings have higher values than the rectangular and inverse Zimmerman wings for $AR = 1.50$. One possible explanation for the difference between planforms is based on the fact that for wings of high aspect ratio the mechanism responsible for lift generation near $\alpha_{C_{L\max}}$ is

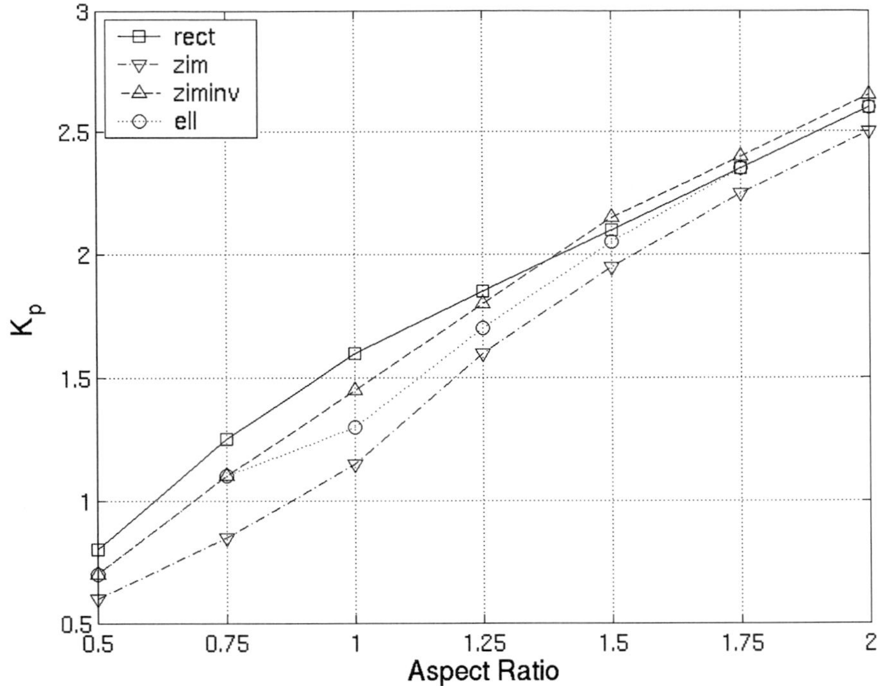

Fig. 2.11 Parameter K_p vs AR and planform.

highly dependent on the shape and planform geometry of the leading edge. The empirical/theoretical analysis proposed in DATCOM [61] was applied to this problem [see [18] for details]. It was concluded that for aspect ratios above 1.50 the predicted values of $\alpha_{C_{L\max}}$ were higher for elliptical and Zimmerman wings than for the other two planforms. It is likely that in the transition region of $C_{L\max}$ and $\alpha_{C_{L\max}}$ the leading-edge planform geometry of elliptical and Zimmerman wings is responsible for delaying the onset of separation. Nevertheless, the structure of the flow near $\alpha_{C_{L\max}}$ for low-aspect-ratio wings at low Reynolds numbers is highly complex. Many flow mechanisms can be interacting in this region.

2.2.6 Center of Lift

One key aspect of low-aspect-ratio wing aerodynamics is the location of the center of lift (or center of pressure) as a function of wing planform, aspect ratio, and angle of attack. The center-of-lift location yields important information about the nature of LAR wing aerodynamics, especially in the nonlinear lift regime.

The location of the center of lift can be calculated by using sting balance data of normal force and pitching moment taken about the $\bar{c}/4$ location of each wing. For this purpose, the parameter h_{CL} is defined as the distance between the leading edge of $\bar{c}$ and the center of pressure, divided by $\bar{c}$ (i.e., $h_{\mathrm{CL}} = x_{\mathrm{CL}}/\bar{c}$). Following the procedure outlined in [18]; a value of h_{CL} is calculated for each wing and aspect ratio at each angle of attack. The uncertainty in the value of h_{CL} was found to vary

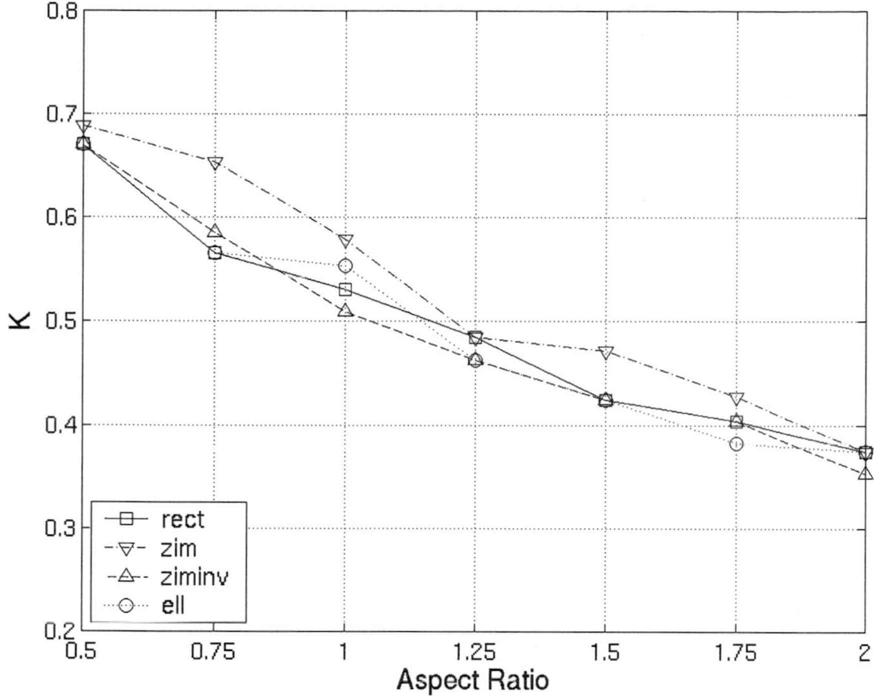

Fig. 2.12 Parameter *K* vs planform.

between 6% of h_{CL} for angles of attack close to 0 deg and 4% of h_{CL} for angles of attack above 15 deg.

Figures 2.16–2.22 provide great insight into the lift-generating mechanisms that operate in the regime of LAR wings at low Reynolds number. The most striking observation that should be made is the drastic shift in the location of the center of lift as angle of attack increases. These large movements of the center of lift with angle of attack present challenges for the trim and longitudinal stability of the subsequent MAV design. As an example, consider the rectangular wing of $AR = 1.00$ shown in Fig. 2.16. For this wing, h_{CL} is close to 0.17 at $\alpha = 0$ deg but increases to 0.40 at $\alpha = 40$ deg. It can be concluded that the chordwise location at which the lift force is acting shifts downstream as the angle of attack increases. This is an expected trend based on the linear-nonlinear theory of LAR wings. For low angles of attack, most of the lift generated by the wing is in the form of circulation lift, as seen in wings of high aspect ratio. Therefore, it is expected that the location of the center of lift will be close to the 25%-chord location at these angles. However, as the angle of attack increases, the lift mechanism shifts to that generated by the wing-tip vortices. These vortices generate lift by creating low-pressure sections on the upper surface of the wing, usually more on the downstream end of the wing. (The wing-tip vortices increase in size and strength as they travel downstream along the upper surface of the wing.) It would be expected that because the lift generated at higher angles of attack is caused by wing-tip vortices (which are stronger near the

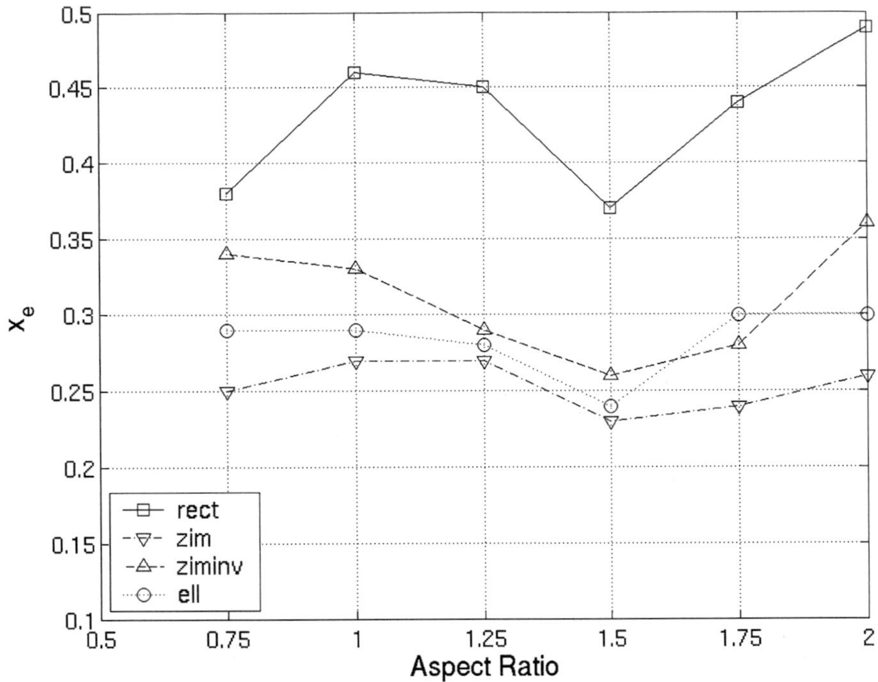

Fig. 2.13 Parameter x_e vs *AR* and planform.

trailing edge) the location of the center of lift would shift downstream. The amount by which the center of lift shifts towards the trailing edge is a direct indication of how much nonlinear lift is being generated by the wing. Wings of $AR = 0.75$ have a much steeper shift of center of lift with α than wings of $AR = 2.00$. If one were to plot h_{CL} vs α for a wing of $AR = 10.00$, this curve would be expected to be a horizontal line centered close to 0.25. The value of h_{CL} would not be expected to change until $\alpha_{C_{L\max}}$ is reached.

Another important conclusion that can be drawn from the center-of-lift analysis relates to the effect of planform and how some planforms are more "linear" than others. Zimmerman and elliptical planforms are seen to have a value of h_{CL}, which is generally lower than that of rectangular and inverse Zimmerman planforms. This is especially true for rectangular planforms, which are seen to have a markedly higher value of h_{CL} than the other planforms. For some high angles of attack, the Zimmerman, inverse Zimmerman, and elliptical planforms have relatively similar behavior. This is probably because Zimmerman and elliptical planforms have a weaker wing-tip vortex system than the one found in inverse Zimmerman and rectangular planforms. As such, it would be expected that the center of lift of the planforms with the weaker vortices would be farther forward than the center of lift of wings with strong vortices. This is indeed the case.

As mentioned earlier, the pitching moment coefficient vs angle of attack for a rectangular wing with $AR = 2.0$ is very sensitive to leading-edge shape. These

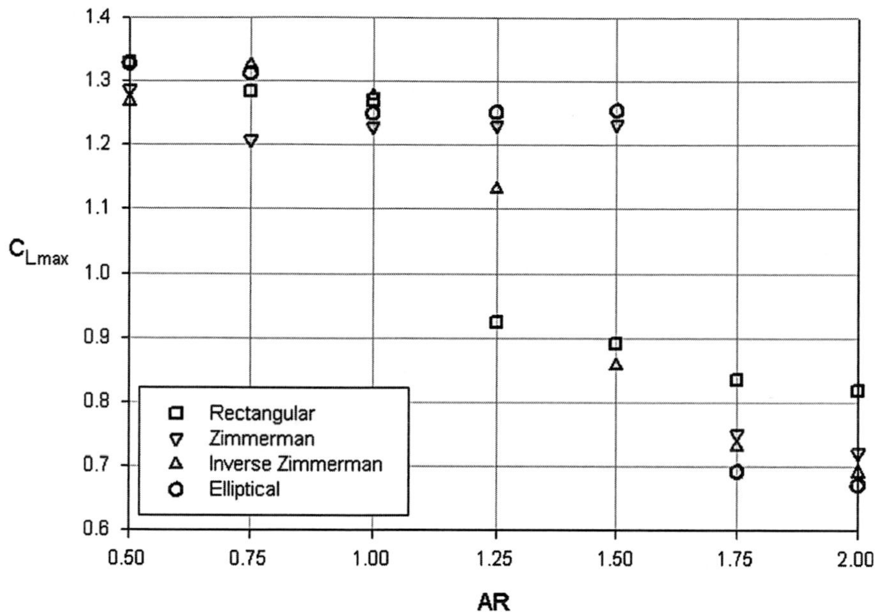

Fig. 2.14 $C_{L\max}$ **of four wing planforms and all aspect ratios for** $Re = 10^5$**.**

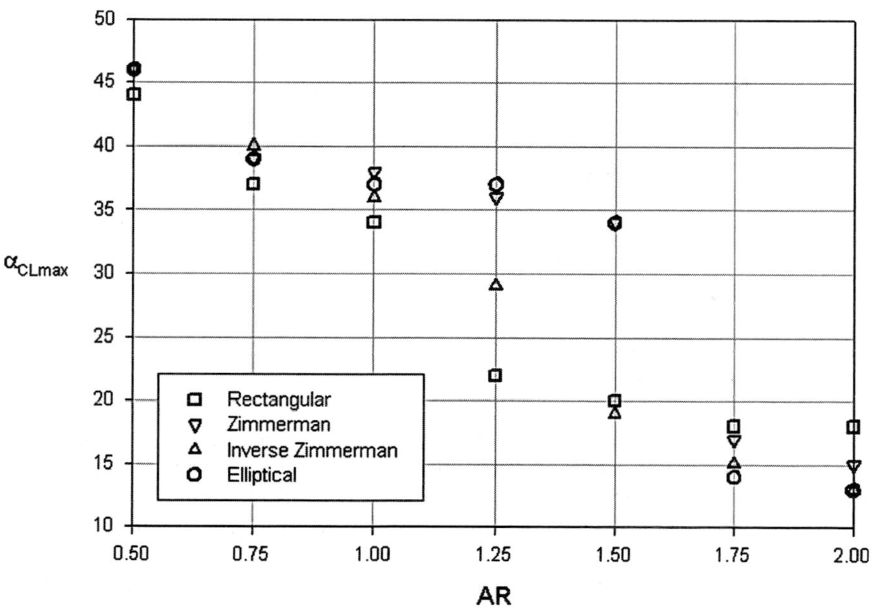

Fig. 2.15 $\alpha_{C_{L\max}}$ **of four wing planforms and all aspect ratios for** $Re = 10^5$**.**

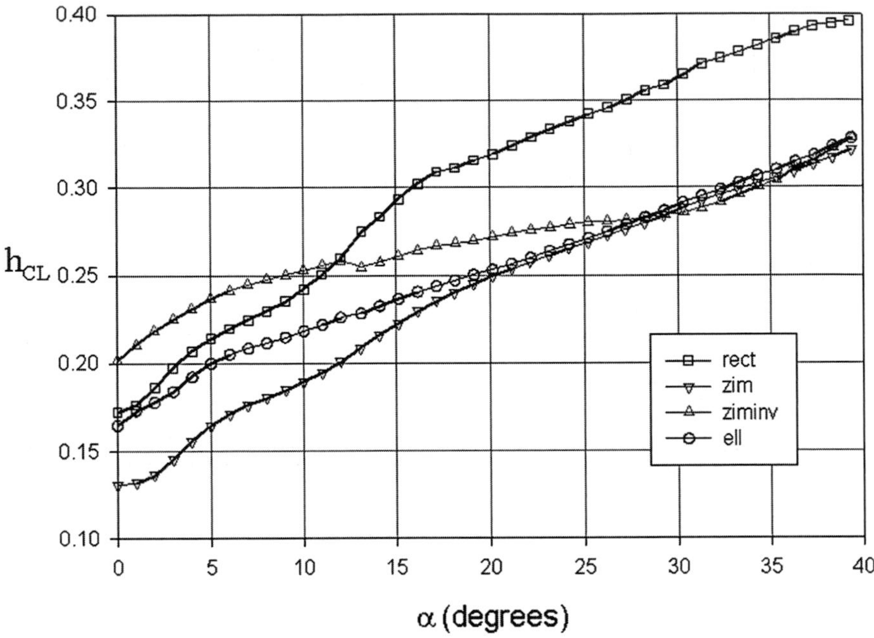

Fig. 2.16 Location of center of lift, with h_{CL} vs α for $AR = 1.00$ and $Re = 1.4 \times 10^5$.

large variations in C_M will produce large variations in the location of the center of lift. Further experiments are needed to clarify this result for lower aspect ratio wings.

2.2.7 Concluding Remarks for Zero Camber

The aerodynamic characteristics of wings of aspect ratio between 0.5 and 2.0 were experimentally determined at Reynolds numbers below 2×10^5. Of primary interest to this study were the effects of wing planform and aspect ratio on the forces and the pitching moment of LAR wings at low Reynolds number. Results showed large nonlinearities in the lift curves, especially for aspect ratios below 1.25. Wings of such low aspect ratios were also found to have high values of maximum lift coefficient and corresponding angle of attack.

Wing planform was determined to have significant aerodynamic effects. The rectangular, inverse Zimmerman, and elliptical planforms were about equally efficient, but the Zimmerman planform at high angles of attack had lower values of C_L/C_D for $AR = 1.0$. For higher values of AR, elliptical planforms were found to yield more favorable characteristics, especially at low to moderate angles of attack.

Additional analysis of the aerodynamic data revealed that analytical expressions for the lift-curve slope are not very accurate over the entire range of aspect ratios and planforms. Comparison of wind-tunnel data with nonlinear equations for lift, drag, and pitching-moment coefficient showed that, up to moderate angles of attack, the nonlinear equations modeled the experimental data quite well, provided that

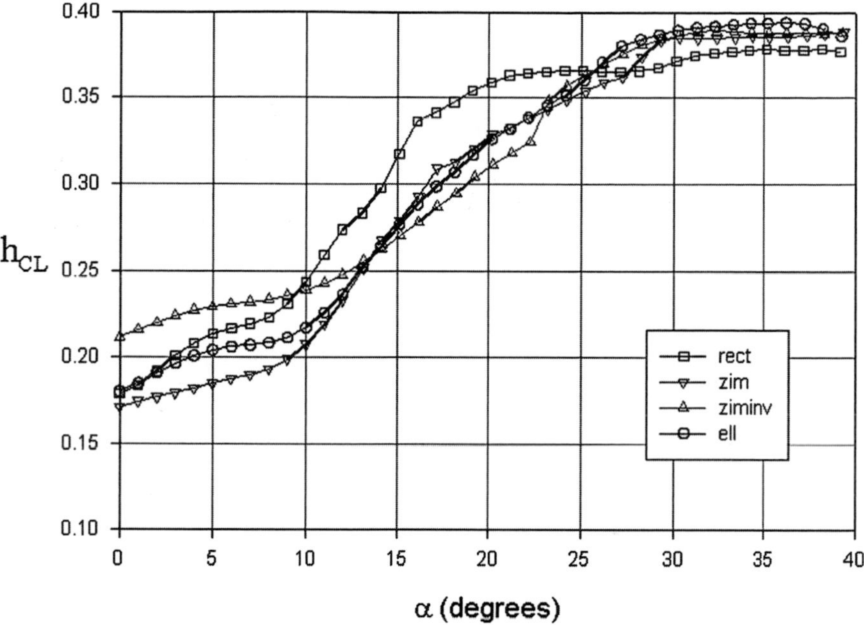

Fig. 2.17 Location of center of lift, with h_{CL} vs α for $AR = 1.50$ and $Re = 1.4 \times 10^5$.

the right equation parameters were used. Optimum values of the parameters to be used in these equations were determined and plotted as functions of aspect ratio and planform.

With respect to performance near $\alpha_{C_{L\max}}$, it was found that a transition exists in the curves of $C_{L\max}$ and $\alpha_{C_{L\max}}$ vs aspect ratio. This transition, which occurs near $AR = 1.25$, provides an indication of the transition between nonlinear and linear lift generation mechanisms at high angles of attack near stall. Also, rectangular/inverse Zimmerman planforms and elliptical/Zimmerman wings have extremely different values of $C_{L\max}$ and $\alpha_{C_{L\max}}$ for an aspect ratio of 1.50. A probable reason for this difference is proposed to be related to the planform geometry of the leading edge of elliptical and Zimmerman wings.

The location of the center of lift as a function of planform, aspect ratio, and angle of attack was determined using wind-tunnel data. This analysis provided great insight into the linear-nonlinear lift character of LAR wings. The location of the center of lift was found to shift towards the trailing edge of the wing as the angle of attack increased. It was proposed that the reason for this shift is related to the nonlinear lift generated by the wing-tip vortices, which are more dominant in the aft section of the wing.

2.2.8 Effects of Camber

A limited number of experiments were performed to obtain the performance trends related to camber [18]. The rectangular, Zimmerman, and inverse Zimmerman planforms with an aspect ratio of 1.0, shown in Fig. 2.3, were modified to

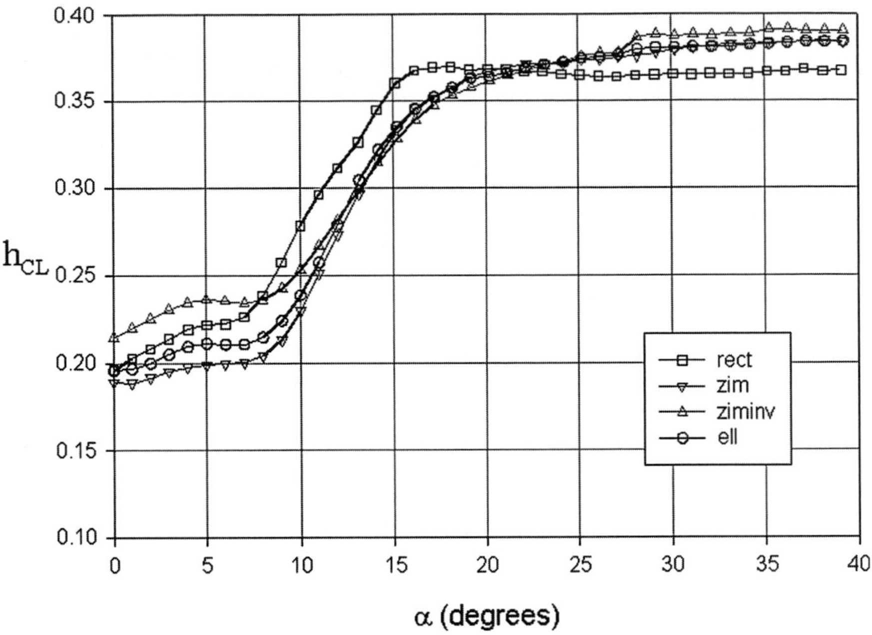

Fig. 2.18 Location of center of lift, with h_{CL} vs α for $AR = 2.00$ and $Re = 1.4 \times 10^5$.

have 4% circular camber. The rectangular and Zimmerman planforms with aspect ratios of 0.5 and 2.0 were also studied. The results indicated the following for a chord Reynolds number of 10^5:

1) Wings with camber have higher lift and lower drag coefficients than those with the same planform but without camber as demonstrated in Figs. 2.23 and 2.24.

2) The values of $C_{L\,max}$ and $\alpha_{C_{L\,max}}$ increase for the cambered wings.

3) The most significant improvement in aerodynamic efficiency using camber occurred for wings with an aspect ratio of 2.0.

Therefore, LAR wings at low Reynolds number benefit from camber as much as high-aspect-ratio wings at higher Reynolds numbers. The effects of 4% circular camber were studied in more detail by Brown [62].

2.2.9 Flow Structure

An increasing number of experimental and numerical studies have improved our understanding of the unsteady flow structure of low-aspect-ratio MAV-type wings [18], [63], and [64]. For example, Jian and Ke-Quin [63] solved the three-dimensional unsteady incompressible Navier–Stokes equations numerically to determine the flow structure over an elliptical wing with an E-174 profile at $Re = 10^3$. They found that for angles of attack larger than 11 deg the flow became bilaterally asymmetric because of a destabilization of the tip vortices. This result might explain the roll instability of most LAR wings. The tip vortices destabilize as a result of their interaction with the separated flow near the center of the wing.

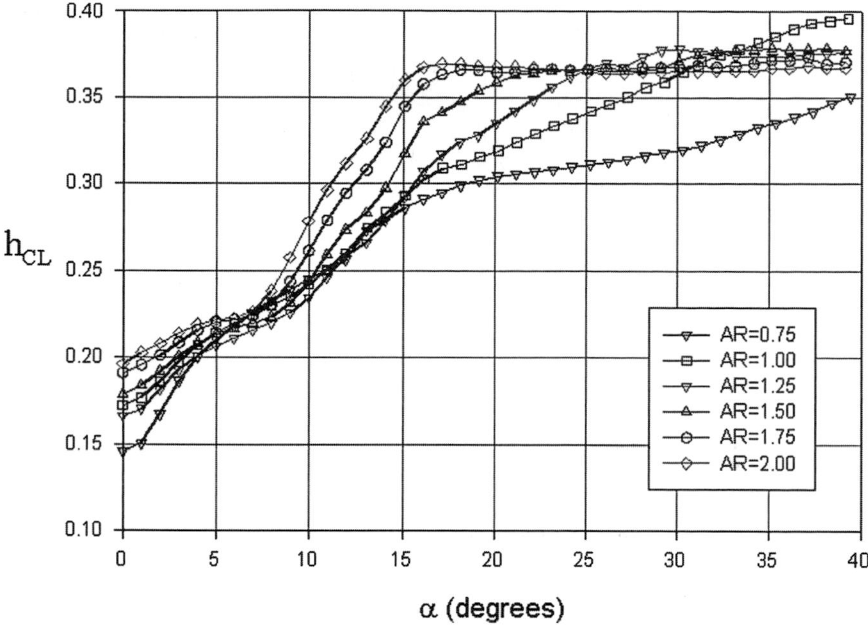

Fig. 2.19 Location of center of lift, with h_{CL} vs α for rectangular planform and $Re = 1.4 \times 10^5$.

Flow-visualization studies in a water channel and the hydrogen bubble technique qualitatively agreed with the numerical results. Similar studies at higher Reynolds numbers are needed in the range that MAVs actually encounter. Another numerical and experimental study was performed to investigate the effect of endplates on the tip vortices of a rigid planform developed for the 15-cm MAV flexible wing designed at the University of Florida [64]. To determine the aerodynamic performance of the MAV wing, the Navier–Stokes equations for incompressible flow were solved for angles of attack of 6 and 15 deg. The numerical results were compared with experiments at Reynolds numbers of 7.1×10^4 and 9.1×10^4. The results indicated that at modest angles of attack the addition of endplates improved the lift to drag ratio by reducing the drag. As the angle of attack became larger, the tip vortices became stronger, and the endplates lost their effectiveness.

2.3 Flexible and Adaptive Wings

Interest in flexible membrane wings was inspired by biological studies of insects [65] and [66], birds [67] and [68], and bats [69] because MAVs are on the same scale as small birds and bats. The development of thin flexible membrane wings for MAVs was the result of both numerical and experimental studies at the University of Florida, Gainesville, Florida, in which it was found that the wings of small birds and bats were thin, flexible, and undercambered [70]. It was further shown that these undercambered wings were more efficient than much thicker rigid wings because they change shape automatically (i.e., passively) in response to the forces

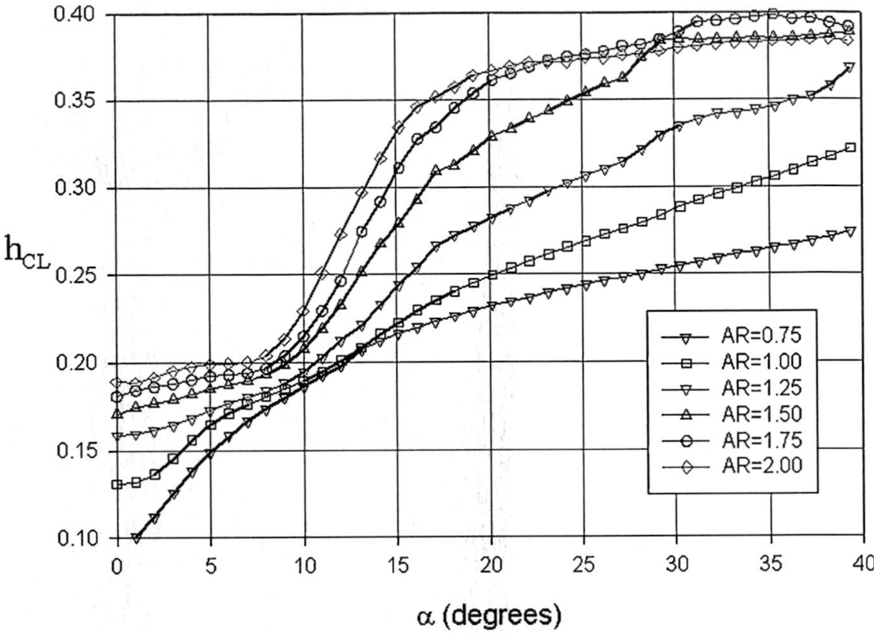

Fig. 2.20 Location of center of lift, with h_{CL} vs α for Zimmerman planform and $Re = 1.4 \times 10^5$.

applied as a result of changes in speed and angle of attack [70] and [71]. This information together with the use of modern composite materials produced a series of MAVs beginning in 1997 [71]. These vehicles won six of the eight Surveillance Competitions (from 1999 to 2006) and three of the five Endurance competitions (from 2002 to 2006) at the annual Micro Air Vehicle competitions as discussed in Chapter 1.

There has long been an interest in changing the geometry of airplane wings (i.e., morphing) in flight to optimize performance for changing missions. It is well known that the Wright brothers used wing warping for lateral control of their gliders and their powered "Flyer" in 1903. Another interesting and old example is the Soviet Union's IS-1 biplane designed in 1938 [72]. The designers' goal was to have a biplane that could take off from a short airfield and once airborne retract the lower wing into the fuselage and become a fast monoplane fighter. The IS-1 biplane's first flight was on 6 November, 1940. Although it flew, its performance was poor because of the added weight of the wing retraction equipment. A modern successful example of a hybrid airplane is the U.S. Navy F-14 swing wing fighter plane. For landing and takeoff the wing is moderately swept, about 20 deg, in order to have good low-speed performance, and once airborne the wing moves into a highly swept, about 68 deg, delta-wing geometry for supersonic flight. It had its first successful flight on 24 May, 1971 [73].

NASA Langley Research Center started a program in 1996 to explore active morphing for different types of airplanes. A recent project supported by NASA

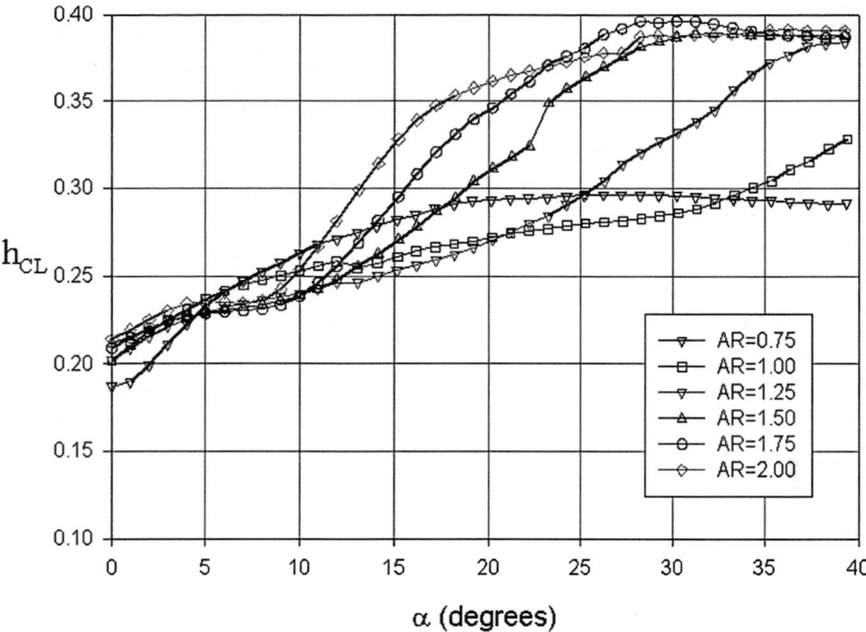

Fig. 2.21 Location of center of lift, with h_{CL} vs α for inverse Zimmerman Planform and $Re = 1.4 \times 10^5$.

resulted in a small vehicle with 61-cm (24-in.) span that mimicked the morphing commonly seen in large birds. This vehicle can vary the gull wing angle to alter its aerodynamic performance and twist its wing tips for roll control [74]. In 2002 DARPA issued a request for proposals (RFP) to investigate "Morphing Wing Structures for Multiple Mission Performance in Unmanned Air Vehicles." Although DARPA was mainly interest in large UAVs, this sparked interest in morphing wings for small UAVs and MAVs so that a single vehicle could achieve both endurance and maneuverability. The international nature of the annual MAV competitions generated a spirit of innovation among university students.

Students and faculty at the University of Arizona began the development of a number of MAVs in 2000 [75] and [76], and they began entering the competitions in 2001. At the Sixth International MAV competition in 2002 at Brigham Young University in Provo, Utah, they introduced a variable camber wing and received the Ingenuity Design award. The camber of the wing could be changed from ground control and therefore was an example of active morphing. A detailed description of a circular and a Zimmerman adaptive wing MAV is presented in the case study in Chapter 6. While continuing their development of adaptive wings, they also received awards for fixed-wing designs as described in Chapter 1.

2.4 Propulsion

Micro-air-vehicle propulsion systems must be as lightweight and efficient as possible. As long as there is no hover requirement, fixed-wing propeller-driven

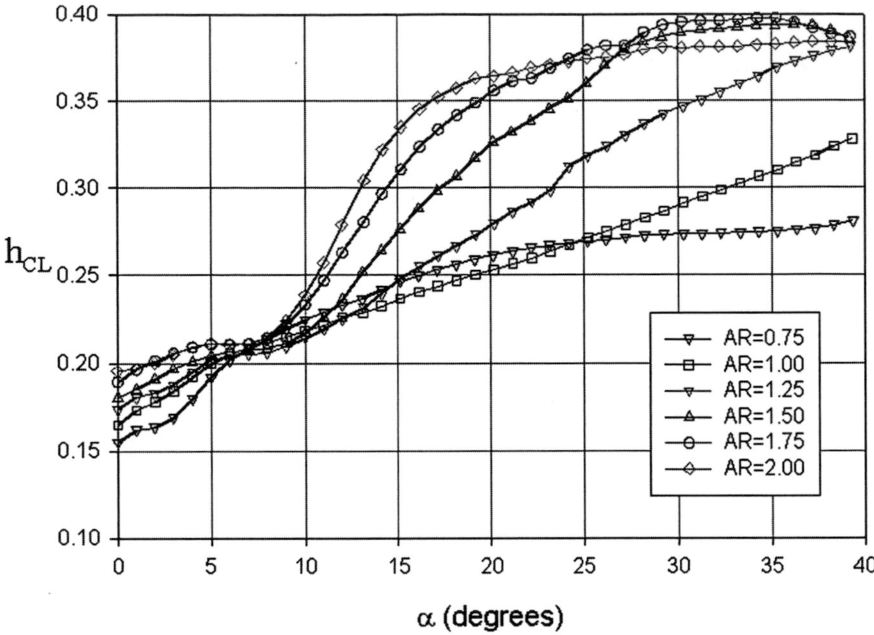

Fig. 2.22 Location of center of lift, with h_{CL} vs α for elliptical planform and $Re = 1.4 \times 10^5$.

MAVs have been found to be the most energy efficient [77]. High energy density and high power density are the most important characteristics. For most MAVs, the power required for both propulsion, controls, and payload is from 2 to 10 W. The energy densities of available storage systems are 47,000 J/g (13.05 W-h/g) for gasoline; 23,000 J/g (6.39 W-h/g) for methanol, 360 J/g (0.10 W-h/g) for lithium batteries, and 100 J/g (0.028 W-h/g) for NiCad batteries [78]. Although gasoline has an energy density more than 100 times that of lithium batteries, internal combustion engines are noisy, difficult to start, and cannot be restarted in flight. However internal combustion engines have been used by early designers especially by those from the model airplane community. Fuel cells might be generally available in the near future [79] and [80]. The first fuel-cell small vehicle (Hornet) was discussed in Chapter 1.

For a propeller-driven vehicle, the power required for level unaccelerated flight is obtained by equating the thrust to the drag and the lift to the weight [81]. This can be written as

$$\frac{T}{W} = \frac{C_D}{C_L} \tag{2.8}$$

$$T_R = \frac{W}{C_L/C_D} = \frac{W}{L/D} \tag{2.9}$$

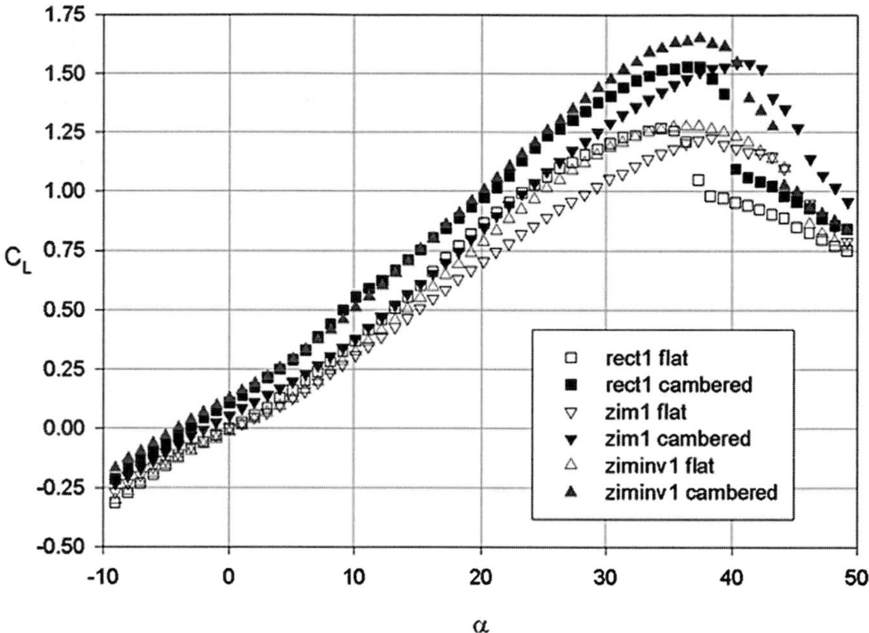

Fig. 2.23 C_L vs α, effect of camber for $AR = 1.00$ and $Re = 10^5$.

and the powered required is

$$P_R = T_R V_\infty \tag{2.10}$$

$$P_R = T_R V_\infty = \frac{W}{C_L/C_D} V_\infty \tag{2.11}$$

Because $L = W = \frac{1}{2}\rho V_\infty^2 S C_L$, therefore

$$V_\infty = \sqrt{\frac{2W}{\rho S C_L}} \tag{2.12}$$

Substituting Eq. (2.12) into Eq. (2.11) gives

$$P_R = \sqrt{\frac{2W^3 C_D^2}{\rho/SC_L^3}} \quad \alpha \frac{1}{C_L^{3/2}/C_D} \tag{2.13}$$

Finally the power required can be written as

$$P_R = W\left(\frac{W}{S}\right)^{1/2}\left(\frac{C_D}{C_L^{3/2}}\right)\left(\frac{2}{\rho}\right)^{1/2}\frac{1}{\eta_P} \tag{2.14}$$

To reduce the power required from Eq. (2.14), it is clear that the total weight and therefore the wing loading, W/S, must be minimized. The wing planform and cross

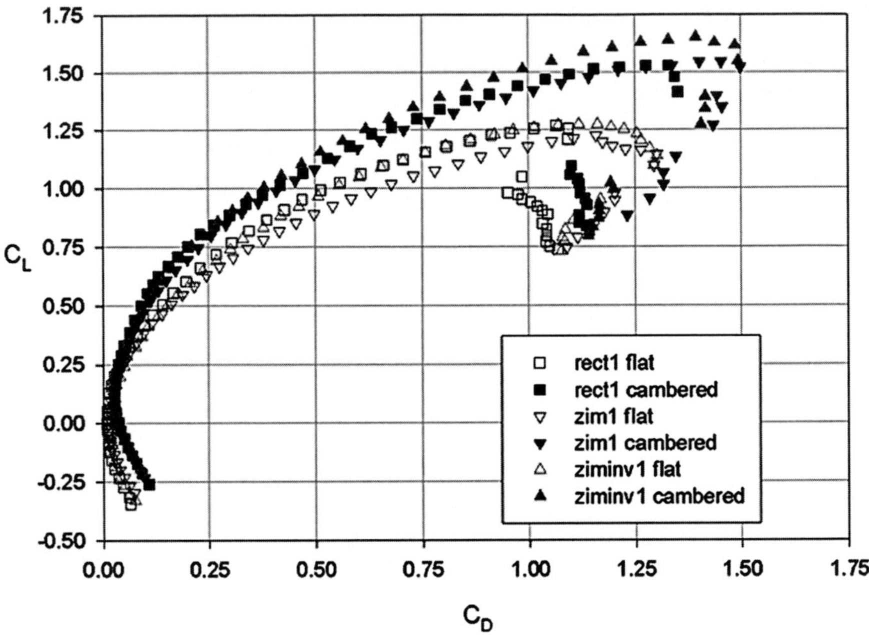

Fig. 2.24 C_L **vs** C_D**, effect of camber for** $AR = 1.00$ **and** $Re = 10^5$**.**

section that produce the largest lift-to-drag ratio must be used, and the propeller efficiency must be maximized. Because micro air vehicles fly at low altitudes, the air density is automatically a maximum.

The power available P_A is the power available to propel the airplane. It is the power delivered to the propeller from the engine/motor shaft P times the propeller efficiency. The propeller efficiency is directly related to the aerodynamics of the propeller and is always less than one. Although numerous MAV designs have used commercially available model airplane propellers, several groups have designed their own in order to improve the performance of their MAV designs [47], [71], and [82]. To maneuver, especially to climb, more power is necessary than indicated in Eq. (2.14). The difference between the power available and the power required for level unaccelerated flight is called the excess power. An approximation to the rate of climb can be obtained by dividing the excess power by the vehicle weight. A more detailed description of vehicle performance can be obtained in [81].

Most MAVs are launched by simply throwing them into the air at a speed that must be at least a little greater that the stalling speed. The stalling speed is the lowest speed at which it can fly in straight and level flight. It is important to make the stalling speed as low as possible. It is clear from Eq. (2.12) that in order to minimize this speed for a given weight and size vehicle this constraint requires the maximum lift coefficient possible [81]. This minimum launching speed can be determined by throwing the MAV with no engine/motor running.

2.4.1 Internal Combustion Engine Propulsion

The smallest internal combustion engine readily available, the Cox® Tee Dee® 0.010, was developed for model airplanes. It has a displacement of 0.164 cm^3 (0.010 $in.^3$) and can produce 20 W of power [80]. It also has a high thrust specific fuel consumption that will limit the MAV range and endurance. A 15-cm (6-in.) propeller-driven MAV with a lift-to-drag ratio of 5 will require 5 W of shaft power for maneuvering and about half that for cruising [80]. Therefore this small liquid fuel engine is overpowered.

For propeller-driven airplanes in level unaccelerated flight with internal combustion engines (i.e., liquid fuel engines), the maximum range depends on the maximum lift-to-drag ratio as shown in Brequet's range equation [81].

$$\text{Range} = \frac{\eta_p}{c_s}\frac{C_L}{C_D} \ln \frac{W_0}{W_1} \tag{2.15}$$

where η_p is the propeller efficiency, c_s is the specific fuel consumption, C_L/C_D is the lift-to-drag ratio, W_0 is the gross weight, and W_1 is the weight, of the airplane without fuel. Thus, the maximum range is directly dependent on the maximum value of (C_L/C_D) at the cruise condition. Brequet's endurance equation for propeller-driven aircraft is

$$\text{Endurance} = \frac{\eta_p}{c_s}\frac{C_L^{3/2}}{C_D}(2\rho S)^{1/2}\left(W_1^{-1/2} - W_0^{-1/2}\right) \tag{2.16}$$

where ρ is the air density and S is the wing area. To maximize endurance, one must maximize $(C_L^{3/2}/C_D)$.

Equations (2.15) and (2.16) do not apply to electrically powered vehicles because their weight remains the same. The case for electrically powered vehicles will be discussed in detail in the next section.

The total drag on the vehicle regardless of the propulsion system is

$$C_D = C_{D0} + \frac{C_L^2}{\pi e(AR)} \tag{2.17}$$

where C_{D0} is the parasite drag coefficient at zero lift and $C_L^2/\pi e(AR)$ includes induced drag caused by lift and the contribution to parasite drag caused by lift. These equations point to the fact that parasite drag, that is, including skin friction and pressure drag, on all of the vehicle's nonlifting parts must be reduced as much as possible.

To reduce $C_L^2/\pi e(AR)$, the aspect ratio (*AR*) can be increased, or the Oswald span efficiency factor e can be increased. Flying at a moderate lift coefficient will also reduce the induced drag (i.e., the drag caused by lift). Because the maximum lift-to-drag ratio usually occurs at angles of attack where the lift coefficient is somewhat lower than its maximum value and because the Oswald efficiency factor e cannot be easily increased, significant reductions in $C_L^2/\pi e(AR)$ are usually accomplished by increasing the aspect ratio.

2.4.2 Electric Motor Propulsion

Electric propulsion has become popular in the MAV and model airplane communities as a result of an increasing number of commercially available small and efficient motors, motor controllers, and advanced technology batteries. Electric propulsion has a number of basic advantages over conventional spark, compression, or glow ignition internal combustion engines. In addition to more rapid and continuing technological improvement, electric power provides 1) high reliability, 2) push-button starting and restarting in flight, 3) precise power management and control, 4) reduced noise and vibration, 5) reduced heat (infrared) emissions, and 6) elimination of volatile fuels and lubricating oil spray. For MAVs that are intended for unobtrusive or low-observability operations, such as close-in military reconnaissance, the advantages of quiet operation and low infrared signature compared with internal combustion engines are particularly important. In addition, all of these advantages remain as MAV size, and power requirements diminish for ultra small vehicles. A variety of reliable electric motor systems, below 1-W and 1-g levels, are currently available and being used in radio-controlled micro model aircraft. The various components of electric propulsion systems, batteries, motors, and propellers are discussed in the following sections, along with some basic technical considerations and analyses that can help guide their selection and use during the design of micro air vehicles.

2.4.2.1 Batteries. The primary disadvantage and limitation of battery-powered flight is the relatively low specific energy (stored energy per unit weight, normally stated in watt hours per kilogram) of batteries. When compared to fuels such as gasoline or alcohol-based glow fuels, the difference in specific energy can be up to two orders of magnitude. This translates to a much higher "fuel" weight requirement for electric-powered vehicles, which made electric aircraft flight impractical prior to the 1970s. Battery technology, however, has improved significantly since that time and continues to improve at an accelerating pace. Small nickel cadmium and nickel metal hydride rechargeable cells used in earlier MAVs had specific energy levels up to approximately 20–50 W·h/kg at useful discharge rates. Modern lithium-ion batteries, such as lithium sulfur-dioxide primary (single-use) cells and lithium-ion polymer rechargeable cells, can provide specific energy levels of 150–200 W·h/kg. The specific power, or delivered watts of power per unit weight, of lithium polymer cells has also increased such that powers of one or more kilowatt/kilogram can be delivered when discharged at nominal MAV rates. Significant further improvements of lithium-ion battery performance in the near term might be unlikely because of basic limits on the battery chemistry, although different chemistries being explored, such as lithium sulfur, might lead to additional improvement. In any case, the specific energy and power of modern lithium-ion chemistry batteries means that electric powered MAVs with useful payloads, power, and endurance are now practical.

Recent flight experiments with fuel cells suggest a possible alternative to battery power. Fuel cells combine a fuel catalytically with oxygen to produce electric power directly, without combustion. Thus, the energy store for a fuel-cell system can be an extremely high specific-energy fuel, generally hydrogen. In 2005 the U.S. Naval Research Laboratory, in cooperation with Protonex Technology

Corporation, flew a small fuel-cell-powered airplane for 3 h, 19 min on a fuel load of 15 g (about 0.5 oz) of compressed hydrogen gas [82], landing when fuel was exhausted. The fuel-cell system supplied a maximum power of about 95 W to the propulsion motor. The airplane, named Spider Lion, weighed 2.5 kg (5.6 lb) and had a wingspan of 2.2 m (7.2 ft), larger than the MAVs considered here. Significant challenges remain, however, in scaling a complete fuel-cell power system down to a size and weight practical for MAV use.

Lithium polymer cells. The most widely used new battery technology for small electric aircraft, offering both high specific energy and power, is the lithium-ion-polymer (LiPo) cell. These became widely available in the early 2000s and now dominate the model airplane and car hobby market. They are rechargeable and offer 5 to 10 times the specific energy of the previously dominant nickel-cadmium (NiCad) and nickel-metal-hydride (NiMH) cells. Average open cell voltage is about 4.1 V, diminishing to an average of about 3.5 V under normal loads, compared to the 1.2 V per cell typical of NiCad or NiMH chemistry. Lithium polymer batteries of one to four cells in series are most commonly used and provide sufficient voltage (3.5 to 15 V) for most electronic and power systems available for MAV use. New, very high-efficiency motors (80% to over 90%) are also becoming available that are specifically designed for LiPo battery voltages, including those for the higher voltages (15 to 30 V) of four- to eight- series cell batteries.

The typical LiPo cell's construction is very different from other rechargeable cells. Smaller cells are normally a thin vinyl pouch or bag, which is vacuum packed with a wound wrapping of thin-film electrode material, in a fluid or gel electrolyte. Shape can vary from flat, paper thin, plate-like discs to rectangular box-like structures. Great flexibility in shape and form is one of the advantages (aside from high specific energy) of the LiPo that make it so popular in portable consumer electronics, including computers and cellular telephones.

Sizing a battery pack for a particular MAV design involves the selection of both capacity (size of cells measured in milliampere hours, or mAh) and voltage (the number of cells in series). The voltage required is determined by the motor being used, because the motor's windings will be tailored to a particular voltage and power range. Details of motor characteristics as related to the battery are covered further in the next section. The "capacity" of the cell or battery is basically a measure of the amount of electricity—number of electrons—the fully charged cell can deliver until it is discharged to a "safe," specified cutoff voltage. Because a cell will slowly drop in voltage as it is discharged, it is important to know what safe cutoff voltages should be used to terminate the cell's discharge. NiCd and NiMH cells are much more robust than LiPo cells in most respects, including voltage cutoff limits. In general, this means they can be totally discharged without suffering serious damage, although it is usually recommended that NiCd and NiMH battery discharge be terminated at about 0.7 to 0.9 V per cell. This will tend to prevent reverse charging weak cells in large series packs of six or more cells. LiPos on the other hand are very sensitive to a minimum cell voltage, similar to a lead-acid car battery. A minimum cutoff voltage of 3.0 V per cell is usually specified by LiPo cell manufacturers. If a cell is allowed to fall below this voltage, damage begins to

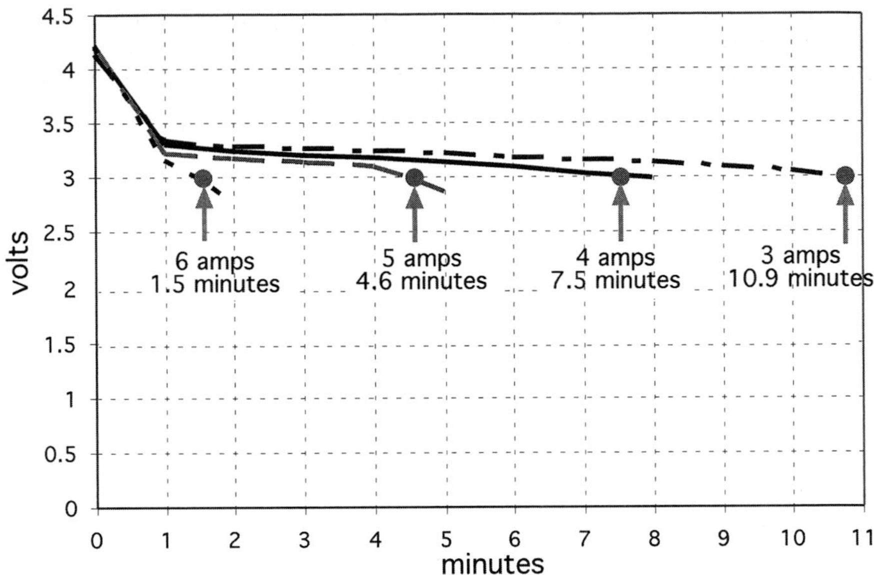

Fig. 2.25 600-mAh LiPo cell discharge profiles.

occur, and the cell's useful life is reduced. Therefore, some means of monitoring and/or managing LiPo battery end voltage is required to avoid battery damage. This cutoff function is often accomplished by the motor electronic-speed-control (ESC) device.

The most revealing signature of a cell's performance and effectiveness is its discharge profile. The usual discharge test is a constant current I discharge of a fully charged cell while the cell's voltage is recorded over time until a prespecified cutoff voltage is reached. Figure 2.25 illustrates typical discharge profiles of a LiPo cell discharged at constant rates of 3 to 6 A. As shown in this case, the initial 4.1 V drops to under 3.5 V within $1/2$ minute; it then slowly flattens out until the 3-V cutoff is reached. In this example, cutoff times vary from 10.2 min at 3 A to 1.5 min at 6 A. From this test we can determine not only the discharge time, but most of the cell's basic performance characteristics. Because both discharge time and average voltage are reduced as discharge rate is increased, it follows that, as more amps are drawn from the cell, the delivered energy (amps $\times$ volts $\times$ time) of the cell will also decline. This performance dropoff with increasing current is primarily caused by the cell's internal ohmic resistance R and the consequent ($I \times R =$ voltage) losses. In addition to the IR voltage drop, internal heating, proportional to I^2, also occurs, which raises the cell's internal temperature and ultimately limits its maximum discharge rate.

Comparing and evaluating batteries. From cell discharge tests as illustrated in Fig. 2.25, the basic cell performance can be calculated as follows. For a fixed discharge rate of I amperes, a cutoff time of t minutes, a cell weight of W g, an average voltage $\bar{v}$ during the discharge (some battery discharge instruments calculate $\bar{v}$ as the discharge takes place; otherwise, find $\bar{v}$ by integration or graphical approximation)

$$\text{Capacity delivered} = I \times v \text{ in ampere minutes} \tag{2.18}$$

$$\text{Delivered energy} = I \times t \times v \text{ in watt minutes} \tag{2.19}$$

Watt minute is a more convenient unit for MAV work than the traditional kilowatt hours.

$$\text{Specific energy} = I \times t \times v/W \text{ in watt minutes/gram} \tag{2.20}$$

$$\text{Specific power} = I \times v/W \text{ in watts/gram} \tag{2.21}$$

It can be argued that, for MAV design purposes, the cell's specific energy is the single most valuable performance measure, because it indicates the amount of energy available per unit of cell weight. This measure, viewed at discharge times of interest (such as mission time), can provide an excellent single evaluation criteria as discussed next.

From the sample discharge plot, notice that both t and v decline as discharge rate increases. This means that capacity or discharge time alone is not a reliable proxy for energy delivered because poor-quality, higher-resistance cells will have a larger voltage drop and hence deliver less energy and power. Note also that specific power increases for a short, high current discharge, while specific energy decreases. In effect, for a given cell there is a tradeoff between power and energy as discharge time varies. Figure 2.26 illustrates this effect for an 8.8-g, 360-mAh LiPo cell. A general method to quantify and evaluate this tradeoff for different cells follows. For any given discharge test, the resulting specific energy E_e and specific power E_p can be plotted as a point on a graph, called a Ragone plot. All possible discharge test points for a particular cell result in a continuous line that quantifies its energy vs power capability. Figure 2.27 shows such curves for five typical LiPo cells having capacities from 135 to 1500 mAh and weighing 3.6 to 31.7 g. Some important characteristics of the Ragone plot are as follows:

1) In all cases the further a point is from the origin, the better the indicated performance.

2) Constant discharge timelines radiate from the origin (in our case with a slope = energy/power × minutes).

3) A cell's energy vs power tradeoff is represented by the slope of its plotted line; the line's distance from the origin indicates its overall capability.

4) When Ragone curves for two cells intersect, the flatter of the two curves represents the cell with higher performance at shorter discharge times (higher specific power), whereas the cell with the steeper curve has better performance at long discharge times (higher specific energy).

5) For a preferred discharge time, the highest E_e cell for that time maximizes overall performance, including power.

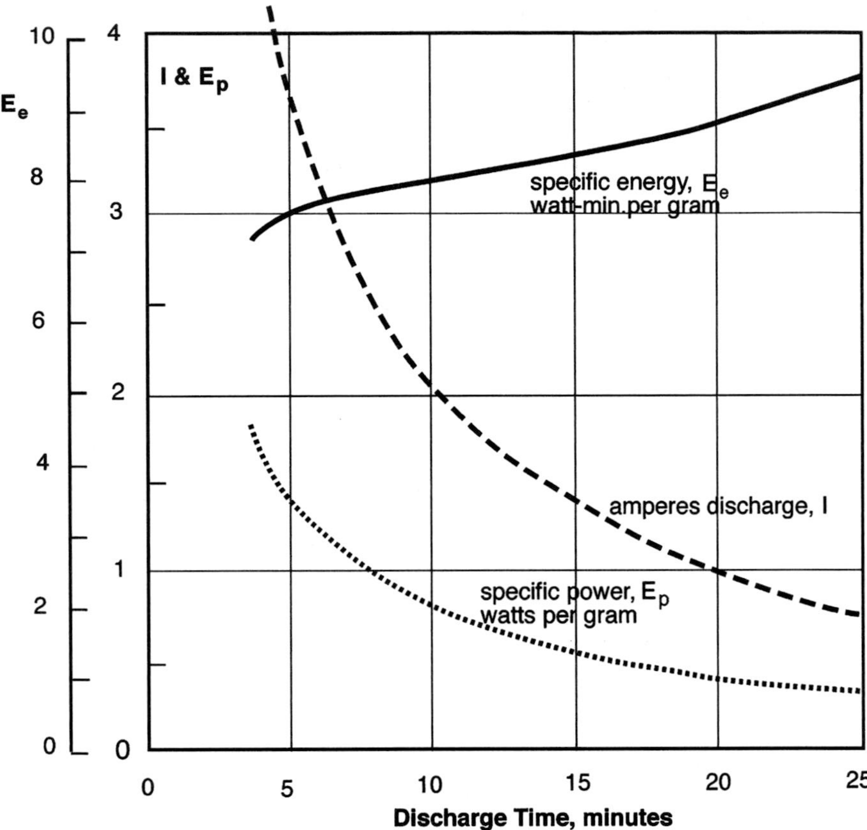

Fig. 2.26 Typical change in specific energy and specific power vs discharge time.

Although high power is beneficial, the highest specific power for a cell can occur at discharge times much shorter than normal MAV operations. In such cases the reduction in specific energy that might have been traded off should be quantified and considered.

The preceding factors suggest that capacity, maximum discharge rate, and/or maximum power in themselves are inadequate measures of merit. A more useful single performance indicator will be a cell's specific energy at useful discharge times.

Cell specification conventions. The approximate, nominal capacity of a cell is generally indicated by its C rating (NiCad, NiMH, and LiPo cells also use this nomenclature). The C rating is expressed in units of milliampere hours (mAh) or ampere hours (Ah). In general, the larger and heavier a cell is, the larger its capacity and C rating will be. Standards have changed over time, but present international standards for LiPo cells define C capacity as follows: a cell has a C capacity rating of n Ah if that cell can be discharged at a constant current of n/5 A from full voltage

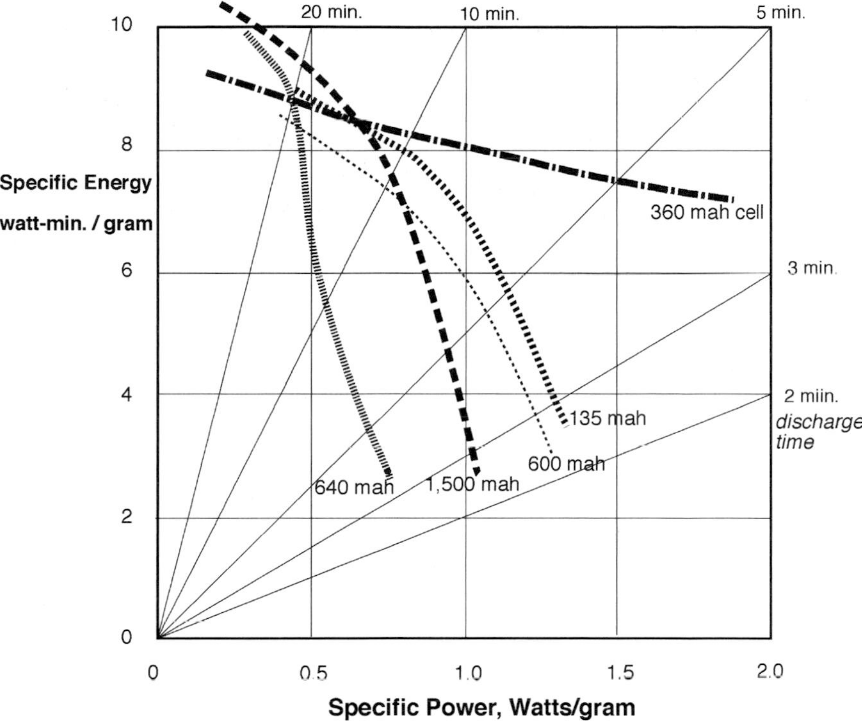

Fig. 2.27 Ragone plot of several small LiPo cells.

to 3 V in 5 h. Its C rated capacity is then n/5 A × 5 h = n Ah, or n × 1000 mAh. As a practical matter, C ratings defined for such low discharge rates over 5 h are only a rough indicator of cell capacity for MAV purposes. Further, rating standards vary widely by manufacturer and are usually not defined or consistently used. It is often improperly assumed, for example, that a cell rated at n Ah should be able to discharge at n amperes for 60 min, or 4 × n A for 15 min, or 10 × n A for 6 min, and so on. As we have seen, this is not the case because, even if normal C rating capacity standards have been used, actual capacity will decline as discharge rate increases. The reduction in delivered capacity depends upon the particular cell's design and quality of manufacture and can vary widely. An illustration of actual, delivered capacity vs discharge rate is shown in Fig. 2.28 for two typical LiPo cells.

In most cases, discharge testing at various discharge rates is the best way to accurately determine actual battery performance that the MAV designer can count on. Ideally, when typical mission power profiles over time can be estimated, battery discharge tests using these profiles will provide the most accurate evaluation of candidate batteries.

Table 2.1 shows the discharge test results for a number of LiPo cells, along with their performance parameters, at various discharge rates and times. Individual cells of identical capacity can be combined into a battery by connecting them in series

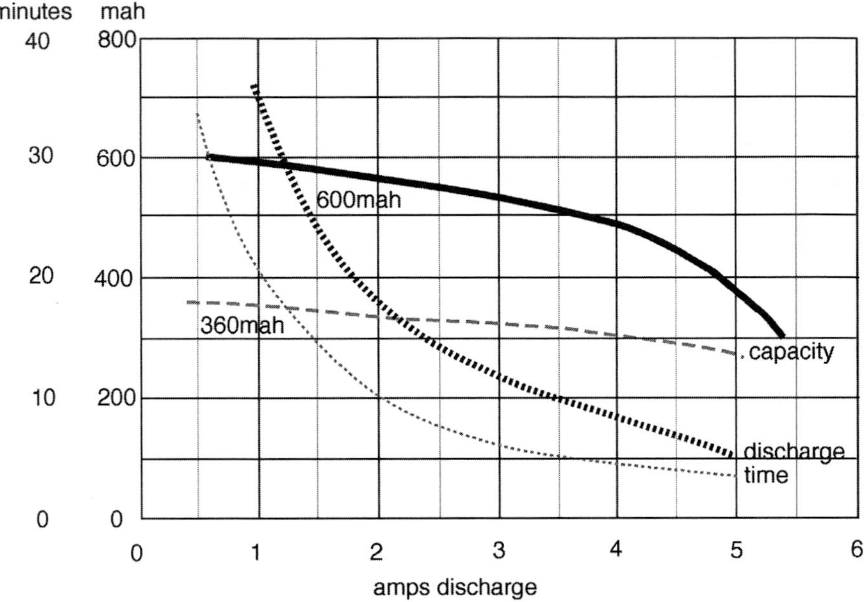

Fig. 2.28 Typical LiPo capacity and discharge time vs discharge rate.

to add voltage or in parallel to increase capacity. Combinations of both series and parallel connections can also be used.

A convenient way to display test data for a particular cell is to plot its discharge time and specific energy vs discharge rate. An example is shown in Fig. 2.29 for three typical LiPo cells of varying weights. They are the 3.6-g, 8.8-g, and 14.3-g cells of 135-mAh, 360-mAh and 600-mAh capacity.

A first step to identify candidate LiPo cells for a particular MAV design is to do the following:

1) Select cell discharge times needed for specific MAV mission profiles.

2) For those times, compare cells by their specific energy E_e. Cells should deliver at least 5 to 8 W·min/g at the desired discharge times. An E_e lower than 3 to 4 W·min/g indicates relatively poor capability—in this case good NiMH or NiCad cells might be an alternative. The higher the E_e, the better.

3) From among the "good" cells of different weights and sizes, final cell/battery selection will be determined by MAV endurance and power requirements (mostly a function of gross weight, wing loading, and L/D) and will be discussed in the next section.

Sizing propulsion batteries. One peculiarity of batteries as a "fuel" is that they are not consumed during flight. This means that—unlike normal fuel that is consumed and constantly reduces aircraft weight during powered flight—battery fuel weight is fixed. In this respect, battery power resembles rubber-band-powered models more than conventionally fueled aircraft. It can be shown that, for the case

Table 2.1 Discharge test results for lithium-ion polymer batteries

Cell	Weight, g	Cutoff, V	Disc. current, A	Cutoff time, min.	Avg. voltage, V	Capacity delivered, mAh	Energy delivered, W.min	Specific energy, W.min/g	Specific energy, kWh/kg	Average specific power, W/g
120 mAh	3.5	0.8	0.7	9.1	1.01	105	6.4	1.8	30.3	0.2
NiMH			1.0	5.6	0.95	92	5.2	1.5	25.0	0.3
"DoubleTime"			1.5	3.4	0.94	84	4.7	1.4	22.6	0.4
300 mAh NiMH	7.6	0.8	2.0	——	——	——	——	——	0.0	——
Tieg 2/3AAA			3.0	5.5	1.02	274	16.8	2.2	36.8	0.4
600 mAh NiMH	12.3	0.8	2.0	18.1	1.09	604	39.5	3.2	53.5	0.2
Tieg AAA			3.0	12.7	1.03	632	39.1	3.2	52.9	0.3
			4.0	8.6	0.99	572	34.0	2.8	46.0	0.3
			5.0	6.3	0.96	522	30.1	2.4	40.7	0.4
135 mAh LiPo	3.6	2.8	0.5	18.0	3.43	150	30.9	8.6	142.9	0.5
Kokam			0.7	12.5	3.37	144	29.1	8.1	134.8	0.7
Powerflite			1.0	8.3	3.24	138	26.8	7.5	124.2	0.9
SYE-301P			1.5	3.1	3.01	76	13.7	3.8	63.5	1.3
350 mAh LiPo	6.38	2.8	1	19.1	3.00	320	57.6	9.0	150.5	0.5
HOMEFLY, Inc.			2	8.3	3.30	274	54.3	8.5	141.7	1.0
			2.5	4.3	3.17	176	33.5	5.2	87.4	1.2
			3	0.6	2.93	26	4.6	0.7	11.9	1.4
360 mAh HD	8.77	3.0	0.4	53.3	3.73	358	80.1	9.1	152.3	0.2
LiPo			1.0	21.1	3.64	352	76.9	8.8	146.1	0.4
Kokam			2.0	10.1	3.5	336	70.6	8.0	134.1	0.8
SO35LC3100...			3.0	6.7	3.44	334	68.9	7.9	131.0	1.2
			3.6	5.3	3.37	320	64.7	7.4	123.0	1.4
			5.0	3.9	3.25	324	63.2	7.2	120.1	1.9
600 mAh LiPo	14.3	3.0	2.0	16.8	3.49	560	117.3	8.2	136.7	0.5
Kokam.FMA			3.0	10.9	3.29	544	107.4	7.5	125.2	0.7
unmarked			4.0	7.5	3.21	500	96.3	6.7	112.2	0.9
prototypes			5.0	4.6	3.14	386	72.7	5.1	84.8	1.1
640 mAh LiPo	13.53	3.0	1.0	37.3	3.59	622	134.0	9.9	165.0	0.3
Kokam, with			1.5	23.4	3.52	586	123.8	9.1	152.5	0.4
terminal board			2.0	11.3	3.45	374	77.4	5.7	95.4	0.5
SO60LB15H....			2.5	7.1	3.38	294	59.6	4.4	73.4	0.6
			3.0	4.3	3.28	214	42.1	3.1	51.9	0.7
850 mAh LiPo	19.2	3.0	2.0	27.0	3.53	902	191.0	10.0	165.8	0.4
ENERLAND,			3.0	16.5	3.43	822	169.2	8.8	146.8	0.5
with heavy wires			4.0	7.8	3.36	522	105.2	5.5	91.4	0.7
			5.0	3.7	3.2	304	58.4	3.0	50.7	0.8
1100 mAh HD	21.9	3.0	3.0	20.7	3.32	1030	205.2	9.4	156.1	0.5
LiPo			4.0	11.0	3.23	730	141.5	6.5	107.7	0.6
Kokam			5.0	6.8	3.15	564	106.6	4.9	81.1	0.7
Lii0LC25....			6.0	1.6	3.14	158	29.8	1.4	22.7	0.9
1.2 Ah LiPo	24.1	3.0	4.0	14.6	3.21	972	187.2	7.8	129.5	0.5
E-Tech			5.0	9.8	3.18	818	156.1	6.5	107.9	0.7
			6.0	5.1	3.12	510	95.5	4.0	66.0	0.8
1.5 Ah LiPo	31.68	3.0	1.5	60	3.66	1502	329.8	10.4	173.5	0.2
Kokam			3.0	30	3.54	1502	319.0	10.1	167.8	0.3
HJKLF18B002			4.0	22.2	3.45	1478	305.9	9.7	161.0	0.4
			5.0	17.4	3.35	1446	290.6	9.2	152.9	0.5
			7.0	11.6	3.23	1350	261.6	8.3	137.6	0.7
			8.0	8.9	3.17	1192	226.7	7.2	119.3	0.8
			9.0	6.1	3.13	910	170.9	5.4	89.9	0.9
			10.0	3.6	3.08	594	109.8	3.5	57.8	1.0
		2.8	10.0	7.8	3.00	1296	233.3	7.4	122.7	0.9
			9.0	9.5	3.11	1416	264.2	8.3	139.0	0.9
			8.0	10.6	3.13	1416	265.9	8.4	139.9	0.8
			7.0	12.4	3.21	1447	278.7	8.8	146.6	0.7
			5.0	18.1	3.37	1508	304.9	9.6	160.4	0.5
			4.0	22.8	3.48	1520	317.4	10.0	167.0	0.4

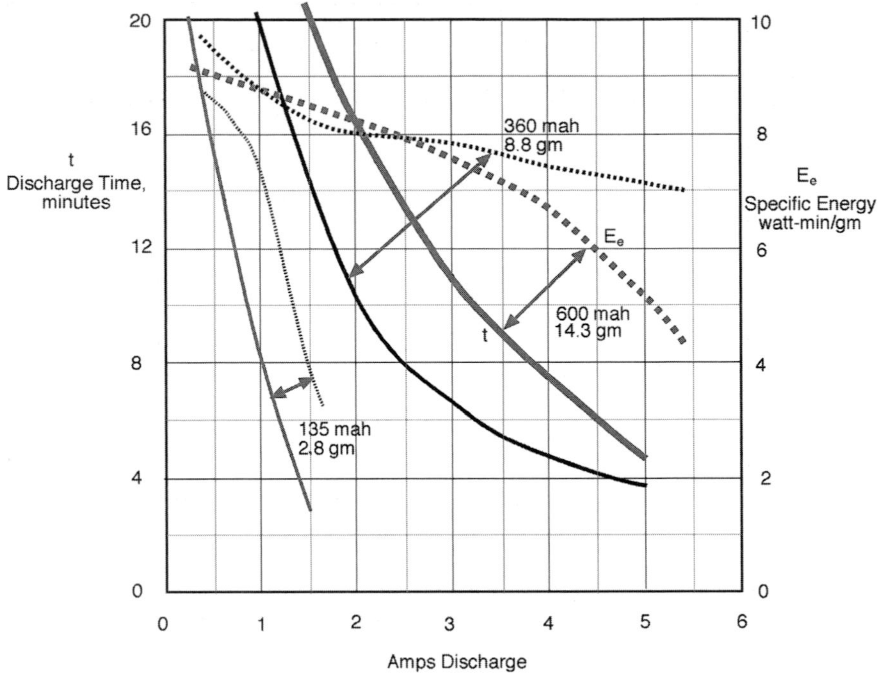

Fig. 2.29 Discharge time and specific energy vs discharge rate.

of battery-powered flight, the endurance of a given aircraft is proportional to

$$\frac{W_B/W_1}{(1 + W_B/W_1)^{3/2}} \tag{2.22}$$

where W_B is battery weight and W_1 is aircraft weight less battery. Keeping W_1 constant, and varying W_B, it follows that endurance is maximized when $W_B = 2 \times W_1$, that is, when battery weight is equal to twice the empty weight of the aircraft, or is $2/3$ of gross weight. This relationship is shown in Fig. 2.30. The maximum region is very flat, and endurance is not reduced much as battery weight falls below $2 \times W_1$. For example, 92% of maximum flight time can be achieved when battery weight is equal to airframe weight, and 71% is achieved with a battery of $0.50 \times$ airframe weight. If endurance or range are important MAV objectives, a battery weight in the neighborhood of at least 25% or more should be considered; below 25%, the endurance potential is declining more and more steeply.

Battery safety. Overcharging of LiPos can lead to rapid cell temperature rise and damage or destruction—even catastrophic, explosive ignition. The Kokam company has found that these effects begin to occur above roughly 60° C internal temperature [83]. Charging protocols and discharge profiles recommended for their cells are intended to keep cell temperatures below this level for improved safety and to maximize cell life. Additionally, many early-model LiPo battery chargers

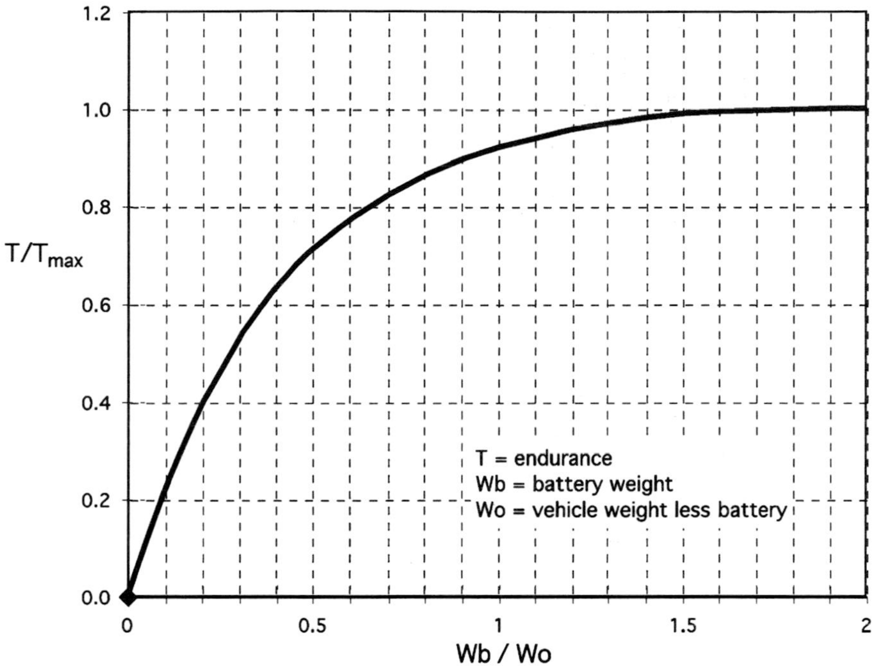

Fig. 2.30 Effect of battery weight on endurance.

use charging protocols that treat the total number of series cells in a pack as a single cell of higher voltage. Because cells age and degrade differently, different cells in a battery pack can become "unbalanced," that is, develop voltage/charge differences among the cells. Continuing to charge and discharge unbalanced packs will increase this imbalance and can lead to significant battery degradation or even catastrophic destruction. Newer, balancing chargers, which sense voltage and charge levels of each individual cell in a series pack, are now considered the safest way to charge LiPo battery packs. Also, because LiPo cells should not be discharged below about 3 V per cell to avoid damage, discharge protection devices should be used to avoid excessive discharge. The larger the number of cells in series in a battery pack, the more important it is to avoid unbalanced cells and to not overdischarge packs. Like all high-power devices, LiPo batteries require careful use. Manufacturer safety guidelines should be understood and observed at all times.

2.4.2.2 DC electric motors. For MAV-size vehicles, dc electric motors are an attractive propulsion option. Small (1 to 20 W) electric motors typically have electrical to mechanical energy conversion efficiencies of 50 to 75%. For larger motors, the electrical efficiency can reach 90% and higher. This compares to less than 10% for the smallest glow-ignition internal combustion engines. For a specific vehicle design, electric motor selection is determined primarily by power requirements and secondarily by considerations of weight, size, and efficiency. Among the bewildering variety of dc electric motors now available, there are three basic types suitable for MAV applications: cored, coreless, and brushless.

Permanent magnet, iron-cored, brushed motors (often called brushed, cored motors) are the most common and numerous of permanent magnet motors available and have been a mainstay of the hobby industry. These motors have permanent magnets fixed within the motor casing (stator) and a rotating armature (rotor) of wire coils wrapped on a core of iron. As the rotor spins, input current to the coils is switched (commutated) between the different poles in the armature through a set of brushes that slide over a ring of rotating switch contacts (the commutator). Ferrite permanent magnets are typically used, though some high-power motors use rare-Earth (e.g., samarium cobalt) magnets because of their ability to withstand high temperatures without loss of magnetic strength. Iron-cored, brushed motors are robust and, because of tremendous worldwide production quantities, are inexpensive and readily available in a wide range of sizes and voltage ratings. For their power, iron cored motors are relatively heavy and are basically low-torque, high-rpm devices. As such, they often need to be geared down to make use of large, reasonably efficient propellers. Their use in MAVs is generally limited to cases where both toughness and low cost are important, or where suitable coreless or brushless motors, with lower weight and higher efficiency, are not available.

Coreless, brushed motors (coreless motors) are most often used in smaller, lighter MAVs, requiring fairly low power. Their operation is similar to the cored motors except that the wire coils of the armature are interwoven without an iron core, usually in the shape of a hollow cylinder. The windings are held intact by a resin coating. The permanent magnet poles often fit inside the hollow armature, which rotates around them on a shaft. Commutation is again through a set of brushes and a commutator ring. The lack of a ferrous core reduces weight and size and improves efficiency by eliminating iron losses caused by magnetic hysterysis and eddy currents in the core. The lighter rotor also has a lower moment of inertia, allowing speed changes to be made faster than in a cored motor. This makes coreless motors particularly useful for servomechanisms. However, without the iron core to act as a support and heat sink for the armature, the coreless motor is more fragile and less tolerant of heating than the more massive cored motors. Costs also tend to be higher for coreless motors than cored motors, but they have lower weight and higher efficiencies. MAV size coreless motors, for example, the 9-g MicroMo DC5-2.4, marketed by Wes-Technik, and the 17-g C. I. Kasei A12C-06-S (aka DC6-8.5), are about 55 to 75% efficient at nominal power levels. Although coreless motor selection overall is limited compared with cored motors, a reasonable range of choices is available. Tiny "pager" type microcoreless motors are often used in the smallest MAV designs (under ~20 g gross weight).

Brushless motors can provide the best specific power (lowest weight for a given power), have the greatest durability, and often have the highest efficiencies of the three types. A brushless motor employs permanent magnet poles in the rotor and fixed, wound electromagnet poles. The stator poles are typically grouped into sets of three that are energized 120 deg out of phase with each other, similar to three-phase ac motors. An electronic controller commutates current to these coils in sequence, producing a rotating magnetic field that spins the permanent magnet rotor. To apply a current pulse to the correct winding at the correct time under varying load conditions, the brushless controller must have a reference to the angular position of the rotor. Sensored brushless motors and controllers do this via magnetic sensors that are built into the motor, reading this information to the controller by hard wiring. Nearly all small brushless motors available at present, however, are sensorless. The sensorless controller has no hard-wired magnetic sensors. It senses the reverse electromotive force (back EMF) induced in the stator coils by the rotor magnets moving past them and with this information uses software to deduce the coil positions.

Until the early 2000s, brushless motors for UAV and hobby use were of the conventional, internal rotor type, in which a small-diameter permanent magnet rotor was inside the wound coils. Their high rotation speeds and low torque, similar to the common brushed motor, usually required a gear reduction system on the output shaft in order to turn larger, more efficient propellers. The arrival of small, external-rotor brushless motors, known as "outrunners," has made brushless motors more practical for aircraft applications. The permanent magnets of an outrunner's rotor are located within a larger-diameter cylindrical external case, spinning outside the fixed, internal stator coils. The increased moment arm of the magnetic rotor provides much more torque than conventional motor geometry. Similar motors, running on ac, have long been used to turn the large, slow moving blades of ceiling fans. The outrunner design is inherently a higher torque/lower rpm device, able to drive a large-diameter propeller without the addition of a gearbox. Though they have only recently been introduced in sizes relevant to the model aircraft community, they are evolving at a rapid rate and now dominate the market for small, high-performance model airplanes. Considering both the electronic controller and the motor itself, brushless motors are currently somewhat more expensive and have a more limited selection than the cored and coreless types, but these differences are diminishing.

In summary, for power levels over 10 W, brushless motors are usually the preferred motor choice considering performance and weight. At smaller sizes their efficiency and weight advantages diminish. They are, nevertheless, more robust, and the outrunner type does not require a gear system. Figure 2.31 illustrates several typical brushed and brushless MAV-size motors.

Electric motor controllers. Brushed motor controllers, often called electronic speed controls (ESCs), are normally smaller, lighter, and less expensive than brushless motor controllers. Their function is to vary the average voltage applied to the motor via pulse width modulation and thereby control the motor's speed and power. Basically, they act as an electronic throttle. They use switching transistors, typically MOSFETs, to chop the incoming dc at a fixed cycle rate, generally 1–100 kHz. The switching transistors are either on or off, providing full voltage or zero voltage to the motor. Within each cycle, the percentage of cycle time in which full voltage is supplied is varied between 0 and 100%, depending on the commanded power level. At these points the ESC acts as an on/off switch and is close to 100% efficient. At partial power settings, however, the efficiency of the controller chopping process is less than 100%, and losses can be significant. Such losses are common to all motor types, brushed and brushless and should be taken into account when estimating overall motor system efficiency.

An additional complexity of brushless motor controllers is that they also perform the essential dc to ac conversion as well as the commutation function. And because

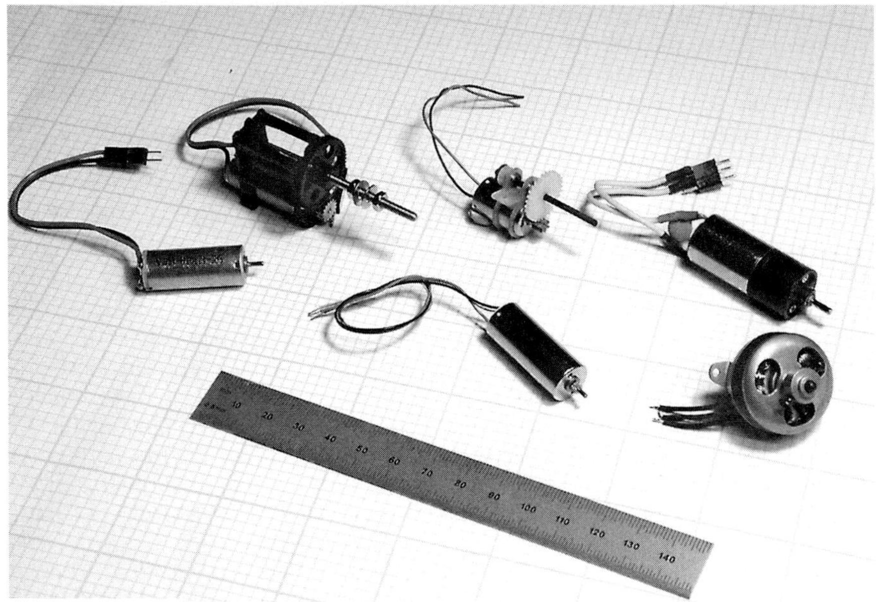

Fig. 2.31 Typical MAV electric motors.

a brushless motor cannot even run without a controller, its controller efficiency must be considered an integral part of the motor's efficiency.

DC motor evaluation, analysis, and selection. How can dc electric motors be quantitatively evaluated, compared, and analyzed? Fortunately, such motors obey a set of rigid, fairly simple relationships among voltage, current, rpm, torque, power, and efficiency. A standard method of characterizing and displaying a dc motor's performance begins with measuring the motor's rpm N and current I as the load on the motor shaft (torque load T_M) is increased from no load T_{M0} to a load that stalls the motor T_{MS}. These measurements are taken at a fixed supply voltage v. Both current and rpm will be linear with torque, as shown in Fig. 2.32. Note that even with no load the motor will draw some current I_0 because the armature windings have a resistance R, and

$$I_0 = v/R \tag{2.23}$$

Also, the rpm at no load N_0 will be proportional to the supply voltage, and

$$N_0 = K_V \times v \tag{2.24}$$

where K_V is a constant, called the speed constant. Two other important characteristics of dc motors are that both stall torque and no load speed are proportional to supply voltage. If voltage is changed, the new rpm vs torque curve is changed as shown in Fig. 2.33. Another defining characteristic of dc motors is that current I is independent of supply voltage, so that the current vs torque line is fixed. From

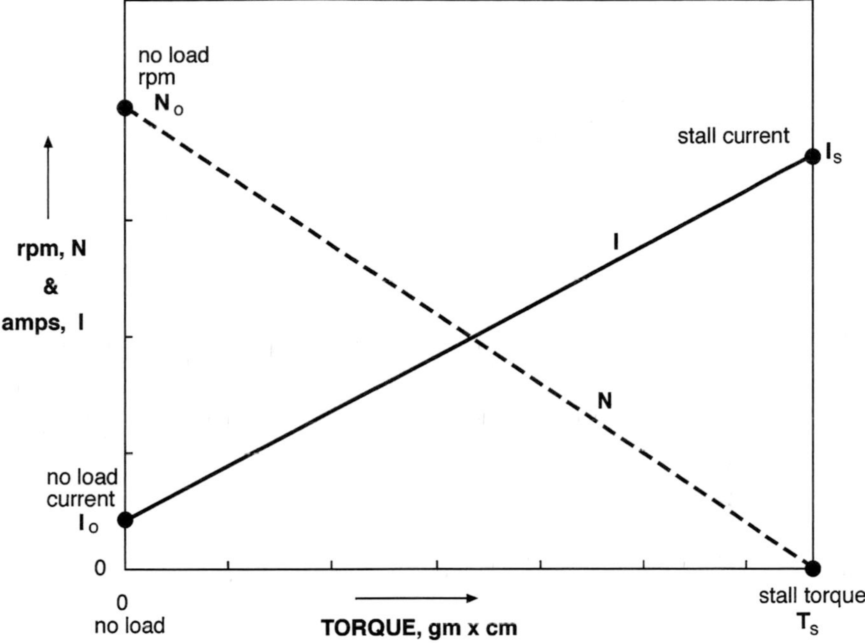

Fig. 2.32 DC motor current and rpm for fixed voltage vs torque.

these rpm and current lines then, virtually all of the important motor performance quantities can be calculated at any value of T_M:

Shaft power:

$$P(\mathrm{W}) = \frac{N \times T_M(\text{g-cm})}{97,500} \tag{2.25}$$

Efficiency:

$$\eta_M = P/v \times I \tag{2.26}$$

Motor heating:

$$H(\mathrm{W}) = v \times I \times (1 - \eta_M) \tag{2.27}$$

It also can be shown that

$$\text{Maximum power } P_{\max} \text{ occurs at } T_M = \frac{T_{\mathrm{MS}}}{2} \tag{2.28}$$

$$\text{Maximum efficiency occurs at } P < P_{\max}, \text{ at current } I = \sqrt{I_0\ I_S} \tag{2.29}$$

where I_S is the stalled motor, or locked-rotor, current at torque T_{MS}. In Fig. 2.34, the calculated quantities of power, efficiency, and heating are added to produce the standard motor characteristic plot. For further details of direct current motor specification, design, and analysis, consult manufacturers' information, for example, Mabuchi Motors Ltd. Catalog # 26, 2000.

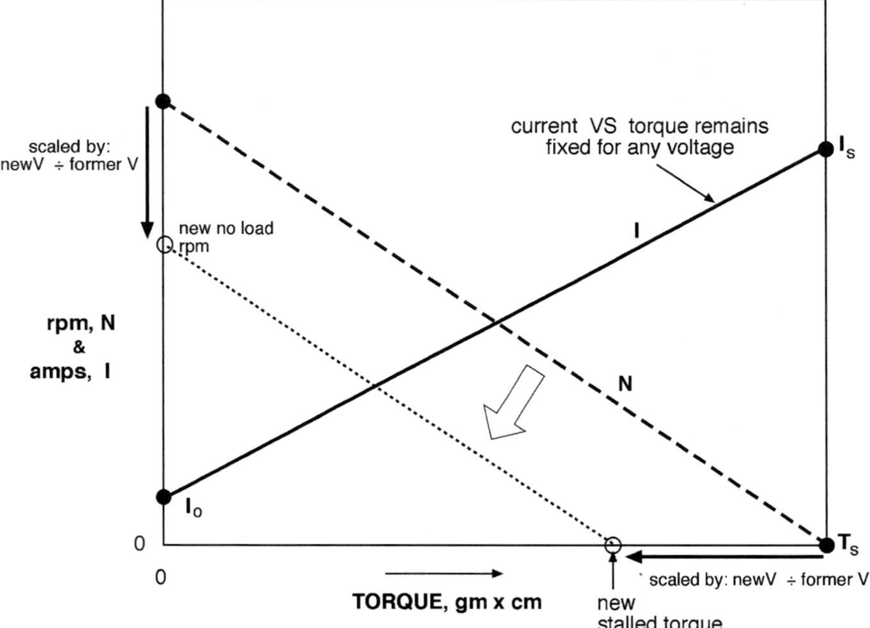

Fig. 2.33 DC motor basic plot scaled for different voltage vs torque.

Motor heating H is not always included in manufacturers' motor plots. Nevertheless, it is an important design item. It often limits the feasible upper operating level of motor power, which is usually below maximum power, and helps define practical applications of the motor. This is especially true for small, relatively low-efficiency motors and very light "pager" coreless motors, which have a low tolerance to overheating. An additional limit to motor performance, especially for coreless motors, is its maximum allowable rpm, above which mechanical failure can occur. Very light loads, and/or excessive voltage, can result in over-rpm and motor damage.

From the typical motor characterization plot of Fig. 2.34, some simple operating principles for dc motors follow:

1) Maximum power occurs midway between no load and stall load.

2) Maximum efficiency occurs below maximum power.

3) Increasing motor loads with larger, higher pitch props will increase current and motor heating and will not necessarily increase power.

4) Reasonable operating loads, current, and rpm are always between maximum power and maximum efficiency.

5) Motor heating can limit maximum useful power.

6) At some point usually below maximum power, a best compromise between power and efficiency exists.

7) If best operating region rpm is too high, gearing down the motor can provide equal power, torque, and efficiency at lower rpm (scaled by the gear ratio), though motor efficiency η_M is reduced by the gearbox efficiency η_G such that new $\eta'_M = \eta_M \times \eta_G$.

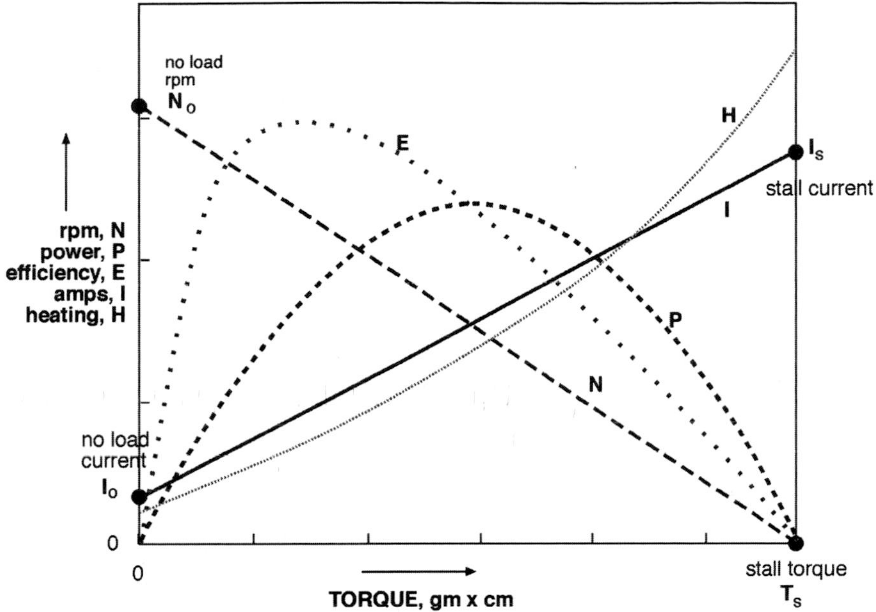

Fig. 2.34 DC motor basic plot for constant voltage.

Generating characterization plots for unknown motors. Because characterization plots completely describe a motor's unique electromechanical characteristics and the fixed relationships among torque, rpm, current, power output, and efficiency, they are necessary inputs to MAV power system design. This includes the evaluation and selection of battery/propeller/motor combinations, as discussed later. Some motor manufacturers (e.g., Mabuchi and MicroMotors) provide detailed engineering specifications for their motors. For many model motors, however, the descriptive specs provided are often quite sparse. A recommended operating voltage and current limits might be given, and at times a power range is indicated, but not always specified as input (battery) power or output (shaft) power. Efficiency data in a usable form are most often missing. In many cases therefore, it will be necessary to develop further information or characterization plots by testing, in order to properly include the motor in the MAV design process. A full dynamometer device that can measure torque, rpm, and electrical input simultaneously is the ideal test instrument. These devices, especially for very small, low torque motors of interest to MAV design, require great precision and are quite expensive because of the low torque levels involved. As an alternative, it is possible to obtain from an accurately instrumented source several standard load propellers for which the rpm vs power relationship is known. By simply measuring motor rpm and input voltage and current (relatively simple measurements) while using a standard load propeller, the motor's output power and efficiency can be determined. Another approach is to use a motor with known characteristics to establish power vs rpm curves for several propellers, which can then be used as standard load propellers on other motors with unknown characteristics.

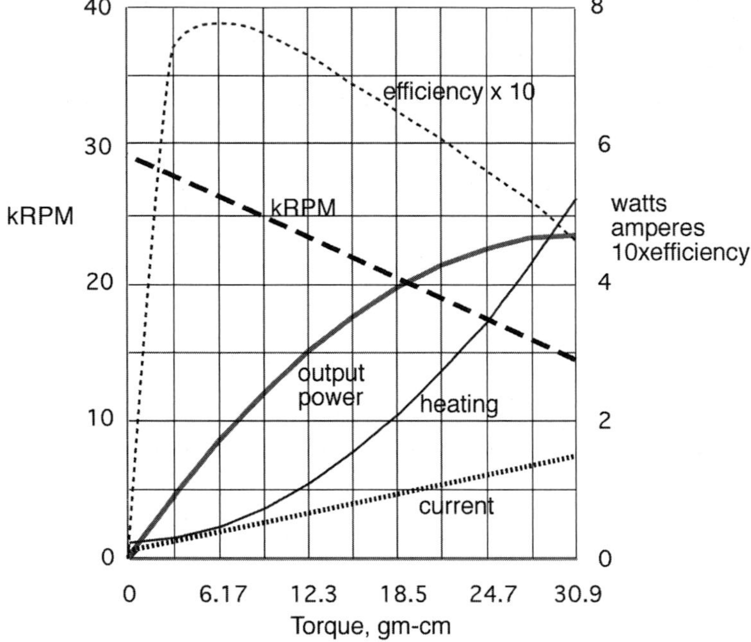

Fig. 2.35 DC5-2.4 performance at 7 V.

Selecting motor/battery/propeller combinations. Figure 2.35 shows the basic characterization plot for the coreless MicroMotor DC5-2.4. The DC5 has been a favorite motor for lightweight RC models and was used in early NRL MITE test vehicles. It will be used to illustrate some of the example design analyses that follow. As shown in the DC5 plot, the maximum power obtainable at 7 V is about 4.7 W. At this point the associated speed is 15,000 rpm, current is 1.4 A, efficiency is 47%, and motor heating is 5.2 W. The maximum efficiency point is 77% at a lower load (torque) value, corresponding to 1.68 W of power and 0.31 A current, with a very low heating of about 1/2 W. Because maximum efficiency always occurs well below the maximum power point, a tradeoff between power and efficiency is necessary in choosing the design power level.

As discussed earlier, it is impractical to operate below the maximum efficiency point or above the maximum power point, because in both cases power and efficiency are falling off. In addition, motor heating can limit the practical upper bound of power at a point below maximum. For the DC5 motor, the manufacturer recommends, and experience has corroborated, that motor heating should be held below about 3 W for improved reliability and reasonable motor life. At 7 V, therefore, a maximum heating of 2.7 W was chosen. The resulting feasible operating region for the DC5, in terms of rpm and current, is shown in Fig. 2.36. The heat-limited maximum output power is 4.3 W, at an efficiency of 61% and motor rpm of 19,000. As can be seen, this is a good tradeoff compared to the maximum power point because motor heating is now cut in half while power is reduced by less than 10%, and efficiency increased by 14%.

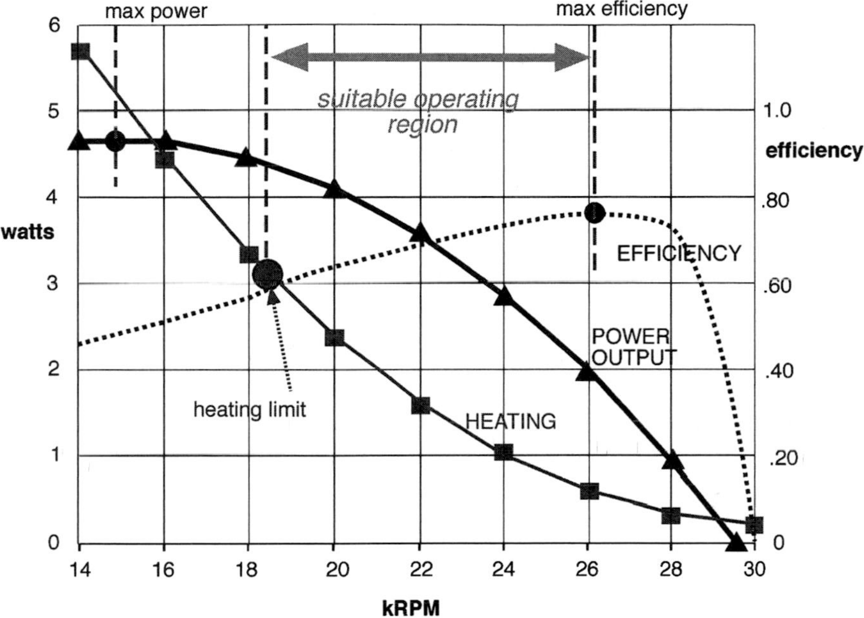

Fig. 2.36 DC5-2.4 operating region, 7 volts.

An example MAV design will be used for illustrative purposes, which is similar to the NRL MITE configurations. It uses *two* outboard mounted motors employing counter-rotating props. If the initial target gross weight is 140 to 200 g (5 to 7 oz), the motor power required can be estimated as follows.

The power required for level, unaccelerated flight was given by Eq. (2.14). A similar, but extended relationship has been developed by W. Hewitt Phillips (Phillips, W. H., correspondence with author, February 1988) in the form of watts of input power required per unit gross weight of the aircraft for various climb angles. Watts per pound is a convenient parameter commonly used by the electric modeling community for initially sizing their electric power systems. For a given vehicle design with known aerodynamic and power system properties such as wing loading, L/D, motor and prop efficiencies, the Phillips' relationship defines the required input power to the motor(s) of an electrically powered vehicle as a function of wing loading and climb angle. It is given by

$$\frac{P_B}{W_0} \approx \left[\frac{1}{L/D} + \sin\phi\right] \times \left(\frac{1}{\eta_P}\right) \times \left(\frac{1}{\eta_M}\right) \times \sqrt{\frac{W_0}{S} \times \frac{2}{\rho\, C_L}} \tag{2.30}$$

where P_B is the motor input power that the battery must supply, W_0 is gross weight, S is wing area, C_L is the lift coefficient, ϕ is the climb angle, η_P is the propeller efficiency, and η_M is the motor efficiency.

Our example MAV design has the following characteristics: gross weight, $W_0 = 170$ g $\Rightarrow$ 1.7 N (6 oz); wing area, $S = 645$ cm^2 (0.69 ft^2); $W_0/S = 26$ N/m^2 (8.7 oz/ft^2); operating lift coefficient, $C_L = 0.7$; $L/D = 5$; motor efficiency,

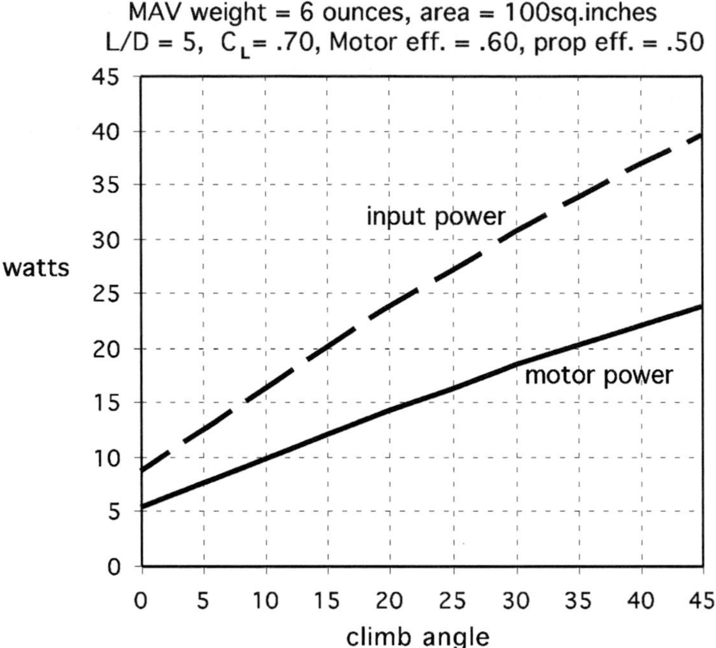

Fig. 2.37 Power required vs climb angle.

$\eta_M = 60\%$; and propeller efficiency, $\eta_P = 50\%$. Using Eq. (2.30) we can compute the battery input power and motor output power required by the example MAV for various climb angles. The results are plotted in Fig. 2.37. In this case, for level flight the power required from the battery is about 5.1 W/N (22.6 W/lb), or 8.5 W total. The motor output power at that input is about 5.2 W total, or 2.6 W for each motor. Because the DC5 motor operating range is from 1.68 W up to 4.3 W of shaft power, total maximum power available is 8.6 W—more than enough for level flight. At maximum power the motors provide an output power margin of 3.4 W, or 65% above the level flight requirement. At full power, the vehicle should be able to sustain a climb angle of about 8 deg—a feasible but fairly low safety margin. It might be adequate for a cruise-only, modest maneuvering vehicle in relatively nonturbulent air. In general, a more conservative, safer design power margin of two or more times level flight power would be desirable. For example, two times level flight power (17 W of input power) would allow climb angles up to about 12 deg and provide additional room for increased power demands when maneuvering, coping with wind and turbulent weather, plus allowances for inevitable payload weight growth. Acceptable power safety margin is a design judgement involving considerations of possible operating environment variations, development contingencies, and future growth requirements. Our MAV example at this stage indicates that weight growth or deterioration in aerodynamic parameters should be monitored and avoided, or higher power motors might have to be selected.

At this point, we can estimate that the 7-V battery must supply 8.5 W, or 1.21 A at 7 V, during level flight, and 1.0 A per motor × 2 = 2.0 A during maximum power operation. If the nominal mission flight profiles, including expected times at maximum power (takeoff, climbout, and maneuver) and at cruise power (transit time) are known, the battery discharge performance can be specified. For example, consider a desired mission time of 15–30 min with a maximum power time of 6 min. The minimum battery capacity required is as follows:

$$\text{Max power capacity} = 6 \text{ min} \times 2.0 \text{ A} = 12 \text{ (A} \cdot \text{min)}$$

$$\text{Cruise capacity} = 1.21 \text{ A} \times 15{-}30 \text{ min} = 18.2{-}36.3 \text{ (A} \cdot \text{min)}$$

$$\text{Total capacity} = 30{-}48 \text{ (A} \cdot \text{min)}, \text{ or about 500 to 800 mAh}$$

If the average voltage during use is about 7.0 V, then the energy supplied by the battery will be

$$\text{Energy} = 7.0 \times (30{-}48) = 210{-}336 \text{ (W} \cdot \text{min)}$$

Because high-quality LiPo cells have specific energies of about 7 W·min/g,

$$\text{Battery weight} = (210{-}336)/7 \approx 30 \text{ to } 48 \text{ g (1.1 to 1.7 oz)}$$

Assuming the example MAV's estimated weight of 170 g includes an allowance for a 43-g (1.5-oz) battery, the battery would be 25% of gross weight, or about $\frac{1}{3}$ the weight of the airframe (vehicle less battery). Figure 2.30 shows that a battery of this weight provides a bit more than half (53%) of the maximum endurance capability of the vehicle, compared to carrying the maximum endurance battery weight of 255 g (9 oz). When battery weight is 25% or less of vehicle gross weight, increasing battery weight can increase endurance significantly, so that in this range it is worthwhile to explore higher capacity, heavier batteries. For example, if a battery weighing 57 g (2.0 oz) were used instead of the 43-g battery, it would be about 31% of gross weight (which would then be 184 g). Compared to the lighter battery, endurance capability would increase more than 15% for a gross weight increase of about 8.3%. In any event, two series-connected LiPo cells of 600 to 700 mAh rated capacity, each weighing about 15–24 g, should meet the design power and endurance requirements.

The next task is to determine suitable propellers for the MAV design that can be matched to the DC-5 motors. Inputs required for this determination include 1) power required to turn propellers at appropriate speeds and 2) suitable propeller pitch and diameter combinations that satisfy flight speed and motor power/rpm constraints. The power to turn a propeller at different speeds can be approximately scaled with the rotation speed cubed:

$$P = K_{\text{Prop}} \times N^3 \tag{2.31}$$

where K_{Prop} is a constant for the particular propeller, depending primarily on diameter, and to a lesser extent upon pitch and blade shape. This approximation has been found useful for typical thin airfoil model propellers of 13–30 cm (5–12 in.) diameter with pitches of 7.6–30 cm (3–12 in.). Peculiar hub and/or blade shape and twist can change the relationship somewhat, including the nominal exponent value of 3, usually within a range of 2.9 to 3.1. K_{Prop} can be estimated

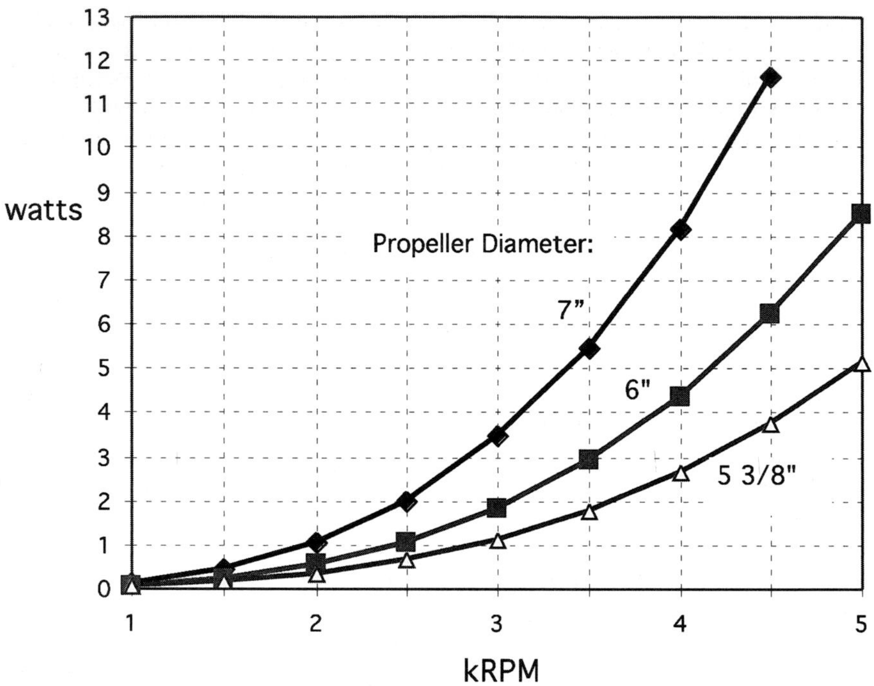

Fig. 2.38 K&P propeller power requirements vs rpm.

from one or two known points of power and rpm for a specific prop and the curve extrapolated to other rpm levels by means of the cubic power law. One can also make use of the large and growing online library of model propeller K_{Prop} factors and exponents measured by contributing electric modelers. Unfortunately, data for MAV props under 8–10 cm (3–4 in.) in diameter are presently rather limited, and aerodynamic theory does not readily predict performance of the tiniest, micro-propellers. Sample curves, developed empirically, of power required vs rpm for typical MAV propellers are shown in Fig. 2.38.

2.4.2.3 Electric propulsion propeller requirements. Having defined motor performance characteristics as well as propeller power absorption curves, the next requirement is to identify specific propellers having characteristics that match the selected motor's power, torque, and rpm profile. Propeller selection, like motor power requirements, depends on desired MAV speed and climb rates, as well as vehicle aerodynamic properties such as drag and lift coefficients, wing loading, and weights. The selection process is an iterative one, and so we will outline one simple approach to arrive at a reasonable first approximation. Ultimately, flight testing will be necessary to finalize the propeller providing the best design solution, generally a compromise between different flight regimes, motor/gearbox performance, and the availability of stock or custom components.

Simple methods for making an initial estimate of suitable power systems for electric model airplanes or MAVs, including propeller parameters, have been developed and described by a number of experienced electric model aircraft designers [84–86]. The methods include both basic aerodynamic theory and the use of empirically derived "rule-of-thumb" performance estimates. Most have been corroborated by extensive flight experience with many types and sizes of electric models. Much of this experience has been with conventional model configurations, often larger and faster than the typical 15–60 cm span MAV weighing 50–350 g. For that reason the estimates should be viewed as first approximations, especially for the very smallest MAVs.

As a first step, speed for level flight V was given in Eq. (2.12) as

$$V = \sqrt{\frac{W_0}{S} \times \frac{2}{\rho C_L}} \tag{2.32}$$

where W_0/S is wing loading expressed in Ns/m^2. For a $C_L = 0.7$, and at sea level, this becomes

$$V\,(\text{m/s}) = 1.53\sqrt{\frac{W_0}{S}\,(\text{N/m}^2)} \tag{2.33}$$

Empirically, it has been found that stall speed V_S for a clean model monoplane is approximately

$$V_S\,(\text{m/s}) \approx \sqrt{\frac{W_0}{S}\,(\text{N/m}^2)} \tag{2.34}$$

Also, a propeller's pitch speed V_P, that is, its nonslip speed of advance, is approximated by

$$V_P\,(\text{m/s}) \approx 0.18 \times \text{rpm (in thousands)} \times \text{prop pitch (cm)} \tag{2.35}$$

To achieve a moderate performance margin beyond level flight (i.e., ability to loop), it has been found that pitch speed should be at least twice stall speed (about 1.3 times level flight speed in our case). More aggressive aerobatic performance, such as performing rolls or WWII fighter-type maneuvers, will require a higher pitch speed, but should be kept less than twice level flight speed. Using these approximations for the example MAV, level flight speed is about 7.8 m/s (26 ft/s) and stall speed about 4.9 m/s (16 ft/s). Based on these results, a reasonable target pitch speed is 10 m/s (33 ft/s).

Several small commercial propellers, available in both right- and left-hand rotation configurations, were analyzed for possible use in the twin-motored MITE vehicle, which requires counter-rotating props. These are the Knight and Purdham Company (K&P) electric props of 15.2 cm diam × 16.8 cm pitch (6 × 6.6 in.), and 17.8 cm diam × 20.5 cm pitch (7 × 8.1 in.). From Eq. (2.34) we find that, to achieve 10-m/s pitch speed, the props must rotate at least 3310 and 2710 rpm, respectively. Comparing the basic DC5 motor plots in Fig. 2.36 to the K&P prop power absorption curves in Fig. 2.38 shows that motor gearing of some sort will be required to reduce motor rpm and increase torque in order to match prop rpm at equivalent power levels. Gear units of 6:1 and 8:1 ratios were selected for evaluation, and therefore shaft rpm is multiplied by 1/6 and 1/8 to produce the power

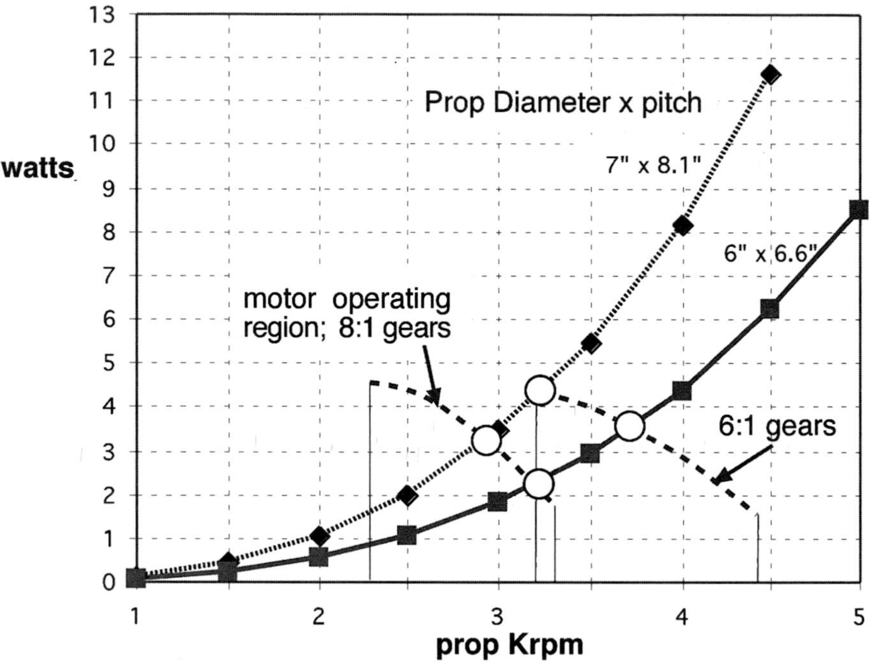

Fig. 2.39 K & P power requirements matched to geared DC5 at 7 V.

vs rpm curves for the geared motors. If gearbox efficiency η_G is known, the power levels P can be reduced by $\eta_G \times P$ to account for these losses.

The geared motor power curves are next overlaid onto the prop power curves to determine coincident design solutions, as shown in Fig. 2.39. The example overlay curves show there are four feasible operating points where prop and geared motor rpm match at equal power levels. These points lie in the suitable motor power operating regions defined earlier, and are listed in Table 2.2. As seen, the 15.2 × 16.8 cm prop with 8:1 gearing provides lower than design pitch speed. This means that this combination might not fly the MAV at acceptable speeds above stall. The

Table 2.2 Possible motor/gear reduction/propeller matches for example MAV

Propeller dia × pitch, cm	Gear ratio	Prop rpm	Motor power, W	Motor, rpm	Motor current, A	Motor efficiency	Pitch speed, m/s	OK pitch speed?
17.8 × 20.5	8:1	2,950	3.1	23,600	0.60	0.73	10.6	yes
17.8 × 20.5	6:1	3,200	4.3	19,200	1.0	0.61	11.5	yes
15.2 × 16.8	8:1	3,200	2.4	25,600	0.4	0.75	9.3	no
15.2 × 16.8	6:1	3,700	3.5	22,200	0.7	0.69	10.9	yes

other combinations will achieve just acceptable speeds, with the 17.8 × 20.5 cm prop with 6:1 gearing providing the highest margin, and at the highest acceptable motor power.

Overall, it was determined that the DC5 power system with 6:1 gear reduction, operating at 7 V with a 17.8 × 20.5 cm prop, could provide modest but adequate performance at efficient power levels and current consumption within the design operating area. A larger variety of gearing ratios and propeller sizes, of course, would allow searching for more design choices within the operating area. Subsequent flight tests of a MITE MAV with an equivalent motor/prop/battery system showed satisfactory performance and verified that initial design estimates were reasonably accurate.

Scaling propellers for gear reduction drive. It is sometimes necessary to estimate what propellers a motor will turn when gearing is added. If a motor has been used to drive a specific propeller on a specific model directly with success, an approximation can be made of how to scale the direct drive propeller for geared operation [86]. To keep the motor load the same with a gear system having speed reduction ratio G and efficiency η_G, the gear-driven propeller's diameter and pitch should be scaled by

$$\text{Scale Factor} = \eta_G^{1/5} \times G^{3/5} \tag{2.36}$$

The reference paper gives this example: assume a direct drive prop of 15.2 × 7.6 cm (6 × 3 in.) is turned at 6000 rpm. If a 4:1 gear system with an efficiency of 95% is added, at the same power and motor rpm the geared motor can turn a prop scaled up by a factor of 2.274, to about a 34.5 × 17.3 cm (13.6 × 6.8 in.) prop. This larger prop will turn at 1500 rpm. Gear systems do add weight and an efficiency penalty, typically 2–7% loss for a good quality gearbox. Nevertheless, they often allow much better matching of the motor's rpm vs torque characteristics to large, efficient propellers.

Propeller design. In some cases a suitable commercial propeller might not be found. Designing and building a propeller might be necessary. A complete analytic treatment of propeller design for minimum induced loss by E. Eugene Larrabee can be found in several publications [87–89], many of which include application to small scale, model propellers. One of the features of the Larrabee minimum induced loss propellers is the unusual blade shape, which has the maximum blade chord at about 40% of the distance from the hub to the tip, with the outboard blade sharply tapered. Diameter to pitch ratios are often about 1.1–1.2 and deviate from constant, helical pitch by reducing pitch somewhat near the blade tips. Sample Larrabee-type blade shapes, which were used in several MITE MAV test vehicles are illustrated in Fig. 2.40. Figure 2.41 shows several MAV props, including the K&P 15.2 and 17.8 cm (6 × 6.6 in.) props. Although current quantitative theory might be of limited use for optimizing the smallest MAV propellers of 5–8 cm (2–3 in.) diam, the detailed analytic procedures and computer code discussed in the references are worth investigating for possible MAV application.

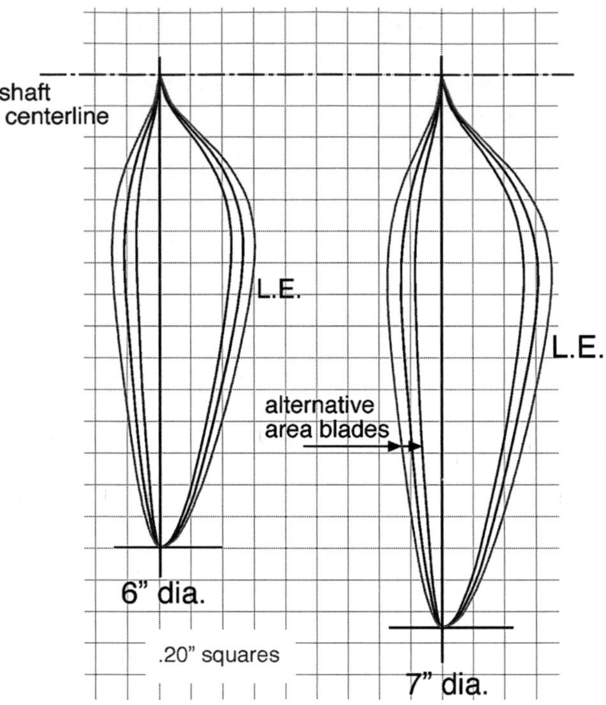

Fig. 2.40 Minimum induced loss Larrabee-type propeller blade shapes for P/D = 1.0 and diameters of 6 and 7 inches.

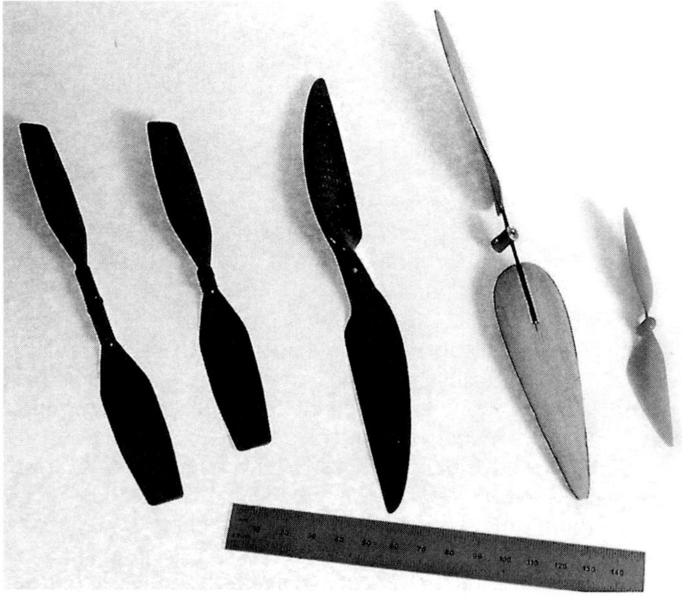

Fig. 2.41 Typical MAV propellers. Left to right: K & P 17.8 cm and 15.2 cm, westech 20 cm, and a 10 cm prop.

2.5 MAV Design Using Experimental Database

The experimental database and empirical relations obtained from the wind-tunnel experiments in the section on rigid wings were implemented within an airplane performance prediction procedure. This procedure was incorporated into a genetic algorithm optimization program that identified near-optimum MAV configurations given certain requirements and constraints. The results obtained revealed that useful design tools could be developed based on the experimental database. [18]

The design example incorporates the requirements and mission objectives that are common to many MAVs. Even though mission objectives vary greatly, many MAV missions share the same characteristics. Torres [18] defined the design problem as "design the smallest air vehicle" that 1) can carry a specified payload consisting of a wireless video transmitter/video downlink, 20 g (0.705 oz), and an additional 20 g (0.705 oz) of payload for optional equipment such as an autopilot, GPS for navigation, other types of sensors, or extra fuel; 2) has an endurance of at least 10 min and a surveillance range of at least 0.8 km (0.5 mi); 3) has a minimum cruise velocity of 40 km/h (25 mph); 4) can fly in vertical gusts of 16 km/h (10 mph); and 5) is maneuverable enough to be able to capture an image of a target on the ground.

The size of the vehicle was defined explicitly as the maximum linear dimension between any two points located on the vehicle while it is airborne. The following **restrictions** were imposed in order to simplify the design process: 1) it should be a fixed-wing tailless airplane (i.e., a flying wing), 2) the wing should be a flat plate with zero camber, a thickness-to-chord ratio of approximately 2%, and 5 to 1 elliptical leading and trailing edges so that it matches the wing models used in the wind-tunnel experiments, 3) the control surfaces shall include a trailing-edge flap and either ailerons or a rudder, 4) the propulsion system shall be predefined as the Cox® Tee Dee® 0.16-cm^3 (0.010-$in.^3$) internal combustion engine (the only small engine of its kind available at the time) and commercially available propeller in the tractor configuration, 5) the nonpayload components shall also be predefined and shall be selected based on the commercially available products (the video downlink payload is also predefined in mass and size), and 6) the structural mass shall be estimated. Taking all of the predefined components (see Table 2.3), structural, fuel, and the 20 g (0.705 oz) of optional mass into account produces an estimated mass of 122.4 g (4.32 oz).

The genetic optimization algorithm was used to identify a group of near-optimum MAV designs from which a design was chosen based on criteria set by the designer. The optimization procedure applied in this manner was found to converge on feasible designs that met all constraints and satisfied all mission requirements and restrictions. The performance of the final wing design with an approximate fuselage (without the engine, propeller, and vertical control surface) was analyzed and compared with wind-tunnel data. The comparison was found to be good, though some discrepancies in pitching moment were found. It was proposed that the differences between predicted and measured performance were caused by an underestimation of the control-surface effectiveness. A prototype of the final design was constructed and successfully flight tested. Flight performance was very good as the airplane was found to be both stable and maneuverable. The final design and arrangement of the components inside this MAV are shown in Figs. 2.42

Table 2.3 Predefined Components

Component	Mass, g	Dimensions, cm	Center of gravity, cm
Engine and propeller	13.6	3.0 × 1.5 × 3.5	0.50 × length
Fuel tank[a]	12.0	4.3 × 1.7 × 1.7	0.50 × length
Video transmitter	7.4	5.5 × 2.0 × 0.75	0.50 × length
Video camera	7.0	1.5 × 2.3 × 1.5	0.50 × length
RC receiver	10.6	4.7 × 2.3 × 1.5	0.50 × length
RC control	5.6	2.0 × 1.2 × 2.5	0.50 × length
Batteries[b]	7.5	3.0 × 0.5 × 2.5	0.50 × length

[a]Holds about 10cc (10 g) of fuel that allows about 10 min of flight.
[b]Batteries power video and RC equipment and can last for about 10 min of flight.

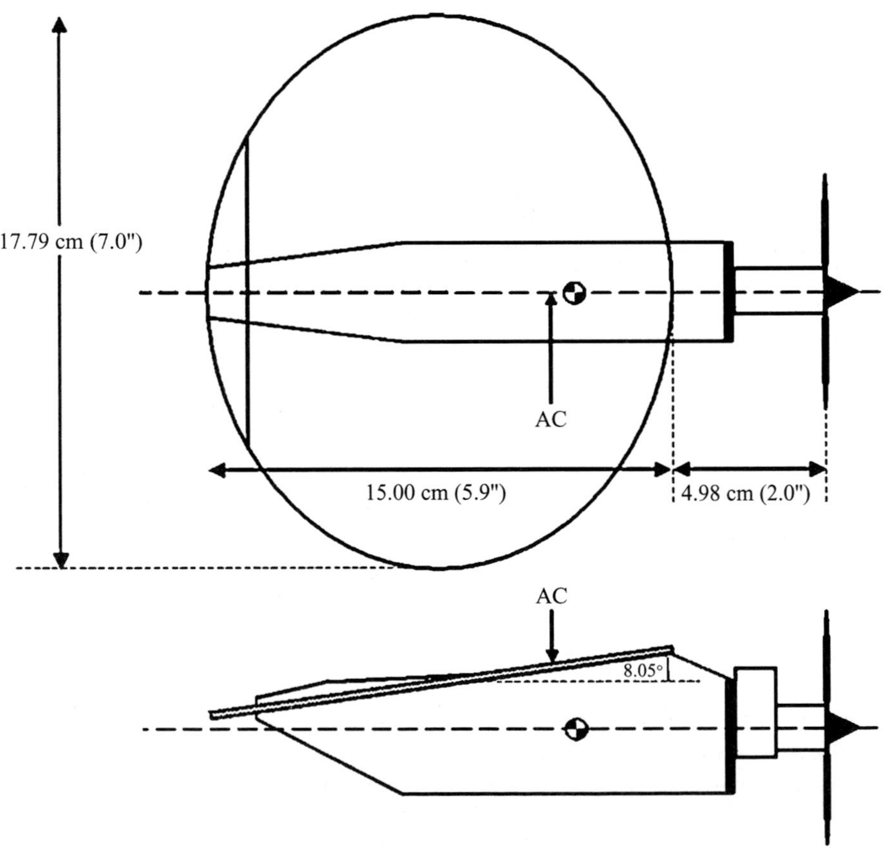

Fig. 2.42 Top and side views of Torres [18] final MAV design.

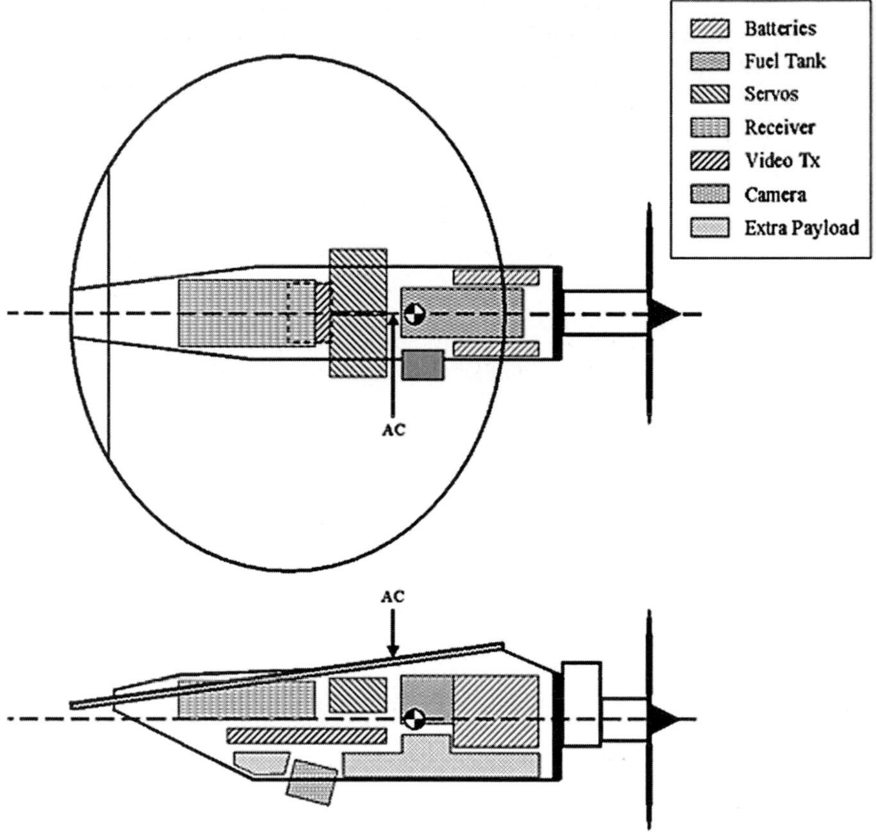

Fig. 2.43 Arrangement of components inside the final design.

and 2.43. Figure 2.44 shows the radio-controlled prototype of the final design. This MAV has an elliptical planform with an aspect ratio of 1.51, a span of 17.8 cm (7 in.), a wing area of 210 cm^2 (32.55 in.^{2}), and a flight speed of about 72.4 km/h (45 min/h). The reason for the relatively high cruise speed can be attributed to selection of the propulsion system. The design requirements and mission objectives impose a restriction on the minimum cruise speed not the maximum value. More details of the design and optimization procedure are in [18].

2.6 Numerical Approach to MAV Design

Recent numerical simulations by Cosyn and Vierendeels [90] produced results that compare well with the two- and three-dimensional experiments performed at Notre Dame on thin rectangular flat plates. They used a commercial CFD program (Fluent) with the Spalart–Allmaras turbulence model and a strip method to study two- and three-dimensional flows over thin rectangular LAR wings with aspect ratios between 0.5 and 2 at a Reynolds number of 10^5. The lift and drag coefficients calculated agreed reasonably well with the experimental data of Pelletier and

Fig. 2.44 Radio-controlled prototype of final design [18].

Mueller [49] up to angles of attack of 10 deg for the two-dimensional cases and up to 15 deg for the aspect ratios of 0.5 and 1.0. As would be expected, the agreement was better for lift than for drag. Lift and drag coefficient results were also obtained for the S5010 reflex profile for the two-dimensional case and for $AR = 0.5$ and 1.0 at Reynolds number of 10^5. Only two-dimensional experimental data were available for the S5010 profile [35] and [44], and the agreement with the numerical results was reasonable at the low angles of attack. The agreement between the CFD and strip method for the S5010 profile for $AR = 0.5$ and 1.0 was good up to angles of attack of about 12 deg.

Cosyn and Vierendeels [91] continued their fundamental research in this area and developed a methodology and design strategies to design a series of fixed-wing micro air vehicles. The approach included the use of a vortex-lattice method, three-dimensional CFD computations, and a numerical propeller optimization method to derive the forces and their derivatives of the MAV and propeller. This information was used for performance and stability-related optimization studies. This approach was used to design an electrically powered 25-cm (10-in.) span flying MAV with an aspect ratio of 1.8 for maximum endurance. A prototype, the UGMAV25 (shown in Fig. 2.45), with a 25-cm (10-in.) span, a wing area of 347 cm^2 (53.78 $in.^2$), and a mass of 58 g (2.0 oz) was constructed and wind tunnel tested. The flight tests were

Fig. 2.45 The 25-cm (10-in.) span UGMAV 25 designed using numerical methods at the University of Ghent, Belgium (reprinted with permission of Jan Vierendeels, University of Ghent, Ghent, Belgium) [90].

successful. Another MAV, UGMAV15 (shown in Fig. 2.46), with a 15-cm (6-in.) span and a mass of 37 g (1.3 oz) was designed, constructed, and successfully flown without any wind-tunnel tests.

2.7 Conclusion

Except for the pioneering research of Schmitz in the 1930s, the most significant research to understand the performance deterioration of low-Reynolds-number airfoils and wings took place in the 1980s. Initially, existing airfoils designed for Reynolds numbers of 5×10^5 or greater were studied at lower Reynolds numbers where the lift decreased, the drag increased, and hysteresis in these forces often occurred. These problems were better understood and solved by using the design methods of Eppler and Drela to design profiles for a specific low-Reynolds-number range. A large number of careful wind-tunnel experiments were used to determine and verify the lift and drag performance and the effects of laminar separation bubbles. The increasing interest in these problems was a result of the desire to improve the performance of jet engines at high altitudes, sailplanes, and small unmanned aircraft. This interest resulted in conferences where active researchers met and shared the results of their research. The challenge involved in the development

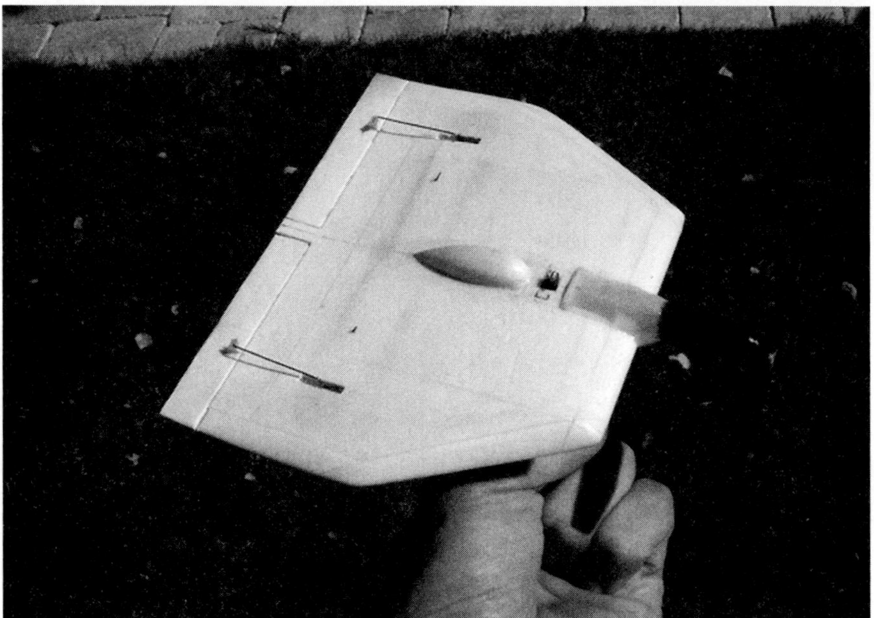

Fig. 2.46 The 15-cm (6-in.) span UGMAV 15 designed using numerical methods at the University of Ghent, Belgium (reprinted with permission of Jan Vierendeels, University of Ghent, Ghent, Belgium).

of efficient micro air vehicles in 1995 combined the problems of low-aspect-ratio wings with those of low Reynolds numbers.

Extensive experimental studies of the effects of planform shape, aspect ratio, and Reynolds number with thin flat-plate wings concluded that aspect ratio and planform shape accounted for greater differences in performance than did Reynolds numbers between 7×10^4 and 1.4×10^5. Experimental results have been useful in the development of numerical methods that can now be used in the MAV design process. Electric motors have significant advantages over internal combustion engines for MAV propulsion. Recent advances in lightweight electric motors and batteries with increased energy densities have made electric propulsion the predominate choice for MAV propulsion systems.

Acknowledgments

The authors would like to thank the following people for their help in supplying information, comments, and suggestions for this chapter: Richard Eppler of Stuttgart, Germany; Michael Selig of the University of Illinois, Urbana, Illinois; Alain Pelletier of Orlando, Florida; James DeLaurier of the University of Toronto, Ontario, Canada; Jan Vierendeels of the University of Ghent, Ghent, Belgium; Sergey Shkarayev of the University of Arizona, and James Kellogg of the U.S. Naval Research Laboratory, Washington, DC.

References

[1]Stallings, R. L., Jr., "Low Aspect Ratio Wings at High Angles of Attack," *Tactical Missile Aerodynamics*, edited by M. J. Hemsch and J. N. Nielsen, Progress in Astronautics and Aeronautics, Vol. 104, AIAA, Reston, VA, 1986, pp. 89–128.

[2]Williams, D. L., II, "Dynamic Lateral Behavior of Low-Aspect-Ratio, Rectangular Wings at High-Angle-of-Attack," Ph.D. Dissertation, Dept. of Aerospace and Mechanical Engineering, Univ. of Notre Dame, IN, April 1996.

[3]Zimmerman, C. H., "Characteristics of Clark Y Airfoils of Small Aspect Ratios," NACA TR 431, May 1932.

[4]Zimmerman, C. H., "Characteristics of Several Airfoils of Low Aspect Ratio," NACA Tech. Note 539, Aug. 1935.

[5]Bartlett, G. E., and Vidal, R. J., "Experimental Investigation of Influence of Edge Shape on the Aerodynamic Characteristics of Low Aspect Ratio Wings at Low Speed," *Journal of Aeronautical Science*, Vol. 22, No. 8, 1955, pp. 517–523.

[6]Wadlin, K. L., Ramsen, J. A., and Vaughan, V. L., Jr., "The Hydrodynamic Characteristics of Modified Rectangular Flat Plates Having Aspect Ratios of 1.00, 0.25, and 0.125 and Operating near a Free Water Surface," NACA TR 1246, 1955.

[7]Neyland, V. Ya., and Stolyarov, G. I., "One Form of Separated Flow over a Low Aspect-Ratio Rectangular Wing," *Fluid Mechanics: Soviet Research*, Vol. 10, No. 6, 1981.

[8]Bollay, W., "A Non-Linear Wing-Theory and Its Application to Rectangular Wings of Small Aspect Ratio," *Zeitschrift fur Angewandte Mathematik und Mechanik*, Vol. 19, 1939, pp. 21–35.

[9]Weinig, F., "Lift and Drag of Wings with Small Span," NACA Tech. Rep. TM-1151, 1947.

[10]Lawrence, H. R., "The Lift Distribution on Low Aspect Ratio Wings at Subsonic Speeds," *Journal of the Aeronautical Sciences*, 1951, pp. 683–693.

[11]Ermolenko, S. D., and Barinov, M. T., "Calculation of the Stream Deflection Angles Behind Rectangular Wings of Small Aspect Ratio at Subsonic Speeds Using Nonlinear Theory," *Soviet Aeronautics*, Vol. 9, No. 3, Fall 1966, pp. 1–4.

[12]Bera, R. K., and Suresh, G., "Comments on the Lawrence Equation for Low-Aspect-Ratio Wings," *Journal of Aircraft*, Vol. 26, No. 9, 1989, pp. 883–885.

[13]Polhamus, E. C., "A Concept of the Vortex Lift of Sharp-Edge Delta Wings Based on a Leading-Edge-Suction Analogy," NASA TN D-3767, 1966.

[14]Polhamus, E. C., "Predictions of Vortex-Lift Characteristics by a Leading-Edge-Suction Analogy," *Journal of Aircraft*, Vol. 8, No. 4, 1976, pp. 193–199.

[15]Rajan, S. C., and Shashidhar, S., "Exact Leading-Term Solution for Low Aspect Ratio Wings," *Journal of Aircraft*, Vol. 34, No. 4, 1997, pp. 571–573.

[16]Hoerner, S. F., *Fluid-Dynamic Drag*, Hoerner Fluid Dynamics, Brick Town, NJ., 1965, Chap. 7.

[17]Hoerner, S. F., and Borst, H. V., *Fluid-Dynamic Lift*, Hoerner Fluid Dynamics, Brick Town, NJ., 1975, pp. 9–8, 9–9, 17–5, and 20–8.

[18]Torres, G. E., "Aerodynamics of Low Aspect Ratio Wings at Low Reynolds Numbers with Applications to Micro Air Vehicle Design and Optimization," Ph.D. Dissertation, Dept. of Aerospace and Mechanical Engineering, Univ. of Notre Dame, IN, April 2002.

[19]Schmitz, F. W., "Aerodynamics of the Model Airplane, Part 1, Airfoil Measurements," Redstone Scientific Information Center, Research and Development Directorate, U.S. Army Missile Command, Redstone Arsenal, AL, 22 Nov., 1967.

[20]McMasters, J. H., and Henderson, M. L., "Low-Speed Single-Element Airfoil Synthesis," *Technical Soaring*, Vol. 6, No. 2, 1980, pp. 1–21.
[21]Lissaman, P. B. S., "Low-Reynolds-Number Airfoils," *Annual Review of Fluid Mechanics*, Vol. 15, 1983, pp. 223–239.
[22]Mueller, T. J., "Low Reynolds Number Vehicles," AGARD-AG-288, edited by E. Reshotko, SPS, Ltd., Loughton, Essex, UK, 1985, 69p.
[23]Carmichael, B. H., "Low Reynolds Number Airfoil Survey," Vol. 1, NASA CR 165803, Nov. 1981, 106p.
[24]Horton, H. P., "Laminar Separation Bubbles in Two and Three-Dimensional Incompressible Flows," Ph.D. Dissertation, Univ. of London, 1968.
[25]Brendel, M., "Experimental Study of the Boundary Layer on a Low Reynolds Number Airfoil in Steady and Unsteady Flow," Ph.D. Dissertation, Dept. of Aerospace and Mechanical Engineering, Univ. of Notre Dame, IN, May 1986.
[26]Brendel, M., and Mueller, T. J., "Boundary Layer Measurements on an Airfoil at Low Reynolds Numbers," *Journal of Aircraft*, Vol. 25, No. 7, 1988, pp. 612–617.
[27]Brendel, M., and Mueller, T. J., "Transition Phenomenon on Airfoils Operating at Low Chord Reynolds Numbers in Steady and Unsteady Flows," *Numerical and Physical Aspects of Aerodynamic Flows IV*, edited by T. Cebeci, Springer-Verlag, Berlin, 1990, pp. 333–344.
[28]Mueller, T. J. (ed.), *Proceedings of the Conference on Low Reynolds Number Airfoil Aerodynamics*, UNDAS-CP-77B123, University of Notre Dame, Dept. of Aerospace and Mechanical Engineering, June 1985, 349p.
[29]*Proceedings of Aerodynamics at Low Reynolds Numbers: $10^4 < Re < 10^6$ International Conference*, Vols. I, II, and III, London, Royal Aeronautical Society, London, 1986, 36 papers.
[30]Mueller, T. J. (ed.), *Low Reynolds Number Aerodynamics*, Lecture Notes in Engineering, Vol. 54, Springer-Verlag, Berlin, 1989, 446p.
[31]Donovan, J. F., and Selig, M. S., "Low Reynolds Number Airfoil Design and Wind Tunnel Testing at Princeton University," (edited by) T. J. Mueller, Lecture Notes in Engineering, Vol. 54, "Low Reynolds Number aerodynamics," Springer-Verlag, Berlin, 1989, pp. 39–57.
[32]Selig, M. S., Donovan, J. F., and Fraser, D. B., *Airfoils at Low Speeds*, Soartech 8, SoarTech Aero Publications, Virginia Beach, VA, 1989, 308p.
[33]Selig, M. S., and Maughmer, M. D., "Generalized Multipoint Inverse Airfoil Design," *AIAA Journal*, Vol. 30, No. 11, 1992, pp. 2618–2625.
[34]Selig, M. S., Guglielmo, J. J., Broeren, A. P., and Giguere, P., *Summary of Low-Speed Airfoil Data*, Vol. 1, SoarTech Aero Publications, Virginia Beach, VA, 1995.
[35]Selig, M. S., Lyon, C. A., Giguere, P., Ninham, C. P., and Guglielmo, J. J., *Summary of Low-Speed Airfoil Data*, Vol. 2, SoarTech Aero Publications, Virginia Beach, VA, 1996, p. 36 and pp. 142–145.
[36]Lyon, C. A., Boreren, A. P., Giguere, P., Gopalarathnam, A., and Selig, M. S., *Summary of Low Speed Airfoil Data*, Vol. 3, SoarTech Aero Publications, Virginia Beach, VA, 1997.
[37]Selig, M. S., Gopalarathnam, A., Giguere, P., and Lyon, C. A., "Systematic Airfoil Design Studies at Low Reynolds Numbers," *Fixed and Flapping Wing Aerodynamics for Micro Air Vehicle Application*, edited by T. J. Mueller, Progress in Astronautics and Aeronautics, Vol. 195, AIAA, Reston, VA, 2001, pp. 143–167.
[38]Eppler, R., and Somers, D. M., "A Computer Program for the Design and Analysis of Low-Speed Airfoils," NASA TM-80210, Aug. 1980.
[39]Eppler, R., *Airfoil Design and Data*, Springer-Verlag, Berlin, 1990, 562p.
[40]Eppler, R., *Airfoil Program System "Profile"*, User Guide, Airfoils, Inc. Port Matilda, PA, website: airfoils.com

[41]Drela, M., and Giles, M. B., "ISES: A Two-Dimensional Viscous Aerodynamic Design and Analysis Code," AIAA Paper 87-90424, Jan. 1987.

[42]Drela, M., "XFOIL: An Analysis and Design System for Low Reynolds Number Airfoils," *Low Reynolds Number Aerodynamics*, edited by T. J. Mueller, Lecture Notes in Engineering, No. 54, 1989, pp. 1–12.

[43]Lennon, A., *Basics of R/C Model Aircraft Design*, Air Age, Inc., Ridgefield, CT, 2002, pp. 111, 135.

[44]Simons, M., *Model Aircraft Aerodynamics*, 4th ed., Special Interest Model Books, Ltd., Dorset, England, UK, 2002, pp. 324–327.

[45]Broughton, B. A., Carroll, C. A., and Selig, M. S., "Wind Tunnel Testing of the S1052 and S1054 Flapped Airfoils for Unmanned Flying Wing Designs," Dept. of Aeronautical and Astronautical Engineering Report UILU ENG 01-0502, Univ. of Illinois at Urbana–Champaign, Feb. 2001.

[46]Kellogg, J., Bovais, C., Dahlburg, J., Foch, R., Gardner, J., Gordon, D., Hartley, R., Kamgar-Parsi, B., McFarlane, H., Pipitone, F., Ramamurti, R., Sciambi, A., Spears, W., Srull, D., and Sullivan, C., "The NRL Mite Air Vehicle," *Proceedings of the 16th International Unmanned Air Vehicle Systems*, 2001, pp. 25.1–25.14.

[47]Grasmeyer, J. M., and Keennon, M. T., "Development of the Black Widow Micro Air Vehicle," *Fixed and Flapping Wing Aerodynamics for Micro Air Vehicle Applications*, edited., by T. J. Mueller, Progress in Astronautics and Aeronautics, Reston, AIAA, VA, Vol. 195, 2001; also AIAA Paper 2001-0127, 2001.

[48]Mueller, T. J., "Aerodynamic Measurements at Low Reynolds Numbers for Fixed Wing Micro-Air Vehicles," *Development and Operation of UAVs for Military and Civil Applications*, RTO-EN-9, Canada Communication Group, Inc, Quebec, Canada, April 2000, pp. 8-1–8-32.

[49]Pelletier, A., and Mueller, T. J., "Low Reynolds Number Aerodynamics of Low-Aspect-Ratio, Thin/Flat/Cambered Plate Wings," *Journal of Aircraft*, Vol. 37, No. 5, Sept.–Oct., 2000, pp. 825–832.

[50]Torres, G. E., and Mueller, T. J., "Aerodynamic Characteristics of Low Aspect Ratio Wings at Low Reynolds Numbers," *Fixed and Flapping Wing Aerodynamics at Low Reynolds Numbers*, edited by T. J. Mueller, Progress in Astronautics and Aeronautics, Vol. 195, AIAA, Reston, VA, 2001, pp. 115–141.

[51]Torres, G. E., and Mueller, T. J., "Low-Aspect-Ratio Wing Aerodynamics at Low Reynolds Numbers," *AIAA Journal*, Vol. 42, No. 5, May 2004, pp. 865–873.

[52]Null, W., Noseck, A., and Shkarayev, S., "Effects of Propulsive-Induced Flow on the Aerodynamics of Micro Air Vehicles," AIAA Paper 2005-4616, June 2005.

[53]Bertin, J. J., *Aerodynamics for Engineers*, 4th. Prentice–Hall, Upper Saddle River, N J, 2002, pp. 163, 168, and 169.

[54]Pankhurst, R. C., and Holder, D. W., *Wind-Tunnel Technique*, Sir Isaac Pitman and Sons, London, 1952, Chap. 8.

[55]Barlow, J. B., Rae, W. H., Jr., and Pope, A., *Low-Speed Wind Tunnel Testing*, J Wiley, New York, 1999, Chap. 10.

[56]Kline, S. J., and McClintock, F. A., "Describing Uncertainties in Single-Sample Experiments," *Mechanical Engineering*, Vol. 75, No. 1, 1953, pp. 3–8.

[57]Kuethe, A. M., and Chow, C-Y., Foundations of Aerodynamics, 5th ed., J Wiley, New York, 1998, pp. 184–187.

[58]Lowry, J. G., and Polhamus, E. C., "A Method for Predicting Lift Increments due to Flap Deflction at Low Angles of Attack in Incompressible Flow," NACA TN 3911, 1957.

[59]Lamar, J. E., "Extension of Leading-Edge-Suction Analogy to Wings with Separated Flow Around the Side Edges at Subsonic Speeds," NASA TR R-428, 1974.
[60]Lamar, J. E., "Prediction of Vortex Flow Characteristics of Wings at Subsonic and Supersonic Speeds," *Journal of Aircraft*, Vol. 13, No. 7, 1976, pp. 490–494.
[61]Hoak, D., and Finck, R. D., "U.S.A.F. Stability and Control – DATCOM," Global Engineering Document, Englewood, Co., Secs. 2.1, 2.2, and 4.8, 1978.
[62]Brown, C. A., "The Effect of Camber on Thin Plate Low Aspect Ratio Wings at Low Reynolds Numbers," Master's Thesis, Dept. of Aerospace and Mechanical Engineering, Univ. of Notre Dame, IN, April 2001.
[63]Jian, T., and Ke-Quin, Z., "Numerical and Experimental Study of Flow Structure of Low Aspect Ratio Wing," *Journal of Aircraft*, Vol. 41, No. 5, 2004, pp. 1196–1201.
[64]Viieru, D., Albertani, R., Shyy, W., and Ifju, P., "Effect of Tip Vortex on Wing Aerodynamics of Micro Air Vehicles," *Journal of Aircraft*, Vol. 42, No. 6, 2005, pp. 1530–1536.
[65]Ellington, C. P., "The Aerodynamics of Hovering Flight,"*Philosophical Transactions of the Royal Society of London*, Vol. 305, No. 1122, 1984, pp. 1–181.
[66]Brodsky, A. K., *Evolution of Insect Flight*, Oxford Scientific Publications, Oxford Univ. Press, Oxford, England, UK, pp. 49–76, 1994.
[67]Tennekes, H., *The Simple Science of Flight – From Insects to Jumbo Jets*, The MIT Press, Cambridge, MA, 1992.
[68]Shyy, W., Berg, M., and Ljungqvist, D., "Flapping and Flexible Wings for Biological and Micro Air Vehicles," *Progress in Aerospace Sciences*, Vol. 35, No. 5, 1999, pp. 455–506.
[69]Aldridge, H. D. J. N., "Kinematics and Aerodynamics of the Greater Horseshoe Bat, Rhinolophus Ferrumequinum, in Horizontal Flight at Various Flight Speeds," *Journal of Experimental Biology*, Vol. 126, 1986, pp. 479–497.
[70]Jenkins, D. A., Shyy, W., Sloan, J., Klevebring, J., and Nilsson, M., "Airfoil Performance at Low Reynolds Numbers for Micro Air Vehicle Applications," *Proceedings of the 13th Bristol International RPV/UAV Conference*, University of Bristol, Bristol, UK, 1998.
[71]Ifju, P. G., Jenkins, D. A., Ettinger, S., Lian, Y., Shyy, W., and Waszak, M. R., "Flexible-Wing-Based Micro Air Vehicles," AIAA Paper 2002-0705, Jan. 2002.
[72]Gunston, B., *Aircraft of the Soviet Union*, Osprey Publishing, Ltd., London, 1983, p. 216–217.
[73]Taylor, J. H., *Jane's Encyclopedia of Aviation*, Crescent Books, New York, 1993, p. 449.
[74]Abdulrahim, M., "Flight Performance Characteristics of a Biologically-Inspired Morphing Aircraft," AIAA Paper 2005-345, VA, Jan. 2005.
[75]Null, W., Wagner, M., Shkarayev, S., Jouse, W., and Brock, K., "Utilizing Adaptive Wing Technology in the Control of a Micro Air Vehicle," *Smart Structures and Materials 2002, Industrial and Commercial Applications of Smart Structures Technologies*, edited by Anna-Maria R. McGowan, *Proceedings of SPIE*, Vol. 4698, 2002, pp. 112–120.
[76]Shkarayev, S., Jouse, W. C., Null, W., and Wagner, M., "Measurements and Performance Prediction of an Adaptive Wing Micro Air Vehicle," *Smart Structures and Materials 2003, Industrial and Commercial Applications of Smart Structures Technologies*, edited by Eric H. Anderson, *Proceedings of SPIE*, Vol. 5054, 2003, pp. 53–65.
[77]Woods, M. I., Henderson, J. F., and Lock, G. D., "Energy Requirements for the Flight of Micro Air Vehicles," *The Aeronautical Journal*, March 2001, pp. 135–149.

[78]Spedding, G. R., and Lissaman, P. B. S., "Technical Aspects of Micro-Scale Flight," *Journal of Avian Biology*, Vol. 29, No. 4, 1998, pp. 458–468.
[79]Ashley, S., "Palm-Size Spy Plane," *Mechanical Engineer*, Feb. 1998, pp. 74–78.
[80]Huber, A. F., II, "Death by a Thousand Cuts—Micro-Air Vehicles in the Service of Air Force Missions," Occasional Paper No. 29, Center for Strategy and Technology, Air War College, Maxwell Air Force Base, AL, July 2002.
[81]Anderson, J. D., Jr., *Introduction to Flight*, 5th ed. McGraw–Hill, New York, 2005, pp. 385–444.
[82]Stroman, R., Kellogg, J. C., and Swider-Lyons, K., "Testing of a PEM Fuel Cell System for Small UAV Propulsion," 42nd Power Sources Conference, June 2006.
[83]"Lithium Polymer Systems," Technical Notes on LiPo safety, The Handbook of FMA, Inc., July 2005.
[84]Phillips, W. H., "Design of Electrically Powered Models for Maximum Climb," *Annual Symposium Report of the National Free Flight Society*, 1978.
[85]Shaw, K., "Electric Sport Scale," *Model Builder*, July 1987.
[86]Groth, C., "Electric Duration by the Numbers," *Annual Symposium Report of the National Free Flight Society*, 2001, pp. 79–84.
[87]Larrabee, E. E., "The Screw Propeller," *Scientific American*, Vol. 243, No. 1, July 1980.
[88]Larrabee, E. E., "Analytical Design of Propellers Having Minimum Induced Loss," *Annual Symposium Report of the National Free Flight Society*, published by NFFS Royal Oak, Michigan, 1977, pp. 90–98.
[89]Larrabee, E. E., "Propeller Design and Analysis for Modelers," *International Symposium Report of the National Free Flight Society*, published by NFFS, Royal Oak, Michigan, 1979, pp. 9–25.
[90]Cosyn, P., and Vierendeels, J., " Numerical Investigation of Low Aspect Ratio Wings at Low Reynolds Numbers," *Proceedings of the 23rd AIAA Applied Aerodynamics Conference*, AIAA, Reston, VA; also AIAA, Paper 2005-4609, June 2005.
[91]Cosyn, P., and Vierendeels, J., "Design of Fixed Wing Micro Air Vehicles," 20th *Bristol UAV Systems Conference*, Bristol, England, UK, April 2005, pp. 25.1–25.15.

3
Autopilot Integration into Micro Air Vehicles

Roman Krashanitsa,* George Platanitis,* Dmytro Silin,† and Sergey Shkarayev‡
University of Arizona, Tucson, Arizona

Nomenclature

$\boldsymbol{A}, \boldsymbol{B}, \boldsymbol{C}, \boldsymbol{D}$	=	state-space model operators
b	=	wing span
C_D	=	drag coefficient
$C_{D^*}, C_{L^*}, C_{m^*}$	=	drag, lift, and pitching-moment coefficient derivatives
C_L	=	lift coefficient
C_l	=	rolling-moment coefficient
C_{l^*}, C_{n^*}	=	rolling-moment and yawing-moment coefficient derivatives
C_m	=	pitching-moment coefficient
C_n	=	yawing-moment coefficient
C_{T^*}	=	derivatives from thrust effects
C_x, C_y	=	longitudinal force coefficients
$C_{x^*}, C_{y^*}, C_{z^*}$	=	force coefficient derivatives in stability x_s, y_s, z_s axes
$\bar{c}$	=	mean geometric chord
F_{A^*}	=	component of the aerodynamic force
I_{xx}	=	moment of inertia about x-axis
I_{xz}	=	product of inertia about x and z axes
I_{yy}	=	moment of inertia about y-axis
I_{zz}	=	moment of inertia about z-axis
K	=	gain constant
K_p, K_d, K_i	=	proportional, derivative, and integral discrete gain coefficients
$K_{p\_c}, K_{d\_c}, K_{i\_c}$	=	proportional, derivative, and integral continuous gain coefficients
L	=	rolling moment
M_{CG}	=	pitching moment about center of gravity
N	=	yawing moment

Copyright © 2006 by Roman Krashanitsa, George Platanitis, Dmytro Silin, and Sergey Shkarayev. Published by the American Institute of Aeronautics and Astronautics, Inc., with permission.

*Research Assistant Professor.

†Graduate Research Assistant.

‡Associate Professor.

p = roll rate
q = pitch rate
$\bar{q}$ = dynamic pressure, $0.5\rho U_1^2$
r = yaw rate
S = wing planform area
T = sampling time
T_d = derivative time
T_i = integral time
t = time
U_1 = steady-state velocity
u = perturbed velocity
v = perturbed side velocity
w = perturbed vertical velocity
x_s, y_s, z_s = stability axes
α = angle of attack
$\dot{\alpha}$ = angle-of-attack rate of change
β = sideslip angle
$\dot{\beta}$ = sideslip angular rate of change
δ_a = aileron deflection
δ_e = elevator deflection
δ_r = rudder deflection
ζ = damping ratio
θ = pitch angle
λ = eigenvalue
ρ = air density
τ = characteristic time
ϕ = roll angle
ω = frequency
ω_d = damped frequency
ω_n = natural frequency
ω_s = sampling frequency

Subscripts

cmd = command or reference value
exp = value obtained using experiment
sim = value obtained using a numerical simulation
θ = pitch channel
ϕ = roll channel

3.1 Introduction

Micro air vehicles (MAVs) are being developed for basic surveillance and reconnaissance missions. They are by definition very small, and they normally operate beyond the line of sight; hence, stability and control are of primary concern for their design. More ambitious missions will require aggressive maneuvering between buildings or trees and the challenge of controlling MAVs around these obstacles increases.

Many MAVs to date need to be flown by a pilot utilizing a visual image relayed to a ground station by onboard electro-optical or infrared video cameras. This method of control proved to be very inefficient and difficult in practice. The angle of view of micro video cameras limits the perception of space, making it difficult for the pilot to assess the MAV's surroundings; his ability to handle the vehicle is diminished by this. Eventually, the quality of video depends on the stability of the airframe. Because of their small size, MAVs demonstrate intrinsically unsteady behavior with high- and low-frequency oscillations disturbing the video image. Thus, an enhanced automatic flight control system is vital for the progress of this technology.

Continual progress in developments in aerodynamic design of low speed and small-size vehicles, emerging microelectronics and micromechanical technologies have reached the state when a fully autonomous MAV has become a reality. A major topic of interest in the present research study is in methods of development and integraion of systems for autonomous flight of an MAV.

Development of fully autonomous MAVs changes priorities in a design process. With no pilot or operator involved in guidance and control, flying qualities of a MAV with related subsystems (autopilot, video system, gimbaled camera, ground station) will have to provide adequate mission performance and acceptable quality of video stream while the handling qualities may have lower priority. Foster and Bowman [1] stated that use of the flying-qualities criteria established for large airplanes (e.g., military specifications [2]) are not acceptable for small unmanned air vehicles. In the study [1], the dynamic stability of several small air vehicles was analyzed using predictive software. Through this extrapolation, flying-quality guidelines were proposed for a small aircraft by scaling down the standards used for a larger aircraft. Note that the standards for flying qualities of piloted aircraft are based on evaluations of handling qualities by many pilots for multiple aircraft under different flight conditions.

Video sequence from a moving source, such as a camera installed on a moving vehicle, contains unwanted inherent vibrations. Several approaches exist to estimate and remove unwanted motions from an image sequence. One approach is to compensate for the undesired motions by controlling motion of the video camera. A dynamic stabilization method [3] was developed to stabilize an image by controlling the motion of a gimbaled camera installed on a submarine. The control law for the camera was based on the prediction of the image motion provided by a constant-acceleration Kalman filter, where first-order forward difference derivative approximations were used.

Computational methods present a software alternative for correction of a video sequence. Jin et al. [4] used a method based on a two-dimensional affine transform to remove the unwanted motion of the video sequence based on detected motion: dollying, tracking, rotational/zooming, or translational types. This method is very sensitive to the motion type; therefore, if detected incorrectly, it can produce the wrong effect on the image sequence. This model can transform only small-amplitude motions of about 4% of the frame height. The computational method in Duric and Rosenfeld [5] is based on required orientation with respect to the horizon. It deals with detecting the horizon lines in the image because far-away lines are not affected by small translations but are affected by the rotation of the image source. The model is based on the three-dimensional affine transform. This approach proved to be successful for image sequences, where the horizon line is detectable and the amplitude of motion oscillation is small ($\sim 10^{-3}$ rad/frame).

Contrary to requirements for piloted aircraft, design requirements for the flying quality of surveillance MAVs are governed by requirements for video stream quality, which in turn depends on quality of transmission, available gimbals for micro video cameras, and software for video correction. Although this problem has been addressed in the recent technical literature, more studies are needed.

Even though MAV design has many ramifications, the development of automatic controls for MAVs still builds upon the fundamentals of aerodynamics and control theory. Flexible-wing MAVs have been investigated by Waszak et al. [6] and Ifju et al. [7]. The authors found that higher angles of attack can be achieved without stalling using a flexible wing that deforms under varying aerodynamic loads, including gust conditions, allowing the wing to see a lower angle of attack at higher pitch attitudes. Also, streamlining the fuselage of the MAV improved the lift-to-drag ratio of the aircraft, resulting in better overall performance. Ongoing investigations in the stability and control of the MAVs are also taking place. Analysis of the static stability derivatives shows the aircraft to be stable in all axes. The nondimensional stability derivatives of the MAV designs were found to be generally larger than those for conventional, piloted aircraft [7].

Several MAVs equipped with automatic control systems have been developed to date. A study by Platanitis and Shkarayev [8] focused on the integration of the MP2028[g] autopilot system for a 90-cm (36-in.) Zagi MAV. The aircraft provides a useful platform for evaluating autopilot integration into MAVs of comparable size. The autopilot uses various feedback loops for navigation and control during autonomous flight. Although one can use empirical approaches (i.e., pick gains, validate in flight tests, then adjust gains) to determine appropriate gains, the motivation of this research was to provide a more systematic approach to determining feedback loop gains. The approach involves determining an analytical model of the aircraft from its structural and aerodynamic characteristics that can then be validated through wind-tunnel experiments and developing the feedback control loops using standard design methods. Failures of the expensive autopilot components during ground tests prevented autonomous flight from taking place.

Taylor et al. [9] developed an attitude stabilization system based on thermal horizon detection. The system is reliable in daytime or nighttime flight, consumes little power, and operates quickly from a cold start. Such a control system even allows an unskilled pilot to fly an UAV.

Hsiao et al. [10] developed a low cost autopilot system with video and autolockup capabilities for autonomous flight and image capturing. The aircraft used was a Monocoupe 90A RC model of 2.44-m (8-ft) wing span and 11-kg (24-lb) takeoff mass. The onboard system consists of a GPS unit to measure position, an air data sensor to obtain speed and altitude data, and a dynamic measuring unit consisting of three microelectromechanical-System (MEMS) gyroscopes and an electric compass for measuring attitude and heading. The entire system provides measurements of the three-dimensional position and orientation of the aircraft and the data are transmitted to an onboard computer to be used by the control laws. Longitudinal control is accomplished by proportional control for altitude hold, with an inner pitch attitude loop that uses proportional-integral-derivative (PID) control. Similarly, lateral-directional control is accomplished by an outer heading hold loop, which uses proportional control, while the inner bank angle control loop uses PID control. Also, an algorithm to calculate the target position of a gimbaled charge-coupled-device (CCD) camera, allowing real-time images to be transmitted to a ground station, uses the aircraft position and orientation data.

Black Widow [11] was the first successful MAV with a 15-cm (6-in.) wing span, an endurance of 30 min, and speed of 48 kmh (30 mph). Black Widow had an automatic control system providing altitude, heading, airspeed hold, and yaw damping. The avionics included a two-axis magnetometer, Pitot tube, and a gyro to measure the turn rate; no GPS unit was installed at that time. This system allowed the aircraft to be flown by an unskilled operator by utilizing video transmitted from the small onboard video camera.

Arning and Sassen [12] presented a 42-cm (16.5-in.) autonomous MAV developed at EADS Dornier. The potential of using MEMS technology to provide size and weight savings, along with reduced power consumption for autopilot hardware mounted on the MAV, is realized. The study provided an overview of the application of sensors for the development of micro flight controls. Despite the current progress in sensor technologies, available microsensors and controllers suffer inaccuracies and instabilities. An alternative solution is in hybridization that combines different sensors into a system.

International micro-air-vehicle competitions have showcased MAVs that demonstrated aircraft flight via an autopilot. Brigham Young University [13] developed their aircraft through an iterative process, which involved a stability analysis of the five aircraft modes (phugoid, short period, Dutch roll, roll, and spiral), until it met functional specifications from various industry organizations. The aircraft had a 60.5-cm (23.8-in.) wing span and a total mass of 264 g (9.31 oz). Implementation of an autopilot provided reliable hands-off control, with capabilities that are competitive with larger UAVs. The design was shown to be competitive with most UAVs that are commercially available. The autopilot interfaced with GPS, Pitot probe, digital communication link and ground station hardware and software. An 8-bit, 29-MHz processor powers the autopilot board and uses pressure transducers, accelerometers, and gyroscopes to measure the airplane's position, attitude, linear and angular velocities, and accelerations. For automatic control, PID loops with feedback from airspeed, altitude, heading, pitch, pitch rate, roll, and roll rate are used to stabilize the aircraft and provide navigation guidance.

A control system for autonomous flight of an MAV for the surveillance mission was developed at RWTH Aachen University [14]. The aircraft used here had a 40-cm (16-in.) wing span and a total mass of 250 g (8.22 oz). An existing autopilot system was modified by the augmentation of a GPS-receiver and telemetry system, which uses a waypoint navigation algorithm. The existing system consisted of three piezoelectric gyros, two pressure sensors (static and dynamic), and a microcontroller to run the control algorithms and invoke control commands. Waypoint navigation algorithms were added to the existing control system, which included altitude and velocity hold, azimuth control, and flight-quality algorithms. Proportional control is used to command heading and proportional-integral (PI) control for altitude and speed hold. Newly developed control laws were integrated into a ground station, allowing gain factors and waypoints to be modified during flight. The system developed is more flexible, and the MAV can navigate pilot independent along GPS waypoints.

Konkuk University [15] made improvements in the flight ability of their MAV entry. A microscale inertial measurement system, the MR01, was developed. The MR01 consists of a one-axis gyroscope sensor and a two-axis accelerometer. When attitude data measured by the MR01 were used as feedback for the servomotor control, longitudinal and lateral stabilities improved. Successful missions have been flown using 13–15-cm (5.1–5.9-in.) wing span MAVs for surveillance in 5-m/s (11-mph) headwinds.

Available off-the-shelf autopilots and microavionics components are expensive, heavy, and of poor quality, especially microsensors. In industry, it leads to proprietary systems preventing a dissemination of knowledge on the development of autonomous MAVs. The viable alternative to this situation is the Paparazzi MAV autopilot research and development project [16] that is founded on the principles of free software [17]. The Paparazzi project was initiated in 2002 by A. Drouin and P. Brisset (ENAC, France) and has since grown into the international project connecting engineers and MAV enthusiasts from many countries.

The Paparazzi autopilot and ground station is a sophisticated set of software and hardware assets, flexible enough to work with various types of flying vehicles. It uses either the Mega or ARM processors and incorporates a GPS unit to determine the current location and an infrared unit to determine the current attitude of the vehicle. Receiver and transmitter units provide communication between the vehicle and the ground station. The autopilot software controls all avionic devices and algorithms for guidance, navigation, and control. It is flexible to accommodate changes in control laws and architecture of the airframe, and can be easily modified. Currently, the project is in the active phase of development and testing of the third version.

A high-quality video can be obtained by the camera with an actively stabilized platform. During the first U.S.–European MAV competition in 2005, several teams demonstrated gimbaled cameras on MAVs equipped with the Paparazzi autopilot. The algorithm synchronizing the actuation of the camera with the rolling motion of the aircraft was utilized to keep the camera pointing to the target on the ground.

Conventional design of larger aircraft is a two-step process, in which the aerodynamic design of an airframe focuses on satisfying performance requirements and then control laws are designed for required stability and controllability. For MAVs, a strong coupling between airframe and control system design was found and investigated in the previous studies [12] and [18]. This coupling necessitates a simultaneous design and optimization of airframe and controls.

The University of Arizona MAV project has been focusing on the development of fully autonomous MAVs since 2003. Several research and development projects conducted at the University of Arizona on the aerodynamics, airframe, and control design of MAVs have been completed, resulting in two exceptional fully autonomous MAVs: 59-cm (23-in.) Zagi and 30-cm (12-in.) Dragonfly. The Zagi is a relatively large but stable (yet very maneuverable) and rugged platform and, therefore, is well suited for experimental investigations of the flight control systems. The Dragonfly is about half the size, but possesses unique flight characteristics, including high maneuverability. This MAV won first place at the first

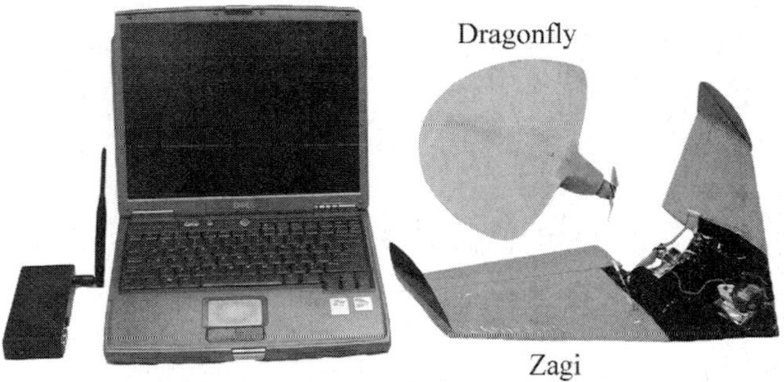

Fig. 3.1 Autonomous MAV system.

U.S.–European Micro-Aerial Vehicle Technology Demonstration and Assessment held in 2005 in Germany.

In this chapter, a method for system integration of autonomous MAVs is a major topic of interest. An MAV-specific approach is presented for the flight-control system design. The process includes the determination of stability and control derivatives using analytical and numerical computational software, simulation of flight dynamics, and closed-loop control design. Telemetry data from flight tests of the autonomous vehicles are used to validate predicted designs. The present study utilizes a control system based on the Paparazzi autopilot. In a series of flight tests of the Dragonfly and Zagi MAVs, telemetry data on control actuation, altitude, attitude, and GPS location of the airplane were collected and transmitted to the ground station.

3.2 MAV Designs and Specifications

Several airplanes were designed, built, and tested during the project on the integration of the miniaturized flight-control system. Two vehicles were chosen for the present study, the Zagi and the Dragonfly (shown in Fig. 3.1), and the Paparazzi autopilot [16] was integrated into both airplanes. With the autopilot onboard, these vehicles possess the following flight characteristics: GPS waypoint navigation by following a specified flight plan, altitude and attitude hold modes, and autonomous climbing and landing. The launching procedure consists of simply a hand throw of the vehicle into the air; the landing can be set along a chosen line. The ground station performs two major tasks: preflight programming of the autopilot and downloading of in-flight telemetry and video; it includes a laptop computer, a radio receiver, and an antenna (see Fig. 3.1). Both airplanes are equipped with video cameras that capture and transmit video to the ground station for analysis, making them very effective for surveillance missions. This section describes the airframe designs of Zagi and Dragonfly.

The 59-cm wing span Zagi is a flying-wing airplane with two winglets added for lateral stability, and elevons are used for longitudinal and lateral control. The total weight of the airplane is 261 g (9.2 oz). The electric brushless motor is mounted aft as a pusher installation. With a throttle setting at 75%, the vehicle has a cruise speed

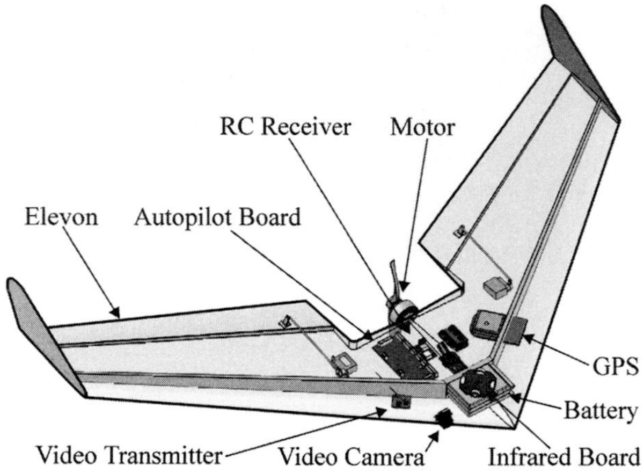

Fig. 3.2 Schematics of the Zagi (SolidWorks™ model).

of 16 m/s (33 mph) and flight endurance up to 15 min. The external design and arrangement of components inside the MAV are shown in Fig. 3.2. Geometrical, mass, and inertia data for the MAV and its components are presented in Tables 3.1 and 3.2. Moments of inertia presented in Table 3.1 are defined with respect to the body-fixed coordinate system with the origin at the center of gravity. They were derived from the SolidWorks™ model of the airplane.

The sweptback wing has an aspect ratio of 3.93, the taper ratio is 0.5, and the leading-edge sweepback angle is 40 deg. The wing design is based upon the MH45 [20] airfoil with a maximum thickness of 9.85% and a maximum camber of 1.63%. The wing was cut from EPP foam and covered with tape. It was reinforced by a

Table 3.1 MAV specifications

Parameter	Zagi	Dragonfly
Wing span, cm	59	30
Length, cm	31	26
Height, cm	6	6
Wing area, cm^2	885	488
Elevon area, cm^2	300	—
Winglet area, cm^2	80	—
Elevator area, cm^2	—	16.7
Fin area, cm^2	—	23
Rudder area, cm^2	—	4.8
CG location from the apex, cm	12.7	3.0
I_{xx},kg·m^2	$3.01 \cdot 10^{-3}$	$2.16 \cdot 10^{-4}$
I_{yy},kg·m^2	$1.33 \cdot 10^{-3}$	$6.38 \cdot 10^{-4}$
I_{zz},kg·m^2	$4.32 \cdot 10^{-3}$	$7.91 \cdot 10^{-4}$
I_{xz},kg·m^2	$-2.70 \cdot 10^{-5}$	$2.45 \cdot 10^{-5}$

Table 3.2 Components of the Zagi

Component	Description	Mass, g
Airframe	EPP foam, Balsa, Kevlar	116
Motor/propeller	Hyperion Z2205-34 [19]/APC 5 × 5 in.	35
Three-cell lithium-polymer battery	Thunder power 730 mAh	46
Autopilot	Paparazzi	30
Video camera/transmitter	MO-S508/BWAV240050	7
Two Microservos	Cirrus-5.4	11
Receiver/speed controller	PENTA 5/Phoenix-10	16
Total	——	261

carbon spar located at 25% of the chord and fiberglass tape along the leading edge. Tapered elevons are fabricated from balsa and are hinged to the trailing edge. Vertical winglets are made from Kevlar™ plastic and glued to the wing tips.

The motor is mounted on a 10-mm-diam (0.39-in.) carbon tube at the cutout in the center of the aft part of the wing. A three-cell lithium-polymer battery with 730-mAh capacity and 11.1-V nominal voltage is used to power the motor and all of the components onboard both the Zagi and the Dragonfly.

The fully autonomous Dragonfly is half the size of the Zagi and, therefore, is better qualified as an MAV category airplane. The external design and arrangement of components inside the MAV are shown in Fig. 3.3. The wing span of the MAV is 30 cm (12 in.), and the total weight is 186 g (6.56 oz). The airplane is controlled by an elevator and a rudder. The vertical tail, installed on the bottom surface of the wing, provides positive coupling between yawing and rolling moments from the rudder. The Dragonfly has a cruise speed of 12 m/s (27 mph) at 50% throttle setting, and a flight endurance of up to 30 min. Table 3.1 shows geometrical and mass specifications of the airplane, and internal components are listed in Table 3.3.

The Zimmerman planform wing of the Dragonfly utilizes a thin, cambered airfoil S5010-TOP24C-REF [21] with a maximum camber of 3% and a maximum inverse camber of 1%. The aerodynamic characteristics of this wing are reported in the case study in Chapter 6. The wing span and aspect ratio of the Zagi are about two times greater than those of the Dragonfly, and the Zagi's weight is 1.4 times greater.

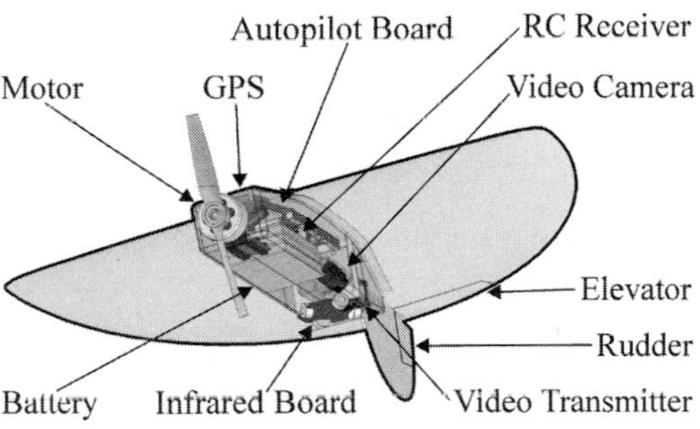

Fig. 3.3 Schematics of the Dragonfly (SolidWorks™ model).

Table 3.3 Components of the Dragonfly

Component	Description	Mass, g
Airframe	Kevlar/Rohacell foam	45
Motor/Propeller	AC-DIYMOT-2207 [19]/APC 4.75 × 4.75 in.	35
3-cell Lithium-Polymer Battery	Thunder Power, 730 mAh	46
Autopilot	Paparazzi	30
Video Camera/Transmitter	MO-S508/BWAV240050	7
2 Micro Servos	Blue Arrow BA-TS	7
Receiver/Speed Controller	PENTA 5/Phoenix-10	16
Total	——	186

Moments of inertia for both airplanes are presented in Table 3.1. Although the moments of inertia in the pitch direction I_{yy} differ by a modest factor of 2.08, the roll I_{xx}, and yaw I_{zz} moments differ by factors of 13.9 and 5.4, respectively. This design feature of MAVs is one major factor affecting the relatively poor stability of MAVs in the lateral direction. Because of the size and weight constraints, there is a lack of ways to control the MAV (only rudder, elevator, and throttle are feasible for this MAV), and this is another factor negatively influencing the airplane's flight dynamics. Note that these issues reflect the current state-of-the-art designs, and many qualities of the MAV will change in the future, first of all, when new, higher energy density power sources become available.

The Dragonfly's wing is molded from a sandwich panel, laminated from Kevlar™ face sheets and a Rohacell foam core. The wing has a constant thickness of 1.5 mm (0.059 in.). To increase its torsional stiffness, the Kevlar cloth is aligned at 45 deg to the wing span direction. The fuselage is also molded from Kevlar cloth and is reinforced by an additional layer of Kevlar around the nose section. The brushless motor is mounted as a tractor installation at the nose of the extended forward fuselage.

3.3 Flight-Control System

In the present project, MAVs were outfitted with the Paparazzi autopilot [16] designed for fully autonomous operation, from launch to recovery. Figure 3.4 shows a diagram of the autopilot components and their connections; their dimensions and masses are presented in Table 3.4. The Paparazzi autopilot includes an autopilot board, a GPS unit, and an infrared sensors board. An onboard antenna, radio receiver, and transmitter (see Tables 3.2 and 3.3) provide a wireless communication link with the ground station.

Based on the flight plan, the autopilot board reads sensor signals, which are fed back through the control laws, and generates control signals to the motor speed controller and rudder/elevator (Dragonfly) or elevon (Zagi) servos (Fig. 3.4). It features two clock-synchronized microprocessors, Atmel Mega8 and Mega128, both running at 16 MHz. The Mega8 supports all radio communications and servo operations, whereas the Mega128 handles the control functions, navigation, data acquisition, and telemetry downloading.

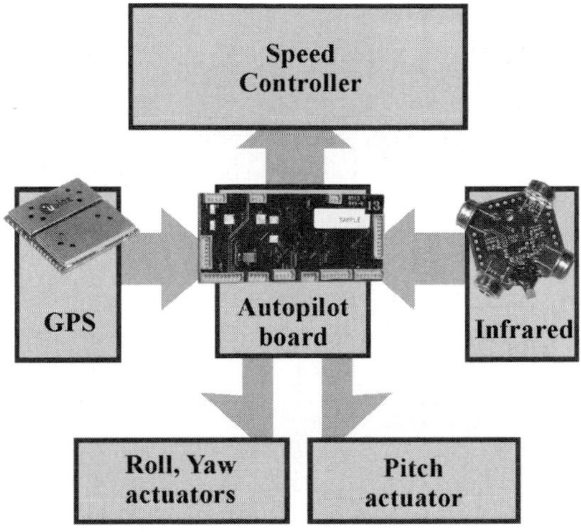

Fig. 3.4 Autopilot principal diagram.

The GPS unit sends coordinates, ground speed, and altitude of the airplane to the autopilot board.

The infrared sensors board, shown in Fig. 3.4, provides the autopilot with the airplane's roll and pitch angles based on measurements of the temperature difference between the sky and ground. The board consists of two pairs of infrared sensors MLX90247 [23] and a two-channel amplifier. Each infrared sensor reads the temperature over its entire field of view and calculates the average; therefore, the main requirement for the infrared sensor mounting is to maximize its field of view. Two pairs of sensors are orthogonally aligned. The temperature difference measured between opposite pairs of sensors allows the roll and pitch of the airplane to be computed.

The flight-control software consists of several modules—configuration files, fly by wire, flight plan, map, autopilot, and GPS tools—and two major parts—the autopilot software onboard the airplane and the ground station software. The autopilot source code is compiled on the ground. In this process, a desired flight plan and the airframe specifications are accommodated. Then, the source code is compiled and uploaded into the Mega128 microprocessor before the flight.

In flight, the autopilot sends telemetry data back to the ground station using the audio/video channel of the transmitter. The data are directed to the sound port of the ground station computer and converted into digital format. Currently, the

Table 3.4 Dimension and mass of the autopilot components

Component	Dimensions, mm	Mass, g
Autopilot board	$61 \times 28 \times 5$	6.6
GPS, U-Blox SAM-LL [22]	$47 \times 32 \times 6$	18.4
Infrared sensors board	$30 \times 30 \times 7$	5.0

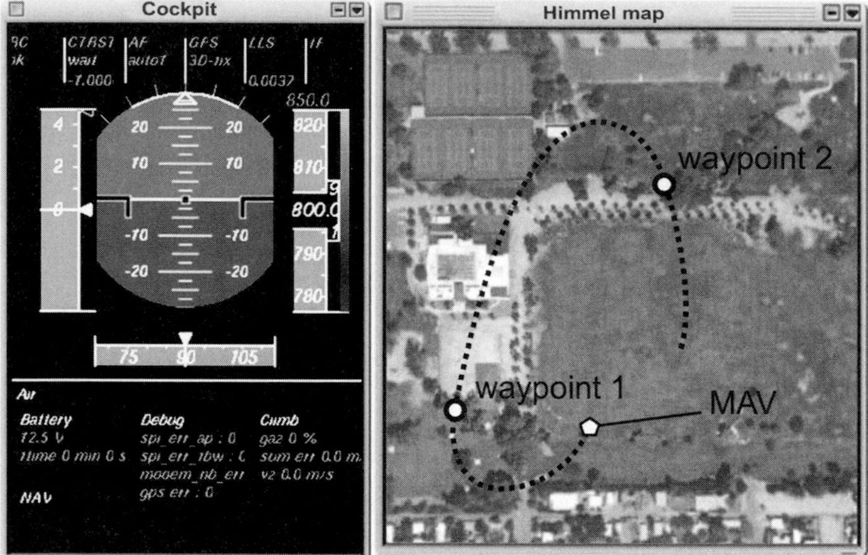

Fig. 3.5 Cockpit and map panels.

telemetry data include GPS-based data, including location coordinates; ground speed, altitude, and climb rate of the airplane; attitude of the airplane provided by infrared sensors; autopilot status data; and position of the control surfaces. The telemetry data are used to display up-to-date information about the location of the airplane on the map, attitude of the airplane, and its altitude above sea level. These data play a major role in performance analysis during the flight tests and adjustments of gains.

The autopilot has two modes of operation: a fully autonomous mode, where the airplane flies according to the specified flight plan, and a stability augmentation mode. The operator can select either of these modes from the ground station. When an airplane is in the fully autonomous mode, a flight plan based on waypoint notation is used. The flight plan consists of one or several blocks executed in a sequential order, where each of the blocks specifies several commands for the autopilot.

When the autopilot is in the stability augmentation mode, the operator of the ground station controls the airplane's desired thrust and roll. Actual commands for servos are computed by the autopilot control loop code. The pitch of the airplane is controlled automatically to ensure stable and controllable flight. Pitch and roll angles are limited to the safe values of the calibrated airplane and are hard coded into the current airframe configuration.

The ground station software consists of the user interface panels and drivers necessary to decode airplane telemetry data. The cockpit user interface panel displays information about altitude, attitude, thrust, power supply, and autopilot parameters of the airplane (Fig. 3.5). The layout of this panel is reminiscent of a conventional airplane cockpit. The upper part of the cockpit panel displays information about the current mode of the autopilot, infrared sensor board, and GPS. The middle part

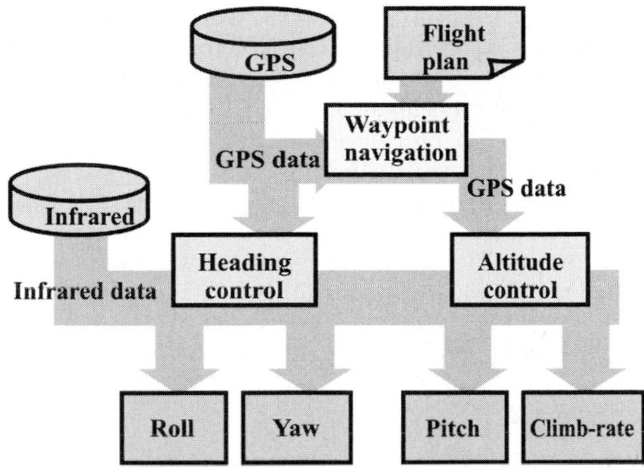

Fig. 3.6 Paparazzi autopilot functional block diagram.

of the cockpit panel houses indicators for the altitude and attitude of the airplane. The bottom part of the cockpit panel shows the condition of the power supply on the airplane and the quality of the data link between the airplane and ground station. The map panel shows the position of the airplane (Fig. 3.5), based on the last received GPS coordinates, and the trajectory of the airplane on the map. It also indicates a current desired position for the airplane flying in fully autonomous mode.

When an airplane flies in a fully autonomous mode, it is navigated by the autopilot, which generates control commands based on the flight plan and airplane location and orientation determined from data measured by the GPS and infrared sensors. A functional diagram of the autopilot is presented in Fig. 3.6. Both waypoint navigation and attitude control are implemented in the autopilot and are realized with the help of navigation and control algorithms for lateral guidance and altitude and attitude control. The algorithm of lateral guidance provides automatic planning of the route. It also checks the current goal conditions, such as distance to the target waypoint and distance to the home waypoint, and determines a bank angle to achieve the desired heading. The attitude and altitude control module implements control laws for altitude hold and attitude hold regimes of the autopilot. Currently, the attitude hold mode is used during the launch phase, and altitude hold is used during the loiter phase of the fully autonomous mission.

3.4 Stability and Control Design for Micro Air Vehicles

To evaluate the effectiveness of control laws for a micro air vehicle, a design process is followed herein, which makes use of the linearized equations of motion of the aircraft. The equations are developed from an evaluation of the various aerodynamic derivatives (stability and control derivatives) in both the longitudinal and lateral-directional motion. These derivatives are then assembled into state-space matrices. The MATLAB®/Simulink [24] program is used to evaluate the open-loop stability of the airplane, as well as to determine control laws for closed-loop stability. The ideal control law is designed for a continuous-time system. Control design, however, is limited by several factors—the fact that measured data from the sensors are sampled, and bandwidth limits of the aircraft dynamics and servomotors have to be considered. Control gains are initially determined using the continuous system model. The discrete equivalents of these gains are evaluated for the related discrete model and adjusted to satisfy stability criteria in discrete time. The gains can then be programmed into an autopilot for flight testing. The following examples will illustrate a control design for determining the necessary gains for use in an autopilot to fly an MAV autonomously.

3.4.1 Dynamic Equations of Motion

Details for determining expressions for the stability and control derivatives can be found in [25], but a brief treatment follows here with features pertaining to MAVs. The longitudinal stability derivatives are determined by first defining the longitudinal forces and pitching moment in the stability coordinate system as follows:

$$F_{Ax} = C_x \bar{q} S, \quad F_{Az} = C_z \bar{q} S, \quad M_{CG} = C_m \bar{q} S \bar{c} \tag{3.1}$$

Assuming small angles, the coefficients of lift C_L and drag C_D can be related to C_x and C_z by

$$C_x = -C_D + C_L \alpha, \quad C_z = -C_L - C_D \alpha$$

Equations (3.1) are differentiated with respect to the nondimensional variables as follows to obtain the force and moment derivatives. A summary of results is shown for u/U_1:

$$\begin{aligned} \left.\frac{\partial F_{\mathrm{Ax}}}{\partial (u/U_1)}\right|_1 &= -\left(C_{Du} + 2C_{D1}\right)\bar{q}S \\ \left.\frac{\partial F_{Az}}{\partial (u/U_1)}\right|_1 &= -\left(C_{Lu} + 2C_{L1}\right)\bar{q}S \\ \left.\frac{\partial M_{CG}}{\partial (u/U_1)}\right|_1 &= \left(C_{mu} + 2C_{m1}\right)\bar{q}S\bar{c} \end{aligned} \tag{3.2}$$

The differentiations must be evaluated at the steady-state flight condition (where all perturbed quantities are zero). Equations (3.1) are subsequently differentiated with respect to α, $\dot{\alpha}\bar{c}/2U_1$, $q\bar{c}/2U_1$, and δ_e. The complete model also includes derivatives caused by thrust forces. For both MAVs presented in this study, the thrust line is assumed to pass through the center of gravity of the aircraft and points horizontally, and the pitching moment caused by thrust is negligible. As such, the thrust will not create any moments about the aircraft center of gravity and will not have any component along the $z$axis. The only component left is in the horizontal direction

$$\left.\frac{\partial F_{Tx}}{\partial (u/U_1)}\right|_1 = -\left(C_{Txu} + 2C_{Tx1}\right)\bar{q}S \tag{3.3}$$

Equations (3.2) and (3.3) form the total speed derivatives in longitudinal direction.

Similarly, stability and control derivatives exist for the lateral-directional dynamics of the aircraft. The lateral-directional forces and moments are given for side force, rolling and yawing moments, respectively, by

$$F_{Ay} = C_y\bar{q}S, \quad L = C_l\bar{q}S\bar{c}, \quad N = C_n\bar{q}S\bar{c} \tag{3.4}$$

Equation (3.4) is differentiated with respect to the dimensionless variables β, $\dot{\beta}b/2U_1$, $pb/2U_1$, $rb/2U_1$, and δ_a.

Once the stability and control derivatives are computed, they are then integrated with the mass, geometric, and inertial properties to obtain the equations of motion in state-space form for each of the longitudinal and lateral motion (see [25] and [26] for details):

$$\begin{aligned}\dot{\mathbf{x}} &= \mathbf{Ax} + \mathbf{Bu}\\ \mathbf{y} &= \mathbf{Cx} + \mathbf{Du}\end{aligned} \tag{3.5}$$

where the state vector for the longitudinal motion is $[u,\ w,\ q,\ \theta]^T$ and for the lateral motion $[v,\ p,\ r,\ \phi]^T$. The state-space form as described in [26] was used for matrices **A**, **B**, **C**, and **D**. Matrix **A** contains the variables for mass, moment of inertia (pitch inertia for longitudinal, roll, yaw and roll-yaw product inertia for lateral), and dimensional stability derivatives for the motion variables. Matrix ***B*** is the control influence matrix containing derivatives for all control surfaces, as well as aircraft mass and inertia variables. Matrices ***C*** and ***D*** are sensor matrices, where ***C*** is the output matrix describing the output signal from the measured state, assumed here as an identity matrix (all state variables measured and available for feedback) and ***D*** is the direct transmission matrix, assumed here as zero [27] and [28]. In general, the output vector **y** can be constructed to suit the requirements of the control objective, such as selecting particular states to be measured. Note that v and w are equivalent perturbed speed variables for β and α, respectively, where for small angles

$$w = U_1\alpha, \quad v = U_1\beta \tag{3.6}$$

for which equivalent dimensional stability derivatives can be determined. This form was used in the development of the state-space matrices in the lateral direction (for the time simulations).

3.4.2 Closed-Loop Control Design for Discrete Systems

Although a data measuring system producing a continuous signal is the most ideal to work with in control design, these systems do not operate this way in reality; instead, the data from a signal are measured at intervals defined by a predetermined sampling rate. This is the case with the Paparazzi autopilot system. All tasks are executed according to the external hardware oscillator on the autopilot board, where a timer is called in the main loop of the autopilot. According to the timer, periodic tasks are executed at 60 Hz. Several periodic subtasks are also executed at varying frequencies. The infrared sensors provide measured data at rates of 20 Hz. The outer navigation loop operates at 4 Hz, while the inner loops controlling roll and pitch can be designed for a 20-Hz sampling rate. The rate at which commands are sent to the servos (deflecting control surfaces) is also 20 Hz. Control laws must therefore be able to maintain the roll and pitch motion stable at this frequency, while not exceeding the dynamics of the servomotors or the roll and pitch dynamics of the aircraft (by sending large oscillatory commands at large frequencies). The servomotor itself has a bandwidth of approximately 8.5 Hz. If a control law actuates the servos near this frequency, it should not be a problem provided that the servos' motion is small (and motion is gradual) at each command step, nor do the servos undergo large oscillations at each step.

With data supplied to the control laws sampled at given rates, it is necessary to design the control laws to maintain the MAV stable for an input signal in discrete time. To accomplish this, the dynamics of the system must be converted to their discrete equivalents. Many textbooks describe discrete control design in great detail, [27] and [28] but some key concepts for defining stability of a closed-loop control design for discrete systems will be presented here.

A continuous signal is transformed into a discrete one by means of the z transform, defined as

$$X(z) = \mathcal{Z}\,[x(t)] = \mathcal{Z}\,[x(kT)] = \sum_{k=0}^{\infty} x(kT)\, z^{-k} \tag{3.7}$$

where z is a complex variable, T is the sampling period, and k is an integer counter for the sampling sequence. It can be shown by using an impulse sampled signal [27] that (via Laplace transform)

$$X(z) = X^{*}(s) = \sum_{k=0}^{\infty} x(kT)\, e^{-kTs} \tag{3.8}$$

where it is seen that we can define $z = e^{Ts}$. The relationship between s and z space just shown can now be used to define the region of stability in the z plane. For a continuous-time system, s is a complex variable defined as, $s = \sigma + j\omega$. Using the definition of z, we get,

$$z = e^{Ts} = e^{T\sigma} e^{j(T\omega)} = e^{T\sigma} e^{j(T\omega + 2\pi k)} = e^{T\sigma} e^{jT(\omega + \omega_s k)}$$

This means that the poles and zeros in the s plane that have frequencies which differ by integer multiples of the sampling frequency $\omega_s = 2\pi/T$ are mapped onto the same location in the z plane. As such, only the first frequency is needed (i.e., $k = 0$). Noting that σ must be negative for stability (located in the left half of the s plane), the magnitude of z for stability is $|z| = e^{T\sigma} < 1$ for asymptotic stability. Thus, the $j\omega$ axis in the s plane corresponds to an angle in the polar coordinate system defined on a unit radius circle in the z plane with the origin at its center. From the oscillatory part of $z(e^{jT\omega})$, it is easy to see that $\angle z = T\omega$. If the frequency ω is held constant and σ is varied, the result is a radial line from the origin of the z plane at an angle of $T\omega$ counterclockwise from the positive real axis. If $\omega = \pm\omega_s/2$, angle of z is $\pm\pi$. On the other hand, if σ is held constant instead, the result is concentric circles in the z plane centered at the origin.

A further useful expression results in constant damping loci in the z plane. The complex variable s can alternatively be written as

$$s = -\zeta\omega_n + j\omega_d, \quad \omega_d = \omega_n\sqrt{1-\zeta^2} \tag{3.9}$$

where the complex conjugate of s will just reflect the point on the real axis in the z plane. Using Eq. (3.9), and the expression for z and for sampling frequency gives

$$z = e^{Ts} = \exp(-\zeta\omega_n T + j\omega_d T) = \exp\left(-\frac{2\pi\zeta}{\sqrt{1-\zeta^2}}\frac{\omega_d}{\omega_s} + j2\pi\frac{\omega_d}{\omega_s}\right) \tag{3.10}$$

where

$$|z| = \exp\left(-\frac{2\pi\zeta}{\sqrt{1-\zeta^2}}\frac{\omega_d}{\omega_s}\right), \quad \angle z = 2\pi\frac{\omega_d}{\omega_s}$$

Now, if the damping ratio is held constant, the preceding expressions allow one to plot constant damping ratio loci in the z plane. Equation (3.10) is useful for mapping the poles of a continuous system to their discrete equivalents. Thus, root-locus design techniques for determining closed-loop gains used for the continuous system can be used for the discrete system as well.

3.4.3 Aerodynamic Parameters

Through the use of analytical software, the aerodynamic derivatives can be determined for a given flight condition and trim setting. The Advanced Aircraft Analysis (AAA) software [29] was used to determine the aerodynamic derivative coefficients for the Zagi from an input of airfoil and geometric data, as well as center-of-gravity location. A flight condition of 16 m/s (36 mph) was chosen, as this is the approximate velocity at which the aircraft would fly in cruise with a trim angle of attack of 2.6 deg and trim elevator setting of −5.8 deg.

Aerodynamic parameters are also obtained for the Dragonfly MAV through numerical simulations conducted with the help of a vortex-lattice computational method and Tornado software [30]. In the analysis, the aircraft was trimmed at an angle of attack of 4 deg and flight speed of 12 m/s (27 mph). In Tornado [30], an approximate model of the MAV was developed, with all lifting surfaces divided into panels. The flight condition was predefined, and the central difference expansion method was used, where aerodynamic coefficients were determined by computing aerodynamic characteristics about small perturbations around the predefined flight condition.

For both the Zagi and Dragonfly, the lateral-directional forces and moments turn out to be very small for $\dot{\beta}$. In most cases, the $\dot{\beta}$ derivatives are negligible for aircraft flying at low speeds [26] and the equations of motion used herein [27] have been derived with these rate derivatives dropped. Similarly, the u derivatives are also small, and the $\dot{\alpha}$ derivatives of the Dragonfly MAV result in small force and moment contributions from them. Therefore, these derivatives are neglected in the model.

Note that assumptions have been made with the Dragonfly model for simplicity in that the thrust line is assumed parallel to the body x axis (z thrust derivatives

Table 3.5 Longitudinal aerodynamic derivatives

	C_D		C_L		C_m	
Variable	Zagi	Dragonfly	Zagi	Dragonfly	Zagi	Dragonfly
1	0.0215	0.0184	0.198	0.0936	0	−0.0105
α	0.117	0.128	3.26	2.426	−0.551	−0.254
q	0	0.1625	2.73	4.51	−1.25	−1.28
δ_e	0.911	0.0214	0.330	0.372	−0.359	−0.228

are zero); propeller moments and propeller wash on the wing have been neglected (although the phenomenon has been investigated, with the effect to be included in future research). A nonzero steady-state moment (at power off) is computed using the trim angle of attack determined from flight tests of the 12-in. MAV used, but this moment is cancelled out by an equal and opposite trim thrust moment.

Tables 3.5 and 3.6 summarize the aerodynamic coefficients for both longitudinal and lateral-directional motion for each of the aircraft. In general, orders of magnitude of the coefficients are similar. Sign changes occur for several moment coefficients (particularly where roll and yaw moments are coupled) because of the vertical tail positioning (for Dragonfly, the tail points below the fuselage), and in some cases, such as the yaw moment caused by sideslip, the relative size of the tail and moment arm affects the size of the dimensionless coefficient. Rolling moments will also be affected by the relative position of the vertical tail aerodynamic center to the MAV's center of gravity. With regard to the Dragonfly aircraft, this vertical moment arm is small, and there is uncertainty as to whether it is above or below the center of gravity (Tornado [30] does not actually calculate it). In particular, rolling moments as a result of yaw motion (where the vertical tail is a major contributor) might be too small to be determined accurately by Tornado.

A close look at the aerodynamic derivatives of both aircraft for pitching moment casued by α and roll and yaw moments caused by β gives an indication of the static stability, as well as the maneuverability of the aircraft. Both aircraft are statically stable in terms of $C_{m\alpha}$, which has a negative sign for both. The Zagi is more stable in pitch compared to the Dragonfly and only requires small changes in angle of attack to produce significant changes in moments. Similarly, the Zagi has a larger, negative $C_{l\beta}$. As such, small amounts of sideslip will induce large rolling moments.

Table 3.6 Lateral aerodynamic derivatives

	C_y		C_l		C_n	
Variable	Zagi	Dragonfly	Zagi	Dragonfly	Zagi	Dragonfly
β	−0.166	−0.234	−0.13	−0.014	0.0501	0.793
p	0.00448	−0.0784	−0.294	−0.17	−0.03	0.0077
r	0.1	0.15	0.0719	−0.0206	−0.036	−0.0614
δ_a	0	——	0.2	——	−0.0101	——
δ_r	——	0.0766	——	−0.00527	——	−0.0363

Table 3.7 Dynamic characteristics of the Zagi

Mode	λ	ω_n	ζ	τ
	Longitudinal dynamics			
Short period	-9.69 ± 28.2j	29.8	0.324	——
Phugoid	-0.0938 ± 0.831j	0.837	0.112	——
	Lateral dynamics			
Dutch roll	-2.27 ± 38.6j	38.6	0.0587	——
Roll	-18.5	——	——	0.054
Spiral	-0.0314	——	——	31.8

The opposite is true for the Dragonfly, which is less stable in this motion variable. Finally, examining the $C_{n\beta}$ values for both aircraft, we see that the Dragonfly has by far a larger yaw moment because the sideslip coefficient is larger than that of the Zagi. Therefore, small sideslip motions will cause the Dragonfly MAV to experience large yawing moments. Relatively speaking, the Dragonfly has a larger moment arm between its center of gravity and the aerodynamic center of the vertical tail than the Zagi. The Dragonfly is more stable in this direction. The centers of gravity of each aircraft are forward of their aerodynamic centers (wing and vertical tail).

3.4.4 Control Design for Zagi

Using the parameters determined analytically for the Zagi, a dynamic model of the MAV is produced, and a control law design example is presented for roll and pitch control. First, an analysis of the open-loop dynamics of each mode is presented. Table 3.7 shows the characteristics of the longitudinal and lateral dynamics in terms of eigenvalues λ, natural frequencies ω_n (rad/s), damping ratio ζ, and characteristic time τ. From the eigenvalues, it can be seen that both the longitudinal and lateral dynamics are stable.

The roll-to-aileron transfer function for the open-loop system in continuous time is summarized in Table A.1, along with the discrete time equivalent at a sample rate of 8 Hz.

The poles of the continuous form of the roll-to-aileron transfer function are already given in Table 3.7 as the eigenvalues of the longitudinal system. Using Eq. (3.10) with a 0.125-s sample time, the discrete time poles are determined and summarized in Tables 3.8 and 3.9, respectively. The zeros of the continuous transfer function are found directly from the roots of the numerator. The zeros of the discrete

Table 3.8 Roll-to-aileron poles for lateral-directional dynamics

Mode	Poles
Dutch roll	0.086 ± 0.747j
Roll	0.0987
Spiral	0.996

Table 3.9 Roll-to-aileron zeros for lateral-directional dynamics

Continuous	Discrete
$-\infty$	-0.439
$-4.63 \pm 34.7j$	$0.105 \pm 0.738j$

transfer function are found directly from the numerator of the discrete transfer function. In this case, the zero-order hold was used to convert the continuous system to a discrete time one. For control design, the zeros corresponding to the z-transform method used (zero-order hold here) must be used in root-locus plots.

For the preliminary design, a control law will be determined for closed-loop stability of the roll-to-aileron transfer function at a sampling time $T = 0.125$ s (or 8-Hz sample frequency). In the autopilot system, the infrared sensors measure the aircraft's attitude at 20 Hz. Actuation commands for the control surfaces must be at 8.5 Hz or less, as this is the approximate bandwidth of the servomotors. The control law must therefore be capable of maintaining stable roll dynamics at this frequency. For time simulation in Simulink of the closed-loop system, a sensor sample frequency of 24 Hz was used and converted to 8 Hz for reading by the control law (24 being an integer multiple of 8). By using frequencies where one is an integer multiple of the other, data quality is preserved during the sample rate conversion by the rate transition block. (This rate transition is considered "safe.)" The control law is designed as if it is seeing an 8-Hz system. Figure 3.7 shows the magnitude vs log frequency of the roll-to-aileron transfer function. Note that the crossover frequency is near 16.2 rad/s (2.58 Hz). This is below the actuation speed of the control law. At the control law rate of 8.5 Hz, the roll-aileron ratio is

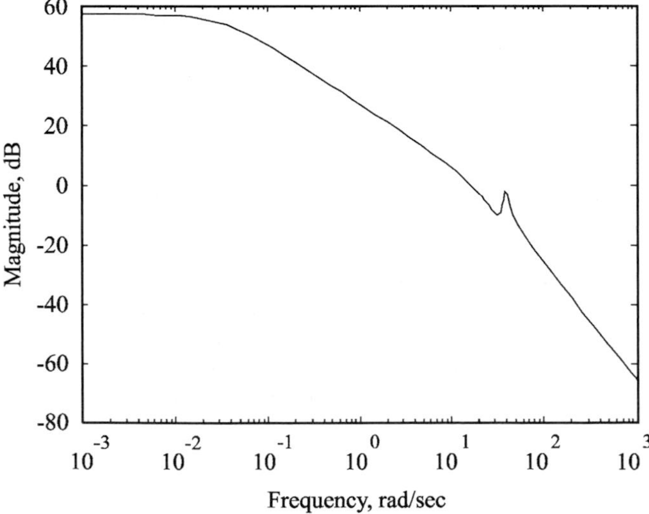

Fig. 3.7 Log magnitude vs frequency for roll to aileron.

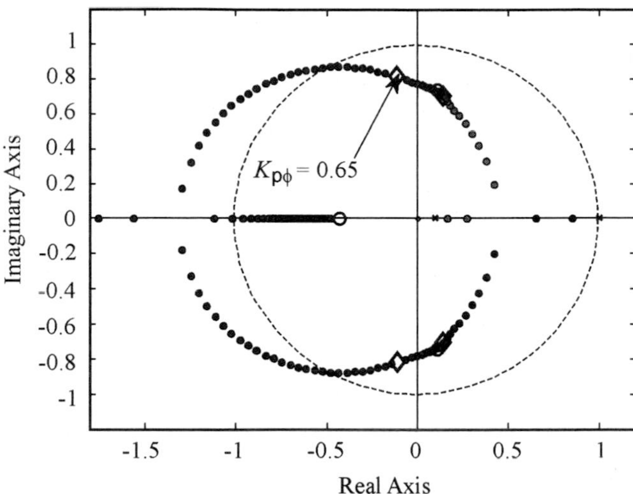

Fig. 3.8 Discrete root locus for roll-to-aileron transfer function for $K_{p\phi} = 0.65$.

about 0.206, which might not be so bad. If at each time step the control actuation is in reasonably small magnitudes without deflections from extreme to extreme (and wind turbulence does not contain significant frequency content in the 8-Hz range or beyond [31]), the MAV should still exhibit a reasonable response. As was already noted by the eigenvalues (Table 3.7), all three lateral modes are stable. The Dutch-roll mode has a damped frequency of 38.6 rad/s (6 Hz). A control law actuating the ailerons at even 8 Hz should still be able to provide a reasonable roll angle response.

The proportional gain previously used in the autopilot for roll control is $K_{p\phi} = 0.65$ and will be used as a starting point in this investigation. Figure 3.8 shows the discrete root-locus plot for the roll dynamics, with $K_{p\phi}$ indicated (diamond marks). The unit circle (stability boundary) is shown dashed. For this gain, the closed-loop gains lie well within the unit circle, but there is a limit as to how much the proportional gain can be increased above 1.05 before the Dutch-roll poles are pushed out of the unit circle into the region of instability. Further increasing $K_{p\phi}$ will push one of those poles to negative infinity along the real axis further destabilizing the closed-loop lateral dynamics.

Roll control in the Paparazzi autopilot also includes derivative control as a further means of stabilizing the roll dynamics. The initial derivative gain set in the autopilot is $K_{d\phi} = 0.65$ (assumed for the discrete system at an 8-Hz sampling rate). It is worthwhile to note here the relationship between the control gains used for the discrete system and the equivalent gains of the continuous system [27] and [28] (without the presence of integral control). This relationship is given by

$$K_{p\phi} = K_{p\phi\_c}, \quad K_{d\phi} = K_{p\phi\_c}\frac{T_d}{T} \tag{3.11}$$

where T_d is the derivative time and the subscript c denotes the equivalent gain for the continuous-time system. The discrete proportional-derivative (PD) control law

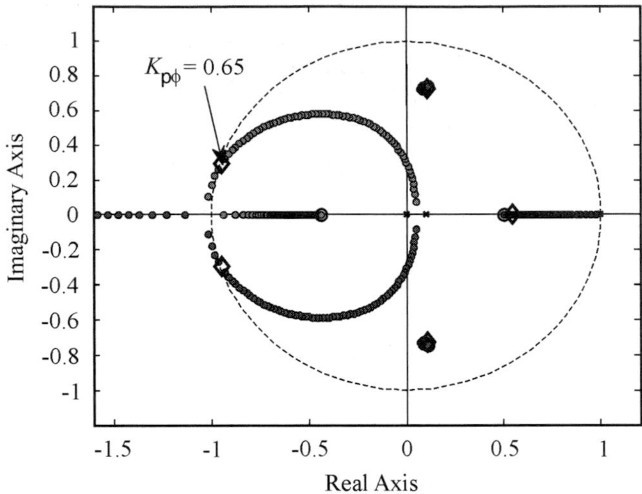

Fig. 3.9 Root locus for roll-to-aileron transfer function for $T_d = T = 0.125$.

is as follows:

$$G_D(z) = K_{p\phi} + K_{d\phi}(1 - z^{-1}) = \frac{K_{p\phi}[(1 + T_d/T)z - T_d/T]}{z} \tag{3.12}$$

With the control law in the forward loop, the modified open-loop transfer function is

$$G(z) = \frac{K_{p\phi}[(1 + T_d/T)z - T_d/T]}{z}\frac{\phi(z)}{\delta_a(z)} \tag{3.13}$$

Using Eq. (3.11), a derivative time of $T_d = T$ will yield $K_{d\phi} = K_{d\phi\_c}$. The proportional gain shown is the same as was used for the proportional only control analysis. This PD control adds a pole at the origin and zero on the real axis at 0.5. The addition of derivative control with the preceding control law parameters made the stability somewhat worse (moving the pole at the origin and the rolling mode pole to the edge of the unit circle—see Fig. 3.9). The derivative time will have to be adjusted and, possibly, the proportional gain so that the closed-loop poles are well within the unit circle.

To improve the stability of the lateral dynamics, the effect of lowering the derivative time is investigated. Using the same roll proportional gain $K_{p\phi} = 0.65$ and changing the derivative time to $T_d = 0.01$s gives the discrete root locus as shown in Fig. 3.10. This time, it is the Dutch-roll poles that migrate towards the unit circle with increased gain, but remain well within the unit circle. However, this is only one indication of the performance of the system. It is necessary to simulate the closed-loop control and examine the time history of the response to a reference roll input.

Figure 3.11 shows the time history of the closed-loop roll control operating at an 8-Hz sampling rate (and control actuation) using output from Simulink. Initially, the aileron deflects to 7 deg, but no more than 3 deg after that. Overall, the aileron

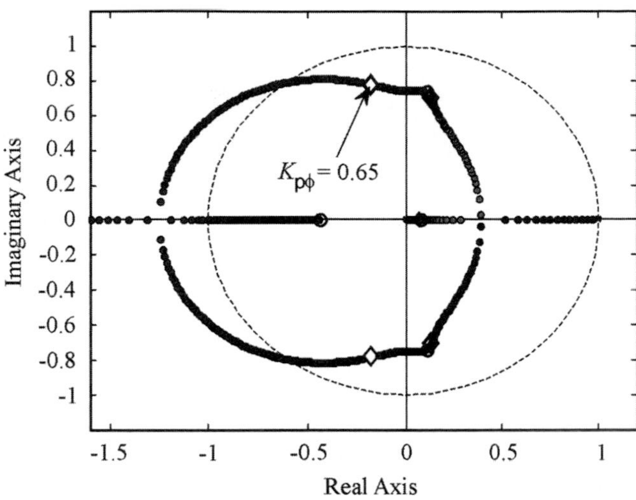

Fig. 3.10 Root locus for roll-to-aileron transfer function for $T_d = 0.01$ and $T = 0.125$.

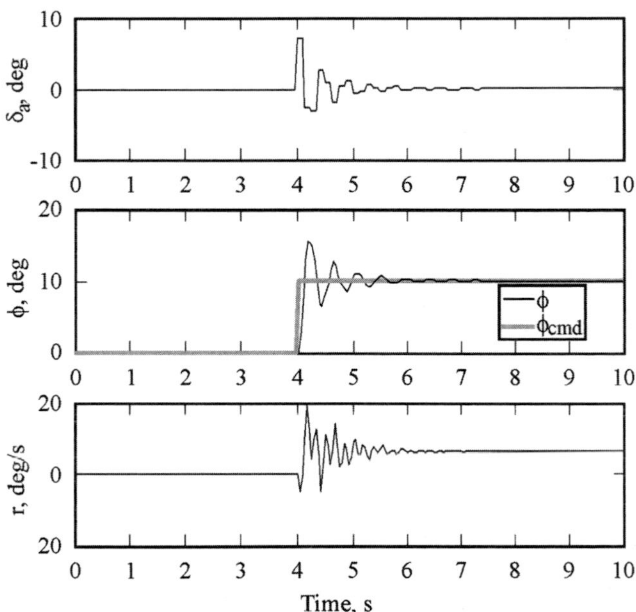

Fig. 3.11 Time response of roll-to-aileron transfer function for $T_d = 0.01$ and $T = 0.125$.

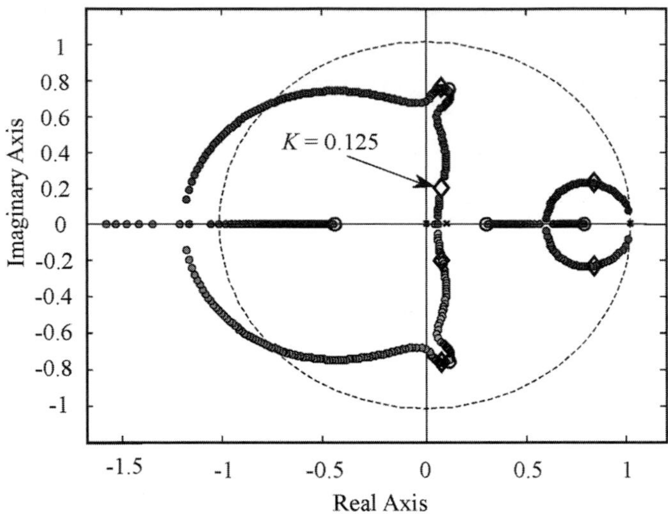

Fig. 3.12 Root locus for roll-to-aileron transfer function for $K_{p\phi} = 0.11$, $K_{d\phi} = 0.04$, and $K_{i\phi} = 0.03$.

actuation appears smooth, with no extreme motion from one command to another (and only oscillates through two cycles in 1 s—well below system and actuator bandwidth and natural frequencies). Although the roll angle could be held by the MAV with this control law, it is evident that other refinements would be necessary to overcome issues such as overshoot. As the aircraft responds to reach the desired roll angle, there is some effect on the other three lateral motion variables caused by coupling. The sideslip angle changes significantly, up to about 6 deg before settling back to zero upon the Zagi reaching its desired roll angle. Also, when the aircraft initiates its roll, there is a large increase in both roll rate p ($\sim$150 deg/s) and yaw rate r (20 deg/s), with much oscillation during this transient phase, until the desired roll angle is reached, where p settles to zero and r is constant ($\sim$7 deg/s). It will be shown that with PID control (and appropriate gain selections) the response of the aircraft improves greatly in all of the lateral motion variables.

From the preceding design procedure, the design iteration should be trivial, where gains can be adjusted to suit response requirements, coupled with iterations on the airframe design. In the flight tests that will be described in Sec. 3.5, the full PID control law is used for roll control, and some gains determined by trial in flight tests of the Zagi are validated here using the preceding design procedure. The following PID gains were used: $K_{p\phi} = 0.11$, $K_{d\phi} = 0.04$, and $K_{i\phi} = 0.03$. At this point, we note the relationship between the continuous and discrete PID control law:

$$K_{p\phi} = K - \frac{K_{i\phi}}{2}, \quad K_{i\phi} = \frac{KT}{T_i}, \quad K_{d\phi} = \frac{KT_d}{T} \tag{3.14}$$

where T_i is the integral time and K is a gain constant (which, incidentally, is the proportional gain for the continuous time control law and variation of this gain will determine the position of the closed-loop poles). For the preceding PID gain

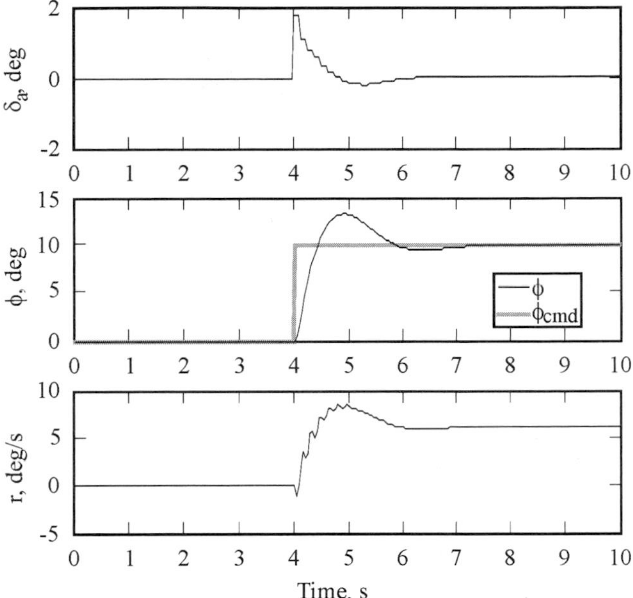

Fig. 3.13 Response for continuous closed-loop roll-to-aileron dynamics with PID control.

values, we have $T_d = 0.04$, $T_i = 0.52$, and $K = 0.125$. Figure 3.12 shows the root locus using this control law, with the position of the closed-loop poles for these gain settings.

Finally, the roll simulation for these gain settings is presented in Fig. 3.13. The response of the closed-loop system with PID is good, with the roll angle settling to the reference value within 2s, and, with the exception of some overshoot initially, oscillation is minimal. The aileron actuation does not exceed 2 deg, with gradual motion at each time step where the aileron is actuated. Also, improvements are seen with the other motion variables over using PD control: β stays within 2 deg at the most, p stays under 40 deg/s, and r stays under 10 deg/s.

Using the same design approach, a control law will be designed for the longitudinal motion of the aircraft using the pitch-to-elevator transfer function. The pitch-to-elevator transfer function for the continuous time system and the discrete system is summarized in Table A.1.

Table 3.10 and 3.11 summarizes the open-loop poles and zeros for the pitch-to-elevator transfer function. Again, the zeros using the zero-order hold transform

Table 3.10 Pitch-to-elevator poles for longitudinal dynamics

Mode	Poles
Short period	-0.275 ± 0.113j
Phugoid	0.983 ± 0.102j

Table 3.11 Pitch-to-elevator zeros for longitudinal dynamics

Continuous	Discrete
$-\infty$	-0.193
-7.86	0.358
-0.929	0.890

method and the root locus for the discrete system with zero-order hold are used. As was already noted by the eigenvalues (Table 3.4), both the short-period and phugoid modes are stable.

Figure 3.14 shows the discrete root locus for the pitch-to-elevator transfer function. The path of the closed-loop poles is shown for increased proportional gain magnitude. Note that by convention, a positive elevator deflection causes a negative pitch response; gains must therefore be negative to ensure proper closed-loop response. As shown, the initial gain does place the closed-loop poles within the unit circle, giving a stable closed-loop system. The most the closed-loop gain can be increased is to around –0.75 before at least one pole migrates outside the unit circle (eventually to negative infinity with further gain increase). Again, using Simulink, the closed-loop system was simulated and the time history plotted for the longitudinal variables and elevator actuation. The results are shown in Fig. 3.15 for sample time period of 0.125 s.

It can be seen that the elevator movement is not very extreme, with generally smooth movements over the course of the automatic actuation. The maximum magnitude of the elevator deflection is only about 2.5 deg, a reasonable amount. It is immediately noted from examining the pitch attitude response that the selected

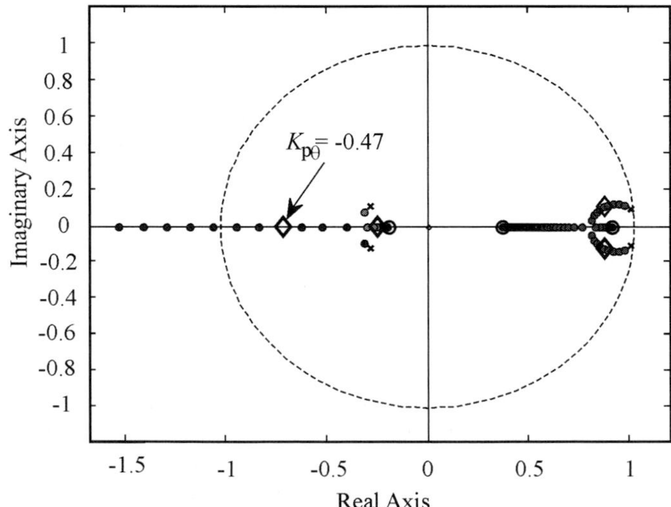

Fig. 3.14 Discrete root locus for pitch-to-elevator transfer function.

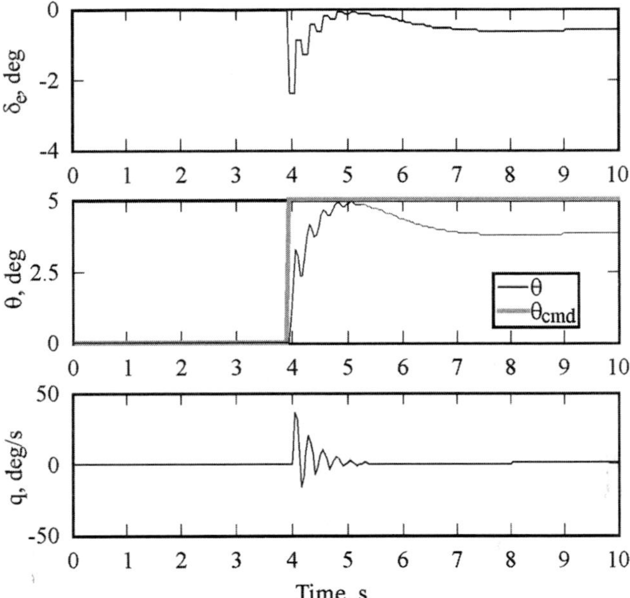

Fig. 3.15 Time response of pitch-to-elevator transfer function for $K_{p\theta} = -0.47$.

gain is insufficient for the closed-loop pitch response. The measured steady-state pitch that the MAV settles to is about 3.7 deg, an error of 26.7%, which is not acceptable. At this point, the easiest solution to this problem is to increase the magnitude of the gain. Additionally, integral control (reduce error and speed up the response) and/or derivative control (reduce overshoot by increasing damping) can be introduced to improve the overall closed-loop performance. At this time, it will be shown what happens when the gain is increased.

Figure 3.16 shows a root-locus plot of the pitch-to-elevator discrete transfer function with a new proportional gain selection ($K_{p\theta} = -0.7$) at 8-Hz sampling frequency. Again, the closed-loop poles still lie within the stable region on the z plane. Figure 3.17 shows the time response of the closed-loop system with the new gain selection. At the new gain selection, the steady-state error has improved somewhat (now it is 25%), but at the expense of greater control actuation. For the longitudinal dynamics of the Zagi MAV, it can be concluded that pure gain control is insufficient at producing acceptable closed-loop performance. Addition of integral and/or derivative control is necessary.

To provide a basis of comparison for the discrete control design presented, closed-loop control for both the roll and pitch of the Zagi is shown for the continuous system. If the data signal from the roll and pitch attitude sensors were continuous, then the equivalent continuous gains would be used in the autopilot. Figures 3.18 and 3.19 show the time history of the closed-loop system to reference roll and pitch signals. As the control law is reading a continuous signal from the sensors in this case, it can be seen that the response is obviously much smoother than if discrete control is used (as should be expected). This observation is illustrated

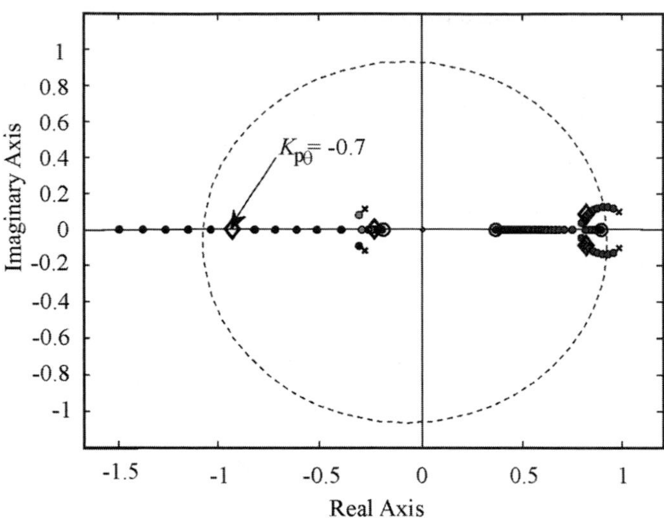

Fig. 3.16 Discrete root locus for pitch-to-elevator transfer function for $K_{p\theta} = -0.7$.

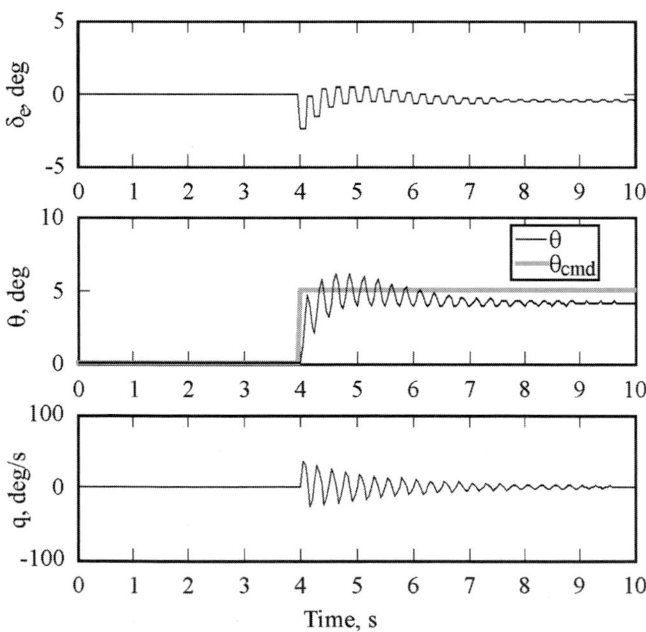

Fig. 3.17 Time response of pitch-to-elevator transfer function for $K_{p\theta} = -0.7$.

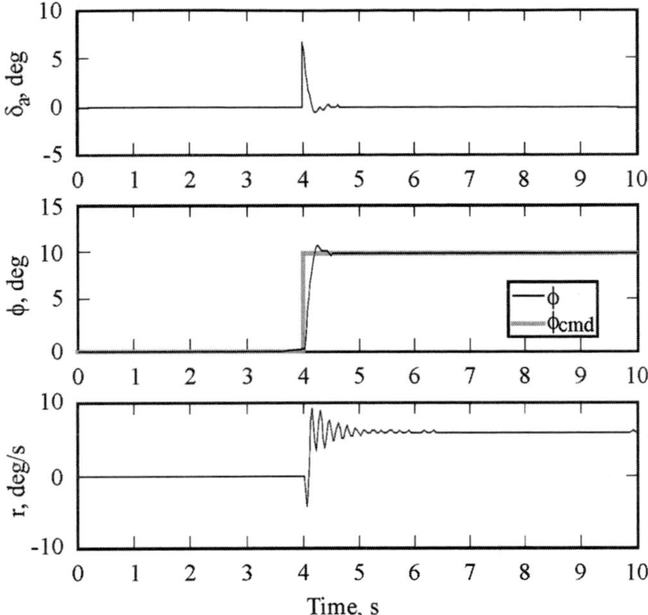

Fig. 3.18 Response for continuous closed-loop roll-to-aileron dynamics.

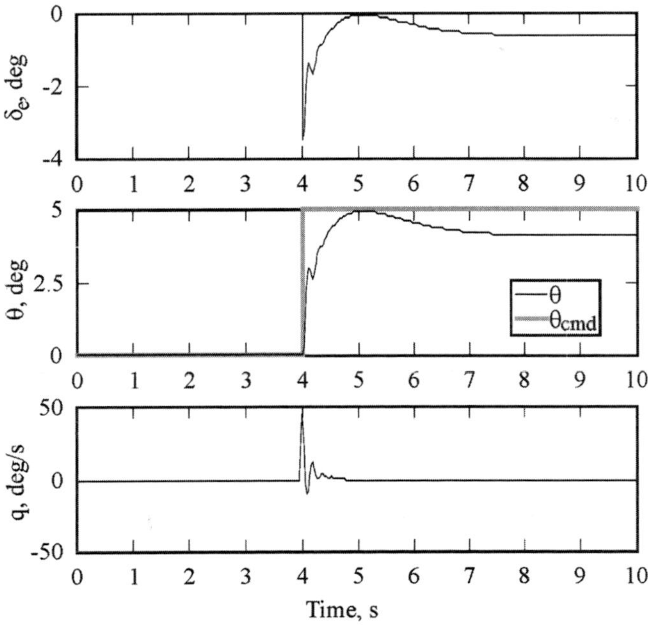

Fig. 3.19 Response for continuous closed-loop pitch-to-elevator dynamics.

Table 3.12 Dynamic characteristics of the Dragonfly MAV

Mode	λ	ω_n	ζ	τ
	Longitudinal dynamics			
Short period	$-7.21 \pm 15.9j$	17.4	0.413	——
——	-0.159	——	——	6.29
——	0.126	——	——	-7.95
	Lateral dynamics			
Dutch roll	$-1.93 \pm 121j$	121	0.0159	——
Roll	-11.1	——	——	0.0897
Spiral	-1.08	——	——	0.926

by comparing Figs. 3.18 and 3.19 to their discrete-time equivalent responses shown in Figs. 3.11 and 3.17. Although there is a smoother response in both cases, it can be seen that with the roll response the roll angle is reached, as it was the case with the discrete-time system (even though it took a second longer for the discrete system to settle). As for the pitch response, again, it can be seen that the pitch attitude does not reach the reference value. The discrepancy is about 20%, only a small improvement over the discrete control design.

Using the continuous-time control analysis provides a useful tool for evaluating the performance of a closed-loop control design. Designing the control law in discrete time is advantageous in that limitations such as system and actuator bandwidth, as well as data sampling times, can be overcome. The performance and stability of a control law designed in continuous time cannot be guaranteed when converted to its discrete-time equivalent; however, a control law that performs well and is stable in discrete-time will maintain that performance in continuous time.

3.4.5 Control Design for Dragonfly

The control design concepts presented for the Zagi can be applied to the Dragonfly MAV as well and will only be described briefly. The Dragonfly uses a rudder control surface to control lateral-directional motion. The autopilot is set such that the rudder will control the roll angle directly, as opposed to the yawing motion that it normally would control. Because roll and yaw motion are coupled, the roll control also affects yawing motion. In this example, the roll and pitch dynamics will be analyzed, and a design example using the roll dynamics will be shown. Candidate gains for each motion will be summarized as well.

Table 3.12 shows the eigenvalues of the Dragonfly at its given trim condition. In the longitudinal dynamics, it can be seen from the eigenvalues that the dynamics are not typical of most aircraft, where the only oscillatory mode present is the short-period one, while the other two eigenvalues are real. Note that one of the eigenvalues shows that the longitudinal dynamics of the Dragonfly are unstable. However, it has been found that there is a range of proportional gains that can provide a measure of stability for the equivalent discrete system. The lateral dynamics, however, are stable. Compared to the Zagi, the Dutch-roll mode is more oscillatory with a lower damping ratio, while roll stability is lower, and spiral stability is higher.

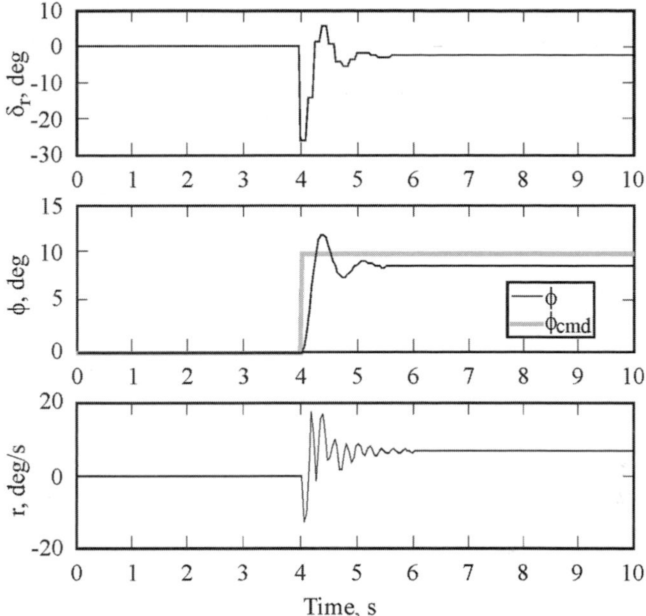

Fig. 3.20 Time response of roll-to-rudder transfer function for $K_{p\phi} = -2.63$, $T_d = 0.001$, and $T = 0.125$.

As was done with the Zagi, control gains were determined in the same way for the Dragonfly using the discrete-time equivalent systems at an 8-Hz sampling rate, with the time responses of the discrete systems shown here. For the gains, the roll response is good (see Fig. 3.20), with little oscillation at the start of control actuation. Between the reference roll angle and the roll angle response, there is a ~10% error that could still be improved upon either by adjusting the PD gains further or introducing integral control as well. Note that as rudder control is being used to control the roll angle, by convention, a negative rudder deflection should lead to a positive roll, which is what occurs here. As the roll and yaw motion are both strongly coupled, actuation of the rudder will also change the heading of the aircraft, evidenced by a somewhat significant change in the yaw rate r, becoming constant when a constant roll angle is reached, indicating a constant rate of change of direction. Roll rate, of course, returns to zero once the MAV reaches the required roll angle.

The PID gains used in the roll control law for the Zagi were also tested with the Dragonfly. Figure 3.21 shows the root-locus plot for this MAV. Again, it can be seen that the gains lie well within the region of stability. Simulation of the closed-loop roll control (Fig. 3.22) shows that the desired roll angle is reached within 4 s of initial command. No overshoot is seen with the roll angle, roll rate is under 5 deg/s during the transient phase, yaw rate is ~8 deg/s, and sideslip motion is negligible. The rudder actuation is quite smooth, with no extreme motion, reaching a constant of −3.5 deg once the desired roll angle is reached, maintaining the aircraft at a constant banked turn at 10 deg. From the simulation, it is evident that these PID gain settings perform well at controlling aircraft roll and holding a desired roll

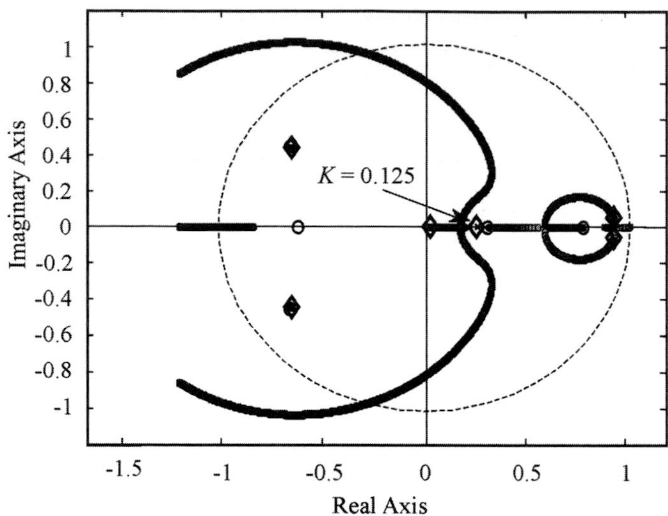

Fig. 3.21 Root locus for roll-to-rudder discrete transfer function with PID control for $K_{p\phi} = 0.11$, $K_{d\phi} = 0.04$, and $K_{i\phi} = 0.03$.

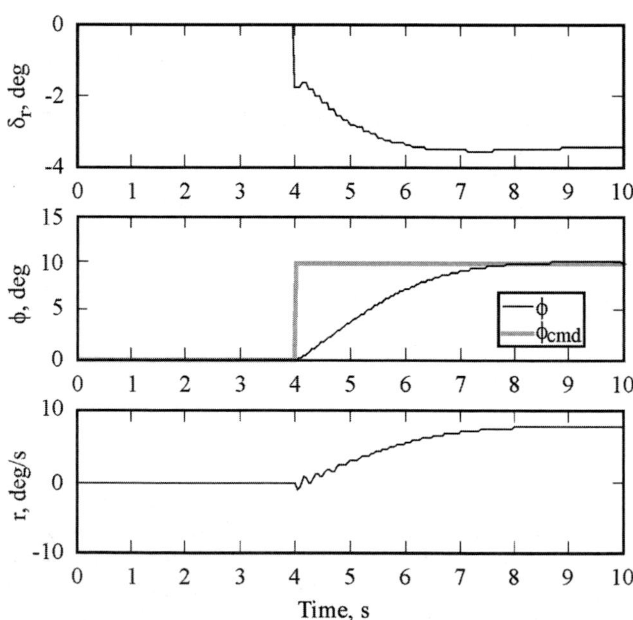

Fig. 3.22 Response for discrete closed-loop roll-to-rudder dynamics with PID control.

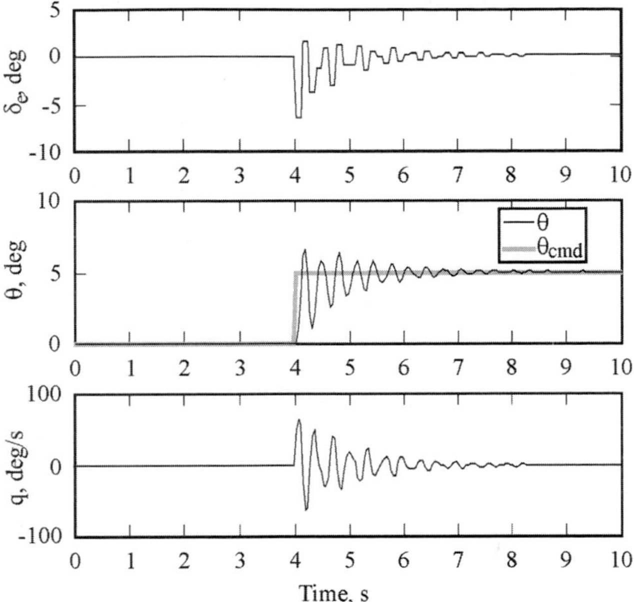

Fig. 3.23 Time response of pitch-to-elevator transfer function for $K_{p\theta} = -1.3$ and $T = 0.125$.

angle. These gains will be further evaluated in flight tests of the Dragonfly MAV and adjusted as is needed to provide adequate roll stability.

Finally, it can be seen from Fig. 3.23 that using proportional control (under the model assumptions made in Sec. 3.4.2) the pitch response is good, with only a 4% error to the reference pitch. Although the response is initially oscillatory in pitch, it only takes about 2 s for the system to reach the reference pitch, with the largest elevator deflection being 6.5 deg in magnitude (occurring at the beginning of control actuation), but much smaller otherwise, being no more than about 3 deg. Although the elevator motion is oscillatory, the amplitude of the oscillation is small. Currently, the autopilot uses proportional control for the closed-loop pitch control, but it might be necessary to add derivative and/or integral control to improve the Dragonfly MAV's response.

3.5 Flight Testing

The goal of the flight tests of the Zagi and Dragonfly MAVs was twofold: 1) to validate and fine tune the designed PID controllers and 2) to investigate flight performance characteristics of autonomous MAVs. Telemetry data, including command input and aircraft response, were the main data source for the investigation of PID controllers. Although all controllers were analyzed and adjusted (roll, yaw, pitch, and climb rate controllers), only details concerning the roll controller will be presented here. Flight tests included straight level dashes, prescribed banked turns, and control pulses simulating wind gusts. By using the stability augmentation

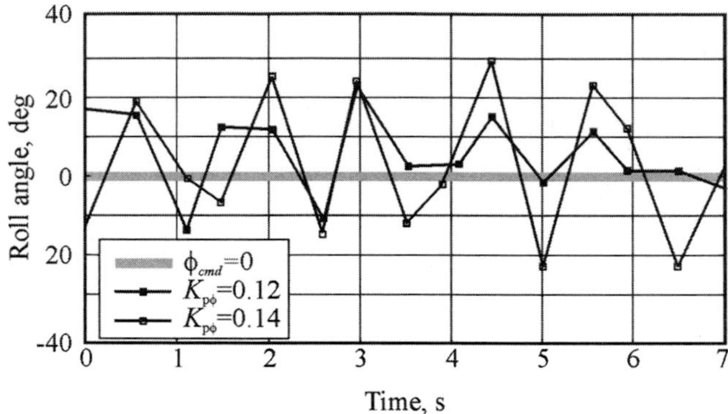

Fig. 3.24 Roll response in a steady level flight.

mode of the autopilot, a pilot introduced the desired initial conditions, after which the roll controller of the autopilot held the desired roll. Finally, fully autonomous flights were performed in order to demonstrate the accuracy and robustness of the flight-control system.

3.5.1 Investigation of the Dynamic Stability of the Zagi

In a series of flight tests, the roll response of the Zagi to prescribed control pulses was investigated. The stability augmentation mode of the autopilot is better suited for this study because the pilot can accurately introduce the desired value of roll ϕ_{cmd}. Then, the autopilot generates control commands for the elevons to hold the roll of the airplane at the desired value. Because navigation and guidance algorithms are inactive in the stability augmentation mode, the overall aircraft response provides necessary information for the evaluation of the quality of the roll controller. The quality of the roll control was studied for a range of roll gain coefficients.

The angle range for the elevons was 30 deg in roll and 35 deg in pitch. The roll response data were measured by infrared sensors and transmitted to the ground station for further analysis. Messages were received at the rate of two to three messages per second for the roll parameters, with a total of about two to three messages per second for all control parameters.

First, the airplane's roll response was studied for steady level flight, that is, the desired roll angle $\phi_{\text{cmd}} = 0$. From the results of simulations presented in Sec. 3.4, two values for the proportional gain coefficient were chosen, $K_{p\phi} = 0.12$ and 0.14 and $K_{d\phi} = K_{p\phi}$ and $K_{i\phi} = 0$. Roll response was measured with the help of the infrared sensors. As can be seen from Fig. 3.24, the roll angle oscillates around the desired roll angle $\phi_{\text{cmd}} = 0$. For $K_{p\phi} = 0.12$, the maximum amplitude of the roll angle is 34 deg and the oscillations decay with time. For $K_{p\phi} = 0.14$, the flight-control system overshoots the desired angle with the maximum amplitude of 51 deg and remains approximately the same for the duration of data recording. The frequency of the oscillations varies in the range from 0.5 to 1.0 Hz for both values of the proportional gain coefficient.

In the second series of tests, the roll transient response was studied after a step-function command changed the desired roll angle, ϕ_{cmd} from 0 to 38 deg. The

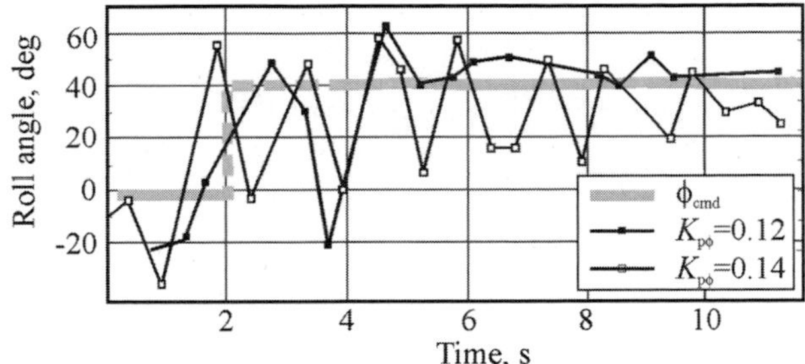

Fig. 3.25 Response of the airplane to the roll command in the form of a step function.

same set of gains as in the previous tests was used in this experiment. The roll response is shown in Fig. 3.25 for two values of $K_{p\phi}$. The response of the airplane was mostly nonoscillatory when $K_{p\phi} = 0.12$ was used, except for the transition period of about 5 s when the maximum roll angle of 61 deg was found. After the first 5 s, the steady state was reached with minimal oscillations.

The experiments show that a higher proportional gain coefficient makes the airplane respond faster, but it overshoots the desired value and causes oscillations. A lower proportional gain coefficient produces a slow, but smooth response. Note that these results were expected based on the study in Sec. 3.4.4 and confirm the adequate response of the airplane control system and robustness of the measurement system.

Another way to assess the quality of the controller is to apply a sharp disturbance to the vehicle and then observe its reaction. A triangular impulse of duration 1.5 s and maximum value of 22 deg was applied to the Zagi in a flight test and in simulations. PID gains of $K_{p\phi} = 0.11$, $K_{d\phi} = 0.04$, and $K_{i\phi} = 0.03$ were used in both the experiment and analysis. Experimental responses, ϕ_{exp} and numerical results of the simulations ϕ_{sim} are shown in Fig. 3.26, together with the desired roll change ϕ_{cmd}.

It can be seen from Fig. 3.26 that the computer-simulated response was faster than the response of the actual airplane, and no overshooting occurred in either case. Although some fluctuations in the roll angle were observed in the experiment,

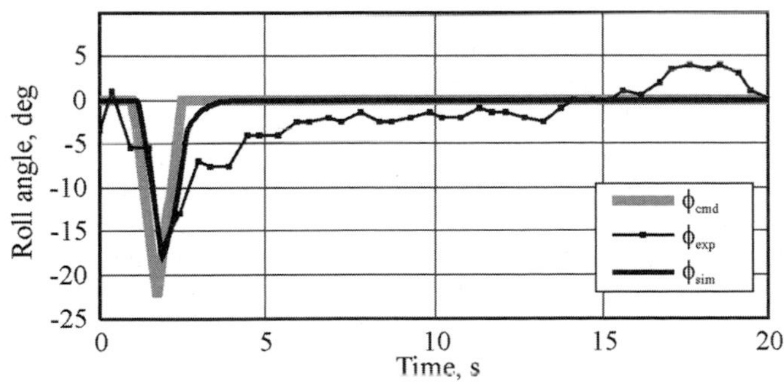

Fig. 3.26 Experimental and numerical time response to a short roll impulse.

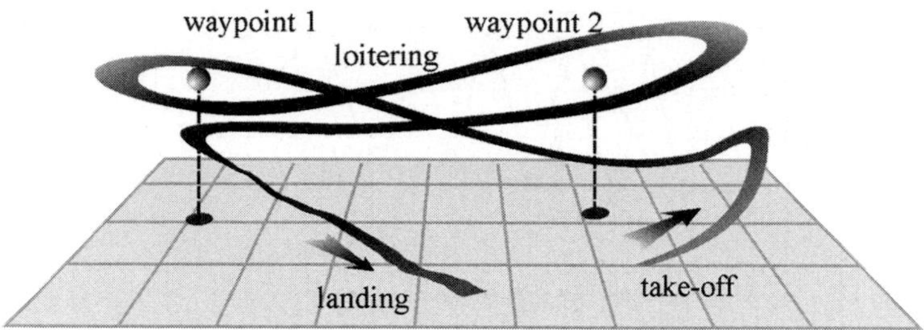

Fig. 3.27 Typical test-flight trajectory.

they were not present in the simulations, because the simulations did not include the effects of atmospheric turbulence. Overall, the results of the numerical simulations agree well with the response of the actual airplane in a roll.

3.5.2 Performance of the Dragonfly MAV in Fully Autonomous Flight

A typical test-flight trajectory of the Dragonfly equipped with the Paparazzi autopilot is presented in Fig. 3.27. During autonomous flight, the autopilot performs waypoint and altitude navigation based on the flight plan uploaded into the MAV. This test flight was designed to validate guidance, navigation, and control algorithms implemented in the autopilot and to demonstrate the potential of the proposed technology. Telemetry data on control, navigation, and location of the airplane were collected and transmitted to the ground station.

This fully autonomous flight lasted for about 30 min and consisted of launch, loitering and landing. Initial values for the PID gain coefficients were introduced using the results given in Sec. 3.4.5. They were adjusted based on the analysis of the acquired telemetry data.

After launching and achieving the altitude specified by the flight plan, the airplane performed a number of figure 8s while loitering between two waypoints 250 m (820 ft) apart. The flight plan required the airplane to perform a large number of banked turns, providing a better evaluation of its flying qualities.

Launching is a high-risk phase for an MAV because of the airplane's sensitivity to wind gusts at low speeds. The launch in fully autonomous mode is performed with a throttle setting at 100%. The airplane holds the pitch attitude prescribed in the flight plan until it reaches the specified altitude, which was 60 m (197 ft) in the conducted flights. Two climb angles (6 and 18 deg) were studied, and both worked reliably. It was determined from GPS measurements that the climb rate was held constant at 2 m/s (4 mph) and 6 m/s (13 mph) for the lower and higher climb angle, respectively. The high climb rate is explained by a high thrust-to-weight ratio of 0.7 (based on the static thrust measurements), enabling aggressive maneuvers.

The maximum speed achieved at the 100% throttle setting during a steady level dash was estimated based on GPS data to be about 22 m/s (49 mph). During the loitering portion of the flight, the throttle setting was at 50%, with a measured cruise speed of about 12 m/s (27 mph). For the current autopilot, it was found that altitude control using the throttle is simple to implement and works reliably. The roll controller limits the maximum bank angle to 40 deg, ensuring stability of

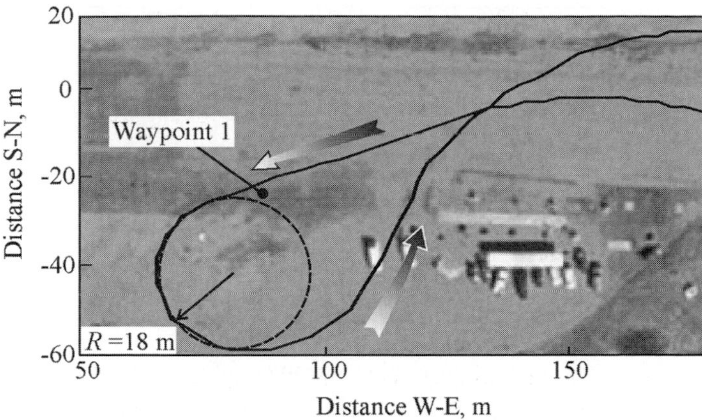

Fig. 3.28 Flight trajectory near the waypoint.

the airplane in turns. In fully autonomous mode with the airplane flying at cruise speed, the radius of turn was slightly less than 18 m (59 ft), as shown in Fig. 3.28. The waypoint shown in the figure denotes a position in the flight area that the airplane should fly through. After the airplane has flown into the vicinity of the waypoint, which is currently set to be within 10 m (33 ft) of the waypoint, the navigation system commands the airplane to head to the next waypoint and computes the desired heading and roll angle.

In one of the test flights, an abrupt stall of the MAV occurred. The airplane was flying at a 50% throttle setting and average speed of 12 m/s (27 mph) when it was hit by a violent wind gust. At that moment, an angle of attack of 29 deg was determined based on recorded GPS and pitch sensors data. A stall angle of attack of 23 deg was reported for the MAV's wing in the case study in Chapter 6. In fact, for the wing with running propeller, the stall point is delayed to a higher angle of attack [32], supporting the hypothesis of stall occurrence. Close investigation of the telemetry data also revealed that overshooting takes place in the climb rate control. After gains were adjusted, no stall occurred in subsequent flights, and the Dragonfly demonstrated a capability of operating in winds of up to 5–8 m/s (11–18 mph).

A transient roll response of 100-s duration is shown in Fig. 3.29. Two time histories are presented on the plot—one for the desired roll angle ϕ_{cmd}, computed by the navigation module, and one for the measured roll angle of the airplane ϕ_{exp}. A proportional-integral-derivative control was used for the roll controller, with the gain coefficients obtained as a result of the design procedure presented in the preceding sections. The maximum error of the measured roll angle is about 11 deg and the average error is about 5 deg. The overall quality of the roll tracking can be regarded as good, and it is believed that it can be improved by further optimization of the flight control system.

The landing sequence is programmed into the autopilot by specifying a value for the throttle and the landing altitude. Through a number of flight tests using a "touch-and-go" procedure, it was found that the airplane performs a gliding descent at a safe angle and lands reliably with a 20% throttle setting. With the elevator maintaining a constant speed, from the altitude of 30 m (98 ft), the airplane glides

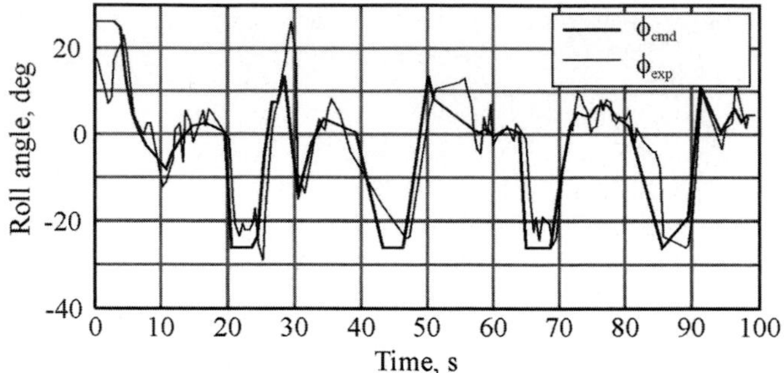

Fig. 3.29 Roll angle of the airplane during a fully autonomous flight.

for about 15 s and travels a distance of about 120 m (394 ft). It corresponds to a gliding ratio of 4:1 and a sink rate of 2 m/s (4 mph).

By flying the trajectories of a moderate difficulty, the performance of the designed PID controllers for the Dragonfly has been validated, and the Dragonfly MAV has performed outstandingly.

3.6 Conclusion

Because of their small size, MAVs are prone to unsteady behavior with high-frequency oscillations, limiting the usefulness of their application. The reduced stability of MAVs leads to greater maneuverability, yet control laws are needed to maintain the aircraft on its flight path, and to provide the stability necessary for proper function of onboard sensors. This is especially needed for smooth flight through wind gusts and turbulence. At the same time, the control laws must allow the aircraft the ability to make aggressive maneuvers. Thus, an enhanced automatic flight-control system is needed for the progress of MAV technology. In this chapter, a systematic approach for the integration of an autopilot into an MAV was the major topic of interest.

The Paparazzi autopilot was integrated into two MAVs: 59-cm (23-in.) Zagi and 30-cm (12-in.) Dragonfly. This chapter described the features of the aerodynamics and structural designs of the MAVs. Both airplanes were equipped with video cameras that capture and transmit video to the ground station, making them very effective for surveillance missions.

Because measured data from the sensors were sampled, and bandwidth limits of the aircraft dynamics and servomotors had to be considered, a discrete time control design was developed to overcome these limitations in that the frequency and speed of control actuation was kept within these limits. Based on the linearized equations of motion of the aircraft, control laws were designed and evaluated. The equations were developed from an evaluation of stability and control derivatives in both the longitudinal and lateral-directional motions that were then assembled into state-space matrices. The MATLAB®/Simulink program was used to investigate the open-loop stability of the airplane, as well as to determine control laws for closed-loop stability. The discrete equivalents of gains were determined and adjusted to satisfy the stability criteria in discrete time. These gains were tested in closed-loop simulations and then were used as initial values by the autopilot.

In a series of flight tests of the Dragonfly and Zagi MAVs, telemetry data on control actuation, altitude, attitude, and GPS location of the airplane were collected. Based on the data collected, the PID controllers were adjusted, and flight performance characteristics were determined for autonomous MAVs. The developed flight-control system was tested in all flight phases, including straight level dashes, prescribed banked turns, control pulses simulating wind gusts, waypoint navigation, fast climbing, and autonomous landing.

Future research will focus on increasing the robustness of the flight-control system and widening the flight envelope. Enhanced control laws will be developed to satisfy the needs for flying very aggressively: the MAV will be capable of sharp turns and pull-outs, steep climbs and descents, and spiraling up and down in narrow spaces. System miniaturization and improvements in control effectiveness and the agility of MAVs will enable a further decrease in the size of MAVs and allow them to play a major role in the future developments of swarms of unmanned air vehicles and other useful applications.

Appendix A. Transfer Functions

Table A.1 Transfer functions for continuous and discrete systems for the Zagi MAV

Transfer function	Time form	Value
$\frac{\theta}{\delta_e}$	Continuous	$\frac{-543.4s^2 - 4778s - 3971}{s^4 + 19.57s^3 + 896.5s^2 + 180.9s + 624.9}$
$\frac{\theta}{\delta_e}$	Discrete	$\frac{-1.339z^3 + 1.413z^2 - 0.1039z - 0.08258}{z^4 - 1.415z^3 - 0.01734z^2 + 0.3637z + 0.08659}$
$\frac{\phi}{\delta_a}$	Continuous	$\frac{506.4s^2 + 4692s + 6.226 \times 10^5}{s^4 + 23.11s^3 + 1581s^2 + 2.777 \times 10^4 s + 8715}$
$\frac{\phi}{\delta_a}$	Discrete	$\frac{1.819z^3 + 0.4151z^2 + 0.842z + 0.4446}{z^4 - 1.267z^3 + 0.8534z^2 - 0.6371z + 0.05568}$

Table A.2 Transfer functions for continuous and discrete systems for the Dragonfly MAV

Transfer function	Time form	Value
$\frac{\theta}{\delta_e}$	Continuous	$\frac{-246.9s^2 - 921.3s - 84.3}{s^4 + 14.45s^3 + 304.2s^2 - 9.801s - 6.072}$
$\frac{\theta}{\delta_e}$	Discrete	$\frac{-0.9719z^3 + 1.035z^2 + 0.2708z - 0.3403}{z^4 - 1.67z^3 + 0.5104z^2 - 0.004867z + 0.1643}$
$\frac{\phi}{\delta_a}$	Continuous	$\frac{-30.41s^2 + 601.7s - 5.135 \times 10^5}{s^4 + 16.09s^3 + 14660s^2 + 1.787 \times 10^5 s + 1.759 \times 10^5}$
$\frac{\phi}{\delta_a}$	Discrete	$\frac{-0.1696z^3 - 0.3269z^2 - 0.2443z - 0.06614}{z^4 + 0.171z^3 - 0.6157z^2 - 0.4127z + 0.1339}$

Acknowledgments

Ft. Huachuca Battle Command Battle Laboratory, Grant number GS09505DN-C650, sponsored this research and development work. The authors would like to acknowledge with pleasure the discussions and suggestions from Jason Denno and Julie Fulmer. Professors T. Vincent and T. J. Mueller read the manuscript and made very helpful comments and suggestions. This project was sponsored, in part, by the grant from US ARO (project manager Dr. S. Sampath). We also would like to thank the rest of the Micro Air Vehicle Project team at the University of Arizona for their contributions to this work: Bill Null, Bret Becker, Daniel Bradley, Anton Kochevar, Jeremy Tyler, Gary Tang, and Bo Han.

References

[1]Foster, T. M., and Bowman, W. J., "Dynamic Stability and Handling Qualities of Small Unmanned-Aerial-Vehicles," AIAA Paper 2005-1023, Jan. 2005.

[2]"Flying Qualities of Piloted Airplanes," Military Specification, Dep. of Defense Military Specifications and Standards, U.S. Dep. of Defense, MIL-F-8785C, Philadelphia, PA, 1980.

[3]Cretual, A., and Chaumette, F., "Dynamic Stabilization of a Pan and Tilt Camera for Submarine Image Visualization," *Computer Vision and Image Understanding*, Vol. 79, No. 1, 2000, pp. 47–65.

[4]Jin, J. S., Zhu, Z., and Xu, G., "Digital Video Sequence Stabilization Based on 2.5D Motion Estimation and Inertial Motion Filtering," *Real-Time Imaging*, Vol. 7, No. 4, 2001, pp. 357–365.

[5]Duric, Z., and Rosenfeld, A., "Image Sequence Stabilization in Real Time," *Real-Time Imaging*, Vol. 2, No. 5, 1996, pp. 271–284.

[6]Waszak, M. R., Jenkins, L. N., and Ifju, P., "Stability and Control Properties of an Aeroelastic Fixed Wing Micro Aerial Vehicle," AIAA Atmospheric Flight Mechanics Conference and Exhibit, Montreal, Canada, Aug. 6–9, 2001. AIAA Paper 2001-4005.

[7]Ifju, P. G., Jenkins, D. A., Ettinger, S., Lian, Y., Shyy, W., and Waszak, M. R., "Flexible-Wing-Based Micro Aerial Vehicles," AIAA Paper 2002-0705, Jan. 2002.

[8]Platanitis, G., and Shkarayev, S., "Integration of an Autopilot for a Micro Air Vehicle," Infotech@Aerospace, Arlington, Virginia, Sep. 26–29, 2005. AIAA Paper 2005-7066.

[9]Taylor, B., Bil, C., and Watkins, S., "Horizon Sensing Attitude Stabilisation: A VMC Autopilot," 18th *International UAV Systems Conference*, March 2003.

[10]Hsiao, F., Chien, Y., Liu, T., Lee, M., Chang, W., Han, S., and Wang, Y., "A Novel Unmanned Aerial Vehicle System with Autonomous Flight and Auto-Lockup Capability," AIAA Paper 2005-1050, Jan. 2005.

[11]Grasmeyer, J., and Keennon, M., "Development of the Black Widow Micro Air Vehicle," 39th Aerospace Sciences Meeting and Exhibit, Reno, NV, Jan. 8–11, 2001. AIAA Paper 2001-0127, Jan. 2001.

[12]Arning, R. K., and Sassen, S., "Flight Control of Micro Aerial Vehicles," AIAA Paper 2004-4911, Aug. 2004.

[13]Flake, J., Frischknecht, B., Hansen, S., Knoebel, N., Ostler, J., and Tuley, B., "Development of the Stableyes Unmanned Air Vehicle," 8th *International Micro Air Vehicle Competition*, The Univ. of Arizona, Tucson AZ, April 2004, pp. 1–10.

[14]Quix, H., and Alles, W., "Design and Automation of Micro Air Vehicles," 1st *European Micro Air Vehicle Conference and Flight Competition*, July 2004.

[15]Chung, D. K., Ryu, J. H., Nam, I. C., Jo, K. Y., Yoon, K. J., Huang, H. C., and Kim, J. H., "Development of Fixed Wing MAV "Batwing"," 9th *International Micro Air Vehicle Competition*. Konkuk University, Seoul, Korea, May 2005, pp. 120–126.

[16]Drouin, A., and Brisset, P., "PaparaDzIY: Do-it-Yourself UAV," 4th European Micro-UAV Meeting, ENSICA, SUPAERO, Toulouse, France, Sept. 2004.

[17]Dixon, R., *Open Source Software Law*, Artech House, Boston, MA, 2004, pp. 287, 288.

[18]Kajiwara, I., and Haftka, R. T., "Simultaneous Optimum Design of Shape and Control System for Micro Air Vehicles," 40th AIAA/ASME/ASCE/AHS/ASC Structures, Structural Dynamics, and Materials Conference and Exhibit, St. Louis, MO, Apr. 12–15, 1999. AIAA Paper 99-1391, June–July 1999.

[19]"Air Craft," Brushless Motors Data Sheet, Oroshi Danchi 24-10 Tokuzen Iizuka-Shi, Fukuoka, Japan, 2006.

[20]Selig, M. S., Guglielmo, J. J., Broeren, A., and Giguère, P., *Summary of Low-Speed Airfoil Data*, Vol. 1, SoarTech Aero Publications, Virginia Beach, VA, 1995, pp. 292–293.

[21]Null, W., and Shkarayev, S., "Effects of Camber on the Aerodynamics of Adaptive-Wing Micro Air Vehicles," *Journal of Aircraft*, Vol. 42, No. 6, 2005, pp. 1537–1542.

[22]SAM-LL Programmable GPS Smart Antenna, U-Blox AG, Zurcherstrasse 68, CH-8800, Thalwil, Switzerland, 2006.

[23]"*MLX90247 Discrete Infrared Thermopile Detectors Data Sheet*," Melexis USA, 41 Locke Road, Concord, NH, 2004, pp. 9, 10.

[24]MATLAB Version 7.0.1 (R14), Simulink ver. 1.1, The MathWorks, Natick, MA, June 2004.

[25]Roskam, J., *Airplane Flight Dynamics and Automatic Flight Controls. Part I*, Design, Analysis and Research Corporation, Lawrence, KS, 2003, p. 576.

[26]Etkin, B., and Reid, L. D., *Dynamics of Flight Stability and Control*, 3rd ed., J. Wiley, New York, 1996, p. 382.

[27]Ogata, K., *Discrete-Time Control Systems*, 2nd ed., Prentice–Hall, Englewood Cliffs, NJ, 1995, p. 768.

[28]Franklin, G. F., Powell, J. D., and Workman, M., *Digital Control of Dynamic Systems*, 3rd ed., Addison Wesley Longman, Menlo Park, CA, 1998, p. 850.

[29]*Advanced Aircraft Analysis AAA Version 3.0 User's Manual*, Design, Analysis and Research Corp., Lawrence, KS, 2005, p. 195.

[30]Melin, T., *User's Guide, Reference Manual, Tornado 1.0, Release 2.3*, Royal Inst. of Technology (KTH), Stockholm, Sweden, 2001, p. 42.

[31]Roskam, J., *Airplane Flight Dynamics and Automatic Flight Controls. Part II*, Design, Analysis and Research Corp., Lawrence, Ks, 2003, p. 362.

[32]Null, W., Noseck, A., and Shkarayev, S., "Effects of Propulsive-Induced Flow on the Aerodynamics of Micro Air Vehicles," 23rd AIAA Applied Aerodynamics Conference, Toronto, Ontario, June 6–9, 2005. AIAA Paper 2005-4616, June 2005.

4
Case Study: Micro Tactical Expendable Rigid-Wing Micro Air Vehicle

James C. Kellogg*
Tactical Electronic Warfare Division, Naval Research Laboratory, Washington, DC

Nomenclature

AR	=	aspect ratio
b	=	wing span
C_D	=	drag coefficient, $D/\frac{1}{2}\rho V_\infty^2 S$
C_{Di}	=	induced drag coefficient
$C_{\mathrm{D\ laminar}}$	=	friction drag coefficient for laminar flow
$C_{\mathrm{D\ turbulent}}$	=	friction drag coefficient for turbulent flow
C_{D0}	=	profile drag coefficient at zero lift
C_L	=	lift coefficient, $L/\frac{1}{2}\rho V_\infty^2 S$
C_{MX}	=	roll-moment coefficient
C_{MZ}	=	pitching-moment coefficient
C_{MZac}	=	pitching-moment coefficient about the aerodynamic center
c	=	wing chord
c_{root}	=	root wing chord
D	=	drag
e	=	Oswald span efficiency factor
K_1	=	profile drag constant for airfoil at low Reynolds number
K_2	=	constant
K_3	=	constant
L	=	lift
P_B	=	power supplied by the battery
Re	=	Reynolds number based on wing root chord, $\rho V_\infty c_{\mathrm{root}}/\mu$
S	=	wing area
V	=	airspeed
V_P	=	propeller pitch speed
V_S	=	stall speed
V_∞	=	freestream velocity
W	=	weight

This material is a work of the U.S. Government and is not subject to copyright protection in the United States.

*Head, Subsystems for Mobility Unit, Code 5712.2, Vehicle Research Section, Offboard Countermeasures Branch, Tactical Electronic Warfare Division.

W_0 = gross weight
x_{ac} = chordwise location of aerodynamic center
α = angle of attack
β = sideslip angle
η_G = gearbox efficiency
η_M = motor efficiency
η_P = propeller efficiency
θ = elevon deflection angle
μ = air viscosity
ρ = air density
ϕ = climb angle

4.1 Introduction

This chapter describes the development of the micro tactical expendable (MITE) fixed-wing micro air vehicle developed at the U.S. Naval Research Laboratory in Washington DC from 1996 to 2002. The purpose of this case study is to present the different facets of the MITE's origin, with particular emphasis on aerodynamics and propulsion, as well as the programmatic context that shaped the work overall. As with any development process, the MITE program required a mixture of creative concepts, applied theory, experimental research, and practical experience, operating within the limits of time and budget. This case study therefore takes a mixed approach in attempting to detail the significant factors—theoretical, experimental, and practical—that influenced the MITE design.

4.2 NRL and DARPA MAV Programs

Micro-air-vehicle (MAV) development at the Naval Research Laboratory (NRL) began in October 1996, concurrently with the Defense Advanced Research Projects Agency (DARPA) MAV research effort. The two programs were complementary, intended to broaden the scope of MAV development by exploring different aspects of vehicle and systems design, payloads, and possible missions. Beyond the distinct technical paths chosen for each effort, the major difference was one of development philosophy. The DARPA program was intended to stimulate growth in new technologies by requiring participants to rise to the challenge of a "DARPA hard problem." Formidable size and weight limits for the vehicle were set as goals: a maximum dimension of 15 cm (6 in.) and a maximum mass of 90 g (3 oz). NRL did not adopt strict limits for its MAV; its objective was to develop "the smallest flight vehicle capable of performing a valuable Navy mission." [1]. In fact, NRL never defined a specific size or weight to distinguish a micro air vehicle from a small unmanned air vehicle (UAV). These different philosophies reflect the different missions of DARPA and NRL. NRL MAVs tested ranged 25–61 cm (10–24 in.) in span and 130–350 g (4.6–12.4 oz) in gross weight.

More specifically, the DARPA-sponsored MAV was envisioned as a tool for the individual infantryman, to gather local reconnaissance in potentially dangerous areas without exposing himself to those dangers. The DARPA program was therefore geared toward the creation of systems ready to be transitioned into operational military equipment. Its tiny size was necessary because the modern foot soldier already carries such a heavy load of equipment that anything new must be exceptionally small and light. On the other hand, NRL's program, funded by the Office of Naval Research (ONR), was tasked with developing and demonstrating new technologies that would be useful to a broader range of MAV applications. It was a 6.2-level program, the Navy's classification for exploratory development and feasibility demonstration, as compared with 6.1 (basic research) or 6.3 (advanced technology/mission demonstration). In 6.2 research, the developed technologies need not be fully integrated into a single package. For example, a mission payload can be flown in a radio-controlled vehicle, while an autonomous navigation system is demonstrated in another vehicle without a payload aboard. This frees the researcher to experiment with a range of different systems, leaving the often daunting task of system integration "as an exercise for the student," in the form of a follow-up 6.3 program.

As the NRL- and DARPA-sponsored work had to deal with the same research areas of low-Reynolds-number aerodynamics, energy storage/propulsion systems with high specific energy and specific power, operational requirements, miniaturized sensing, flight control, navigation systems, etc., some overlap of the programs was inevitable. This overlap included the intended MAV mission as a foot soldier's reconnaissance tool. Because the U.S. Navy encompasses the U.S. Marine Corps (USMC), supporting the infantryman was a relevant Navy mission for the NRL MAV. Nevertheless, the programs were coordinated from the outset to avoid duplication of effort. NRL concentrated on single-use, expendable systems, necessitating a very low unit cost. DARPA looked towards reusability, allowing a more sophisticated system with a relatively higher cost. DARPA explored global-positioning-system (GPS) navigation systems, liquid-fueled propulsion systems, and micro imager payloads while NRL worked on optically based navigation, electric propulsion, and other payloads. In addition to the portable recon tool mission, NRL also experimented with the one-way delivery of micropayloads, such as ground sensors or electronic warfare packages, to specific locations.

4.3 Development of the MITE Configuration

The NRL developmental MAV was named the MITE, for micro tactical expendable. The MITE was not a single design, but a series of aircraft that differed in size, payload, airfoil, and other details while maintaining the same basic configuration. This configuration was shaped at the outset by Richard J. Foch, head of NRL's Vehicle Research Section, who made the initial design choices based on a tradeoff analysis that included aerodynamic performance, payload interface, launch and recovery methods, and compact storage. Because compactness is fundamental to the MAV concept, the first choice was for a flying wing. This also fit the designer's belief that simplicity and a low part count were best for a tiny expendable airplane. Wing aspect ratio was the next consideration, and it was here that the low Reynolds number regime of MAVs brought about a departure from traditional airplane design experience. At the chord Reynolds numbers considered in the initial MITE design studies, between 5×10^4 and 10^5, a low-aspect ratio wing was found to be most suitable. Although the low aspect ratio increases the induced drag of the wing, the increase in wing chord raises the Reynolds number, which improves the boundary-layer characteristics, hence performance, of the airfoil. Additionally, a low aspect ratio is desirable for an MAV because it concentrates wing area in a

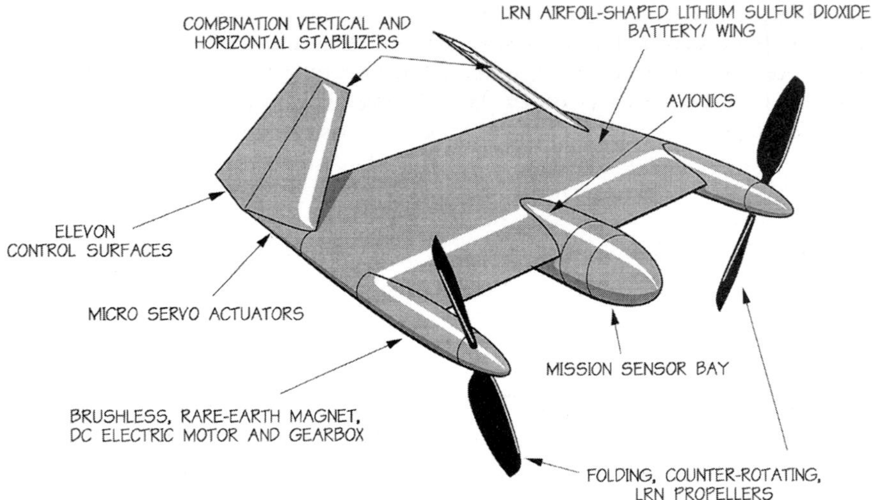

Fig. 4.1 Original MITE concept for a 15-cm micro air vehicle.

compact shape, resulting in a vehicle that is smaller overall. Purely in this regard, the optimum geometrical shape for a wing planform is a circle, though aerodynamic or structural factors might dictate otherwise. Propulsion choices rounded out the vehicle concept. Dual propellers provide slipstream flow over nearly the entire wing for enhanced lift and control at low speeds, and they counter rotate to control torque effects. The initial MITE concept is shown in Fig. 4.1; its origins are described in more detail in the next section.

4.3.1 Wing Aspect Ratio for Minimum Drag at Low Reynolds Number

For a high-Reynolds number wing in level, constant-speed flight, operating at the velocity for maximum range (minimum required thrust), induced drag and parasite drag are equal, each representing 50% of the total drag. When flying at its best speed for endurance (minimum required power), the induced drag component is even greater, two–thirds of the total drag. Consequently, designers working in the high Reynolds number regime of human-carrying airplanes, $Re > 10^6$, must minimize induced drag to maximize range and endurance. Induced drag, a direct consequence of the aerodynamic production of lift, is the result of a spanwise component of the airflow caused by the tips of the finite-span wing. It is proportional to the square of the lift coefficient and inversely proportional to the wing aspect ratio (the ratio of wing span to average chord). Because the optimal cruise lift coefficient is predetermined by the airfoil shape, reducing induced drag requires increasing the aspect ratio. For a given wing loading, this means increasing the wing span and reducing the wing chord. Increasing span increases the wing root bending moment, whereas reducing the chord implies a reduction in wing thickness and possible height of the wing spars. Therefore, increased aspect ratio will reduce wing strength and stiffness unless additional structure, and therefore weight, is added. Generally, the wing aspect ratio is made as large as structural or cost considerations allow.

In the aerodynamic regime of MAV wings, however, increasing the aspect ratio leads to difficulties before the structural limit is reached. The profile drag for a large aircraft's wing is relatively constant with variations in Reynolds number, but this is not the case at the MAV scale. From Hoerner [2], the profile drag coefficient in the range $10^3 < Re < 10^6$ for laminar flow on a flat plate is:

$$C_{\text{D laminar}} = \frac{1.328}{\sqrt{Re}} \tag{4.1}$$

For turbulent flow, empirical studies have shown that the profile drag coefficient for $Re < 10^6$ is found to be

$$C_{\text{D turbulent}} = \frac{0.074}{\sqrt[5]{Re}} \tag{4.2}$$

An MAV wing will experience a combination of laminar and turbulent flow. Again using empirical data, a general expression for the total profile drag coefficient at zero-lift C_{D0} in the Reynolds-number range applicable to MAVs is found to be

$$C_{D0} = \frac{K_1}{\sqrt[3]{Re}} \tag{4.3}$$

where K_1 is a constant determined empirically for a given airfoil. Thus, the profile drag coefficient increases as Reynolds number decreases. At sufficiently low Reynolds number, profile drag becomes the dominant drag component. Therefore the design of an efficient MAV wing requires an optimization between minimizing induced drag without significantly increasing the airfoil profile drag. Total drag is the sum of induced and profile drag:

$$C_D = C_{D0} + C_{\text{Di}} = \frac{K_1}{\sqrt[3]{Re}} + \frac{C_L^2}{\pi e AR} \tag{4.4}$$

where C_L is the coefficient of lift, e is the Oswald span efficiency factor determined by the wing planform and the additional profile drag at nonzero lift, and AR is the aspect ratio. Recall that $Re = \rho Vc/\mu$, where ρ is air density, V is airspeed, c is average chord length, and μ is air viscosity, and that $AR = b/c$, where b is the wing span:

$$C_D = \frac{K_1}{\sqrt[3]{\rho Vc/\mu}} + \frac{C_L^2}{\pi eb/c} \tag{4.5}$$

For a given weight and cruise speed, constants $K_2 = C_L^2/\pi e$ and $K_3 = \rho V/\mu$ are defined to simplify Eq. (4.5) to

$$C_D = \frac{K_1/\sqrt[3]{K_3}}{\sqrt[3]{c}} + \frac{K_2/b}{1/c} = \frac{K_1/\sqrt[3]{K_3}}{\sqrt[3]{c}} + \frac{K_2 c}{b} \tag{4.6}$$

To find the wing chord for minimum total drag, the differential dC_D/dc is set equal to zero:

$$\frac{\mathrm{d}C_D}{\mathrm{d}c} = 0 = \frac{-1/3 K_1 c^{-4/3}}{\sqrt[3]{K_3}} + \frac{K_2}{b} \tag{4.7}$$

Solving for c gives

$$c = \left[\frac{K_1/\sqrt[3]{K_3}}{3K_2/b}\right]^{3/4} \tag{4.8}$$

$$c = \left[\frac{K_1 \pi e b}{3C_L^2 \sqrt[3]{\rho V/\mu}}\right]^{3/4} \tag{4.9}$$

Empirical equation (4.9) was derived by Kevin Ailinger, an aeronautical engineer at NRL, to determine, as a first-order approximation, an optimal tradeoff between induced drag and profile drag for MAVs [3]. This equation indicates why MAVs typically do not lose aerodynamic efficiency at very low aspect ratios as significantly as do larger aircraft. The decreased aspect ratio, hence the increased chord length, raises the chord Reynolds number, producing benefits that limit the increase in profile drag for wings in the MAV regime.

4.3.2 Propulsion

Battery-powered electrical propulsion was chosen for the MITE. This decision was based on many factors, not the least of which were the many drawbacks of the available alternative, internal combustion (IC) reciprocating engines. For MAV-scale IC engines, most of these drawbacks are inherent and would be difficult or impossible to overcome by additional engineering research and development. The high surface area-to-volume ratio of miniature engines causes rapid heat loss, so maintaining the temperatures needed for good combustion is difficult. Fuel/air mixture control is awkward because regulating components, such as carburetors, are so tiny that external contamination and the manufacturing tolerances of the parts significantly affect their operation. As a result, these engines are highly temperamental; they are difficult to start, adjust, and control. This can cause stoppages if the mixture becomes too rich or too lean. Because small IC engines are generally started by an external device, often the operator's finger, there is no chance for a restart in flight. External starting would make the rapid deployment of a large number of MITEs extremely labor intensive from the ground and virtually impossible from an aircraft. An internal starter would be an unacceptable complexity for an MAV, mechanically awkward and adding significant weight even if it doubled as a generator for onboard payload power.

Additional disadvantages of small IC engines are their inherently high noise and vibration levels, and high operating speeds–typically 20,000 to 30,000 rpm. The low observability of a bird-sized aircraft would be shattered by the high-pitched snarl of a tiny piston engine, and some of the MITE's intended missions required unobtrusive operation. Reducing shaft rpm is desirable to increase propeller efficiency through lower disk loading, but the small IC engine's high rotation speed and vibration levels would necessitate an extremely robust and heavy gearbox to accomplish this. Such a gearbox is impractical for MAVs. Typically, small diameter, high-rpm propellers are used with small IC engines, further increasing operating noise.

Driving MAV propellers with electric motors removes these disadvantages. Electric motors are self-starting, and solid-state electronic motor controllers provide reliable, repeatable control over their entire speed range. They are quiet, and their low vibration makes possible the use of lightweight, efficient gear reduction systems. By increasing propeller diameter and reducing rpm, disk loading and propeller noise are reduced while efficiency is increased. Gear ratios can be selected to optimize motor loading for propellers of different diameters and pitches, expanding the parameter space available to the designer.

The major weakness of battery powered flight is the batteries. Batteries have far lower specific energy than fuels, so that a fuel-powered MAV with an efficient engine can have greater flight endurance than a battery-powered MAV of the same launch weight. However, electric motors have higher energy conversion efficiencies than MAV-scale IC engines and lend themselves to greater propeller efficiencies, so that this disadvantage is somewhat mitigated.

Finally, it was felt that electric propulsion for MAVs had far greater potential for improvement than small IC engines and would benefit more from a research and development effort. The booming market for portable electronics in the mid-1990s was driving large investments in battery technology, and improved microelectronics promised smaller, more capable motor controllers, particularly for brushless motors. NRL took advantage of these advances; over the course of the MITE program, from 1996 to 2002, the specific energy of commercially available cells relevant to MAVs more than doubled.

The reliability and ease of control of electric motors made practical the use of twin, contra rotating propellers for the MITE. A small weight price was paid by having two motors instead of one, but many advantages were gained. Balancing the propeller torque reaction gave better lateral stability at low speed, which is particularly important at launch. All of the MITEs flown were hand launched without difficulty. The propellers were placed ahead of the wing leading edge on both sides, so that much of the wing and control surface area was blown by the propeller wash. This increased control authority, increased the chord Reynolds number, and helped energize the flow over the wing, reducing its tendency to separate on the upper surface. The rotation of the propellers was set such that their helical flow worked to counteract the wing-tip vortices on both sides; the right propeller turned clockwise (right hand) and the left counterclockwise (left hand) when viewed from behind. This is an application of the Zimmermann effect [4] for reducing the drag of low aspect ratio wings. Finally, the twin propellers gave a clear field of view forward from the aircraft nose for any payload located there, such as a video camera.

The problem of single-engine-out performance is of great concern in traditional multi-engine airplane design, but was not a problem for the MITE. Electric motor systems have proven to be extremely reliable; because the MITE was an unmanned, expendable vehicle, accommodating the remote possibility of a single motor failure was not considered worth the effort.

4.4 Detailed Design of the MITE

The initial MITE concept shown in Fig. 4.1 was formulated as a 15-cm (6-in.) span vehicle, in line with the DARPA MAV concept. Detailed design studies of this 15-cm vehicle began at the outset of the program, but at that time many of the components needed to build a flight-test aircraft of that size were not available. Nevertheless, early experimental verification of the design was highly desirable and

Fig. 4.2 MITE 2B, with a micro video camera in the nose.

could be accomplished with a radio-controlled engineering test model. Although a 15-cm span aircraft was a major challenge in the mid-1990s, a version of it scaled up by 2.5 times was not. This version was designated MITE 2 (see Fig. 4.2); it had a 36.8-cm (14.5-in.) span and was the first MITE configuration to fly under radio control. MITE 2 was flown without a payload for initial tests and later carried a video camera and transmitter payload. Larger and smaller MITEs were later designed, built, and flown, as will be discussed. In this case study, MITE 2 will be examined in the greatest detail.

In contrast to human-carrying aircraft, experimental MAVs for flight test can usually be built quickly and inexpensively because of their small size. Therefore it is relatively easy to build and fly a model to test a concept or design modification. This practice was used throughout the MITE program. Indeed, most of the engineers who created the MITE and its systems are also flying model aircraft hobbyists, many with considerable success in model aviation design and competition. Their practical experience was a valuable supplement to their engineering skills and was capitalized upon in many aspects of the MITE program. This was not, however, a case of hobbyists who took themselves to be engineers, but rather engineers who are also dedicated hobbyists.

4.4.1 Aerodynamic Design

As a plank-type flying wing, the MITE required a reflexed airfoil for low pitching moments to provide trim flight with reasonable longitudinal stability. At the low

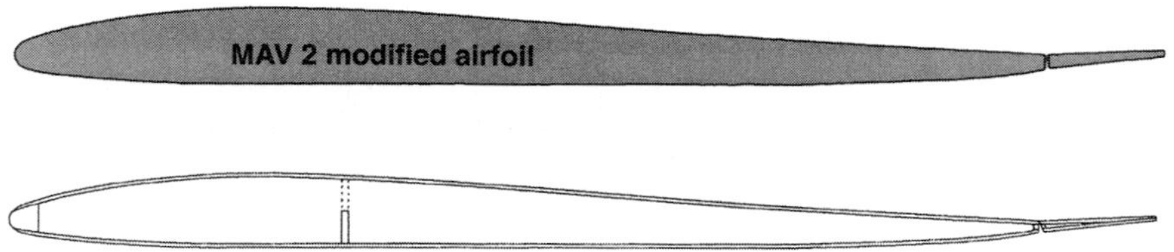

Fig. 4.3 SM-2 low-Reynolds-number reflexed airfoil, as used on MITE 2–MITE 5.

Reynolds numbers involved it was desirable to make the airfoil as thin as practical to minimize profile drag, while retaining enough spar depth to ensure that the structure had sufficient strength. This strength was not dictated by flight loads, which at the MAV scale are quite small, but rather by the handling loads experienced when the aircraft is picked up and worked on by engineers, technicians, and the occasional ham-fisted program manager. Also, because the first MITEs were to be engineering test models, a certain amount of crashworthiness and ease of repair was desirable. This need for ruggedness increases further for an operational vehicle that is to be carried and operated by soldiers or Marines in the field, where conditions rarely resemble those of the laboratory.

Donald Srull, an experienced aeronautical engineer and world-class flying model designer, selected a reflexed, flat bottom airfoil for MITE, similar to those designed and used by flying wing pioneer Alexander Lippisch with success on various lightweight, flying-wing powered gliders and model aircraft. The basic airfoil, called the SM-2, had also been successfully applied, with thicknesses of 6 to 12%, in a number of free-flight and radio-controlled flying-wing model aircraft. The airfoil was thinned to 5.2% for better performance at low Reynolds numbers (see Fig. 4.3). Figure 4.4 shows lift coefficient for the SM-2 plotted vs drag coefficient

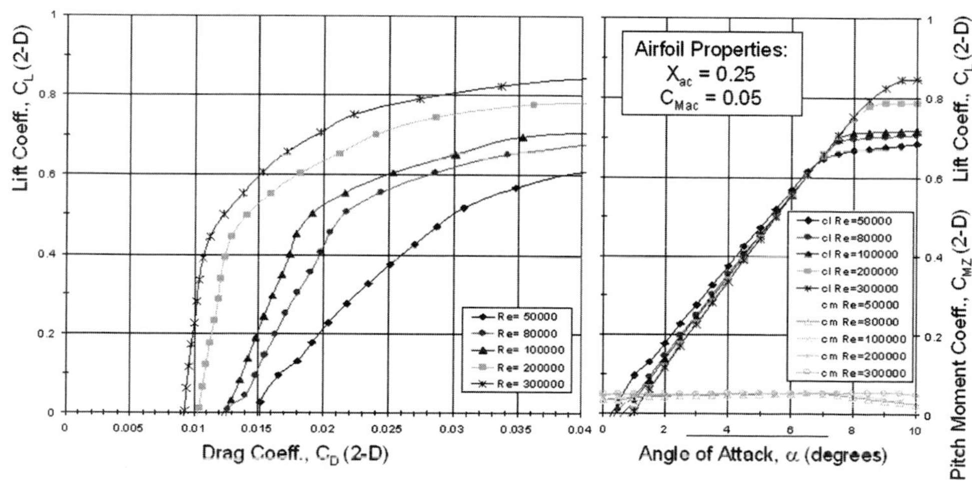

Fig. 4.4 Characteristics of the Srull SM-2 airfoil computed by the Drela XFOIL code.

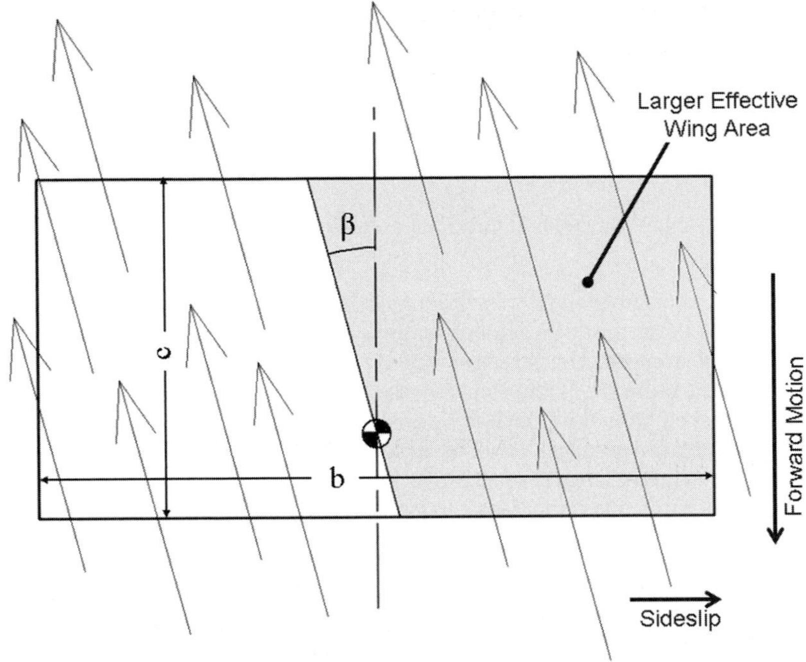

Fig. 4.5 Lateral stability for a low-aspect-ratio wing.

and angle of attack, as well as pitching moment vs angle of attack, as computed by the two-dimensional Drela XFOIL code [5] and [6]. Although a higher lift coefficient could have been achieved with more camber, this also increases the pitching moment and necessitates additional reflex and can lead to additional drag and control difficulties. The low camber shape made for a simpler, more robust wing structure, and the flat bottom simplified airframe assembly. Full-span elevons for roll and pitch control made up the aft 10% of the chord, and fixed vertical stabilizers were added at or near the wing tips. In later MITE experiments, differential thrust between the propellers was used for yaw control.

The MITE's lateral stability benefited from its low-aspect-ratio. Low-aspect-ratio wings have an inherent stabilizing effect similar to dihedral, leading-edge sweepback, or high wing placement. This allowed all of the MITEs to be built with flat, zero-dihedral wings, simplifying construction and improving strength. This stabilizing effect is illustrated in Fig. 4.5. Consider a rectangular wing with span b, chord c, and its center of gravity at the 25% chord position, which is perturbed in roll from level flight. It will begin to sideslip in the direction of the lowered wing tip. The airflow over the wing will be the vector sum of its forward velocity and the sideways velocity (toward the raised wing tip), so that its actual path is along the stability axis at an angle β with respect to the aircraft body centerline. Because aerodynamic moments are taken about the stability axis, this effectively increases the wing area on the side moving into the slip and reduces it by the same amount on the other side, generating a moment that acts to restore the wing to level flight. The area of the lowered wing is increased by

$$\tfrac{1}{2}(3/4c\sin\beta)\times\tfrac{1}{2}(3/4c)-\tfrac{1}{2}(1/4c\sin\beta)\times\tfrac{1}{2}(1/4c) \tag{4.10}$$

$$=\tfrac{1}{2}(9/16c^2\sin\beta)-\tfrac{1}{2}(1/16c^2\sin\beta) \tag{4.11}$$

$$=1/4c^2\sin\beta \tag{4.12}$$

Because the raised wing loses the same area, the total change in wing area is:

$$\Delta S = 1/2c^2\sin\beta \tag{4.13}$$

As a fraction of the entire wing, this change is:

$$\frac{1/2c^2\sin\beta}{b\,c}=\frac{c\sin\beta}{2b} \tag{4.14}$$

and, because aspect ratio $AR = b/c$ for a rectangular wing, the effective change in wing area at slip angle β is

$$\frac{\sin\beta}{2\,AR} \tag{4.15}$$

Therefore, the tendency for the wing to restore itself to level flight rises with decreasing aspect ratio. In practice, this effect becomes noticeable for wings with $AR < 6$. The different MITEs that were flight tested had aspect ratios of $1.0 \leq AR \leq 2.4$, and, because the center of gravity was located at the 15–20% chord position for all the MITEs, the lateral restoring moment was actually greater than that indicated by Eq. (4.15). The first experimental verification of the MITE configuration was an electrically powered free flight model, shown in Fig. 4.6, constructed and flown by Bernard H. Carson, a U.S. Naval Academy Professor of aeronautics, test pilot, and flying model builder. It was inherently stable in all three axes.

MITE 2's twin vertical stabilizers were placed on the upper surface, about 3 cm inboard from the wing tips. On subsequent MITEs the vertical stabilizers were relocated to the wing tips and extended below as well as above the wing surfaces. The stabilizers also doubled as wing-tip plates to possibly impede the wing-tip vortices and reduce induced drag.

4.4.2 Aerodynamic Design Refinement by Computational Fluid Dynamics

Computational fluid dynamics (CFD) was used rather than wind-tunnel testing to examine and refine the aerodynamic characteristics of the MITE. Three-dimensional CFD simulation essentially divides a volume of space into a three-dimensional array, or mesh, of discrete elements, and blows virtual air (or some other fluid) over a virtual model of an aircraft (or other structure) within that volume. The state of the fluid is computed in each of the discrete elements, so that an overall picture of the flowfield and pressure distribution can be created. These simulations require enormous computational power, for two reasons. First, the state of the fluid in each element depends directly on the state in the adjacent elements, which in turn depend on the state of their adjacent elements, and so on.

Fig. 4.6 Electrically powered free-flight aircraft built to test the flight stability of the MITE configuration.

A vast number of iterations is needed, therefore, to converge on a solution. Also, the elements must be small enough to resolve the fine structure of the flow, and the entire volume must be large enough to extend out from the structure to a distance at which the flow is essentially undisturbed. Three-dimensional CFD simulation that comes close to resembling reality is only possible because sufficient computing power is now available, and capabilities should certainly increase in this regard.

CFD has capabilities that cannot be practically achieved in a wind tunnel. The virtual model can be made to move, so that banks, turns, and other dynamic maneuvers can be simulated in a way that would be difficult or impossible in a wind tunnel. The pressure distribution over a surface can be computed at a level of detail that physical instruments on the surface of a wind-tunnel model cannot match. Flow visualization is also far easier in a virtual world. On the other hand, CFD cannot at present simulate areas where a mixture of both laminar and turbulent is present; the governing equations for an element are either laminar or turbulent, and transition points between the two must be defined. Also, it should not be assumed that CFD simulations are inexpensive to run simply because no physical model or facility is needed—high computational power comes at high cost.

CFD for the MITE vehicles was carried out by Ravi Ramamurti, John Gardner, and William Sandberg at NRL's Laboratory for Computational Physics and Fluid Dynamics. The simulations began with three-dimensional CAD models of the

MITE configurations to be studied, which were the original 15-cm MITE concept, MITE 2, MITE 4, and MITE 6. Several techniques were used to reduce the number of discrete elements in the simulation, thereby reducing the computational demand. First, because the MITEs were symmetric along their centerline, only one half of the vehicle was simulated. To give finer resolution where needed, the mesh elements were smaller in locations such as the leading and trailing edges, wingtip, and portions of the upper surface, where the pressure and velocity gradients were large or the flow was likely to be unsteady. The elements were larger in less critical areas. Also, sections of the mesh could move during the simulation to track features such as shed vorticity. Simulating MAVs accurately at low Reynolds number is challenging because the flow is unsteady in some locations, notably in the separation bubbles. Unsteady, turbulent flow is by its nature chaotic, and impossible to fully predict. Turbulent flow models have only been validated for fairly simple geometries thus far and must be extrapolated for more complex problems.

These studies had two objectives: to determine optimal configurations from the aerodynamic performance point of view and to determine the aerodynamic lift coefficient, drag coefficient, pitching moment coefficient, and roll moment coefficient vs angle of attack for various elevon deflection angles. These provided input to a six-degrees-of-freedom flight simulator that was used to evaluate stability, performance, and optimal control laws for MITE autopilot designs. The CFD code used was the Finite Element Flow solver (FEFLO) [7], an unstructured finite element code based on tetrahedral elements. FEFLO was developed by Ravi Ramamurti at NRL and Rainald Löhner at the George Mason University, Fairfax, Virginia. The governing equations are the incompressible Navier–Stokes equations, which are solved in an arbitrary-Lagrangian–Eulerian (ALE) formulation. The equations are discretized in time using an implicit time stepping procedure. The flow solver is able to capture the unsteadiness in the flowfield, if such exists, and is time accurate, allowing local time stepping as an option. The resulting expressions are subsequently discretized in space using a Galerkin procedure with linear tetrahedral elements. For the algorithm to be as fast as possible, the overhead in building element matrices, residual vectors, etc. must be kept to a minimum. This is accomplished by employing simple, low-order elements that have all of the variables (three flow velocity components and the pressure) at the same node location. The resulting matrix systems are solved iteratively using a preconditioned-conjugate-gradient (PCG) algorithm. The code has the options to be for run Euler, laminar flow, or incorporate a Baldwin–Lomax turbulence model. The solver has been successfully evaluated for both two dimensional and three-dimensional, laminar and turbulent flow problems by Ramamurti and Löhner [8] and Ramamurti et al., [9] and has been validated by extensive comparisons between experimental data and computed results.

Figure 4.7 shows the pressure distribution over the original MITE concept vehicle, with a wingspan of 15 cm and an aspect ratio of 1.25. The pressure contours on the body surface are shown. The dark regions indicate stagnation regions of high pressure. A number of configuration changes were examined to find an optimum. The high pressure in the stagnation regions around the wing body junction were reduced by the addition of fairings, and it was found that drag was reduced by mounting the fuselage pod below the wing rather than integrating it into the leading edge.

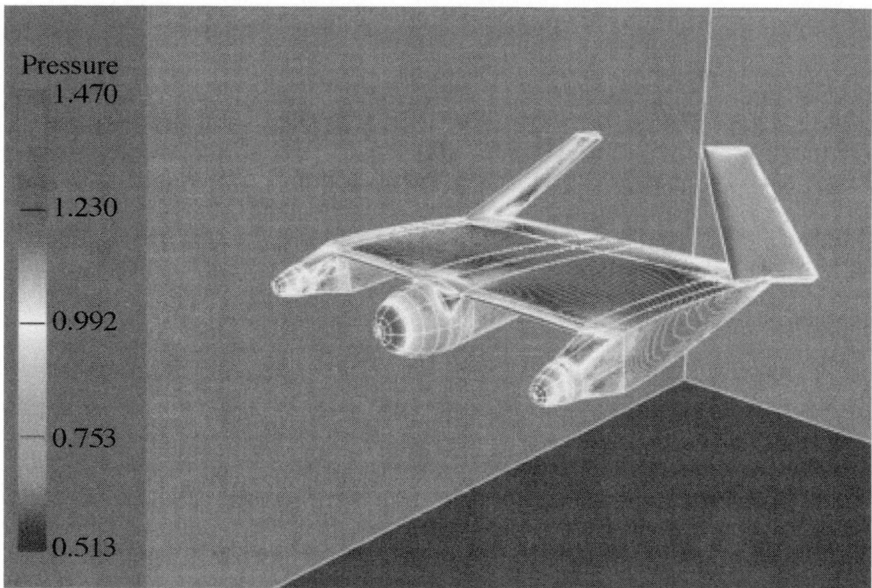

Fig. 4.7 Computational-fluid-dynamics simulation of the pressure distribution on the 15-cm MITE concept. (See color plate following p. 178.)

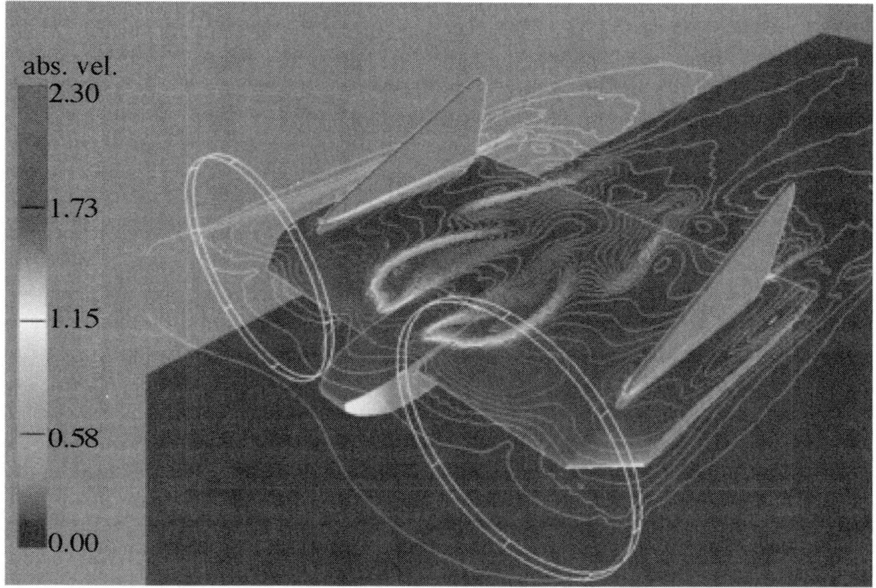

Fig. 4.8 Computational-fluid-dynamics simulation of the flowfield around MITE 2, with the propellers represented by actuator disks. (See color plate following p. 178.)

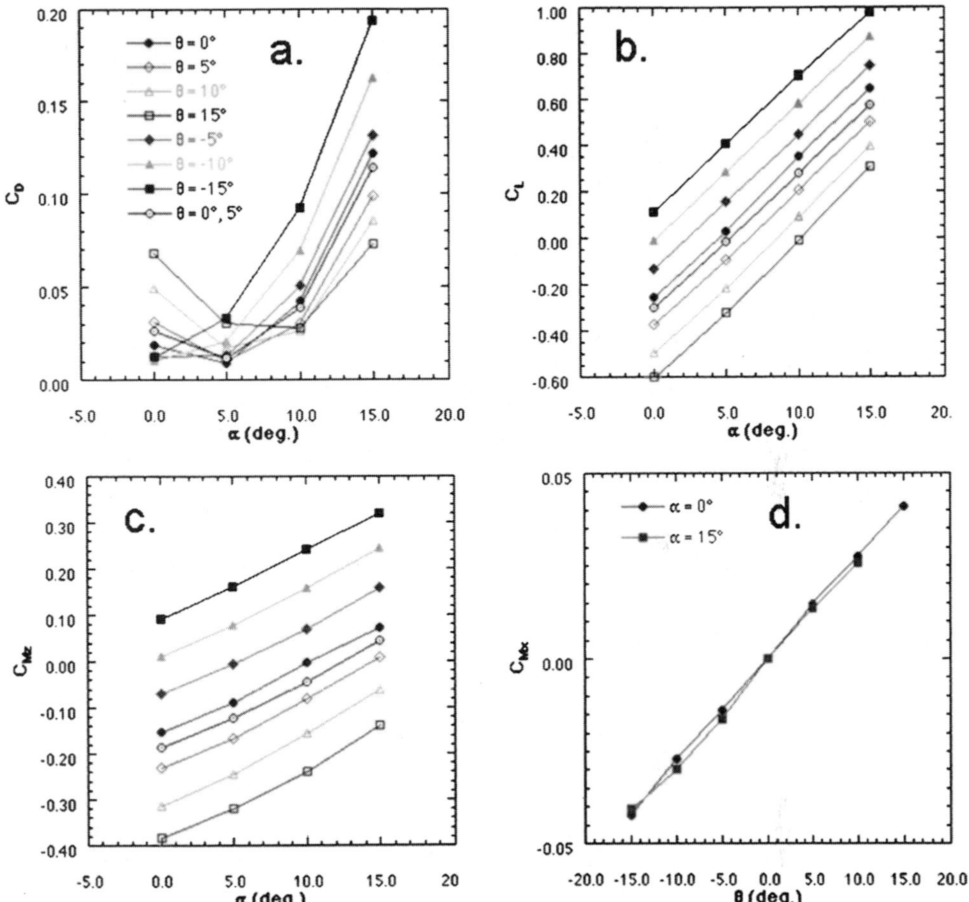

Fig. 4.9 Aerodynamic coefficients for MITE 2 derived by computational fluid dynamics.

Flow over small vehicles is significantly modified by the influence of the propeller. Time-dependent simulation of the propeller would have been very time consuming, and multiple simulations were needed at different vehicle attitudes to determine the aerodynamic coefficients and to investigate several different MITE configurations. Instead, it was found adequate to model the propeller as an actuator disk with a distribution of axial and radial momentum sources that matched the thrust and drag of the actual propeller. Figure 4.8 shows a simulation of the MITE 2 configuration, in which the velocity contours include effects of the counter rotating propellers. This simulation was at an angle of attack of 15 deg. Note the induced separation at the wing-body junction on the MITE 2 at the high angle of attack. The results of the aerodynamic coefficients derived for the MITE 2 configurations are shown in Fig. 4.9 [10]. The drag coefficient C_D, lift coefficient C_L, and pitching-moment coefficient about the center of gravity C_{MZ} are shown vs angle of attack α in Figs. 4.9a–4.9c, respectively, for elevon deflection angle θ of -15 deg $\leq \theta \leq +15$ deg and for differential elevon deflections of 0 deg and 5 deg. Figure 4.9d shows roll moment coefficient C_{MX} vs differential elevon deflection angle.

4.4.3 Propulsion System Design

The final engineering, construction, and initial flight testing of the MITE 2, as well as the later MITEs, were conducted by Donald Srull. MITE 2 had a wing span of 36.8 cm (14.5 in.) and a chord of 24.6 cm (10 in.), except at the wing tips, which were raked back at 45 deg for the outer 2.5 cm of the leading edge. This gave it a wing area of 929 cm^2 (1.00 ft^2) and an aspect ratio of 1.46. Initial flights without a payload, a version designated MITE 2A, were made at a gross weight of 129 g (4.6 oz). MITE 2A used a pair of Faulhaber/MicroMo DC5-2.4 coreless motors, turning Knight and Purdham Company (K&P) propellers of 17.8 cm diam × 20.5 cm pitch (7 × 8.1 in.) through 8:1 gear reduction drives. A micro video camera, transmitter, and a larger battery pack were then added, and the aircraft was redesignated MITE 2B, with a gross weight of 211 g (7.4 oz). Using the procedures described in Chapter 2, the motors, gearing, and propeller selection are examined to determine the effect of the increased weight on propulsion performance for MITE 2B. Estimates for MITE 2B were as follows: gross weight (with payload), $W_0 = 2.1$ N (7.4 oz) (mass of 211 g); wing area, $S = 929$ cm^2 (1.0 ft^2); operating lift coefficient $C_L \approx 0.6$; L/D ratio ≈ 5; motor efficiency $\eta_M \approx 50\%$; and propeller efficiency $\eta_P \approx 55\%$. Level flight velocity V for MITE 2B is

$$V = \sqrt{\frac{2W_0/S}{\rho\, C_L}}, \tag{4.16}$$

where W_0/S is wing loading expressed in Newtons/m^2. So, $V \approx 8$ m/s (26.5 ft/s) at sea level. The estimating relationship in Chapter 2 defines the power that must be supplied by the flight battery P_B of a given aircraft in order to climb at various climb angles ϕ as

$$\frac{P_B}{W_0} \approx V \times \left[\frac{1}{L/D} + \sin\phi\right] \times \left(\frac{1}{\eta_P}\right) \times \left(\frac{1}{\eta_M}\right), \tag{4.17}$$

where P_B is in watts, W_0 in newtons, and V in m/s. Using the preceding estimates, battery power per unit weight is $P_B/W_0 \approx 5.7$ W/N for level flight, giving $P_B \approx$ 12 W and 22 W for a 10 deg climb.

Note that these are the electrical power levels that must be supplied by the battery, not the power output from the motors or propellers. This differs from IC powered aircraft design, in which shaft power to the propeller is the usual standard. In a battery-powered aircraft, the battery is the most critical component of the power system. The selection of cells available that are suitable for flight propulsion is much more limited than the range of electric motors available. In general, a battery pack with sufficient specific power and specific energy for the application is selected first, and then a motor is found to fit the battery.

For MITE 2B to achieve acceptable flight performance, the battery pack was required to deliver about 12 W for level flight and 17–20 W peak to the twin motors,

Table 4.1 Discharge test results of batteries available for MITE 2

Cell	Weight, g	Discharge rate, A	Energy capacity, W·min	Energy density, W·min/g	Av. voltage, V	Delivered capacity, mAh	Cutoff voltage, V	Cutoff time, min
430 mAh Tadiran	12.5	0.7	59.2	4.73	2.62	376	2.2	32.2
Li/MnO_2		1.0	44.9	3.59	2.53	294	2.2	17.7
rechargeable		1.5	32.0	2.56	2.47	216	2.2	8.6
		2.0	18.3	1.47	2.36	130	2.2	3.9
800 mAh Tadiran	18.3	1.0	93.0	5.08	2.55	608	2.2	36.5
Li/MnO_2		1.5	82.6	4.50	2.56	538	2.2	21.5
rechargeable		2.0	71.7	3.92	2.51	476	2.2	14.3
		2.5	60.2	3.29	2.39	420	2.2	10.1
		3.0	46.7	2.55	2.40	324	2.2	6.5
750 mAh	11.3	0.7	82.9	7.15	2.30	595	1.6	51.0
CR2 Li/SO_2		1.0	57.1	4.92	2.20	425	1.6	25.0
primary		1.5	34.6	2.98	2.10	280	1.6	10.3
50 mAh	3.6	0.5	4.7	1.31	1.14	35	0.8	4.1
Sanyo NiCd		1.0	1.8	0.50	0.99	31	0.8	1.8
rechargeable		1.5	1.1	0.29	0.99	27	0.8	1.1
80 mAh	5.73	0.5	8.7	1.53	1.17	63	0.8	7.5
Sanyo NiCd		1.0	3.3	0.57	1.09	50	0.8	3.0
rechargeable		1.5	1.8	0.32	1.03	44	0.8	1.8
110 mAh	7.3	0.5	6.7	0.91	1.19	94	0.8	11.3
Sanyo NiCd		0.7	6.7	0.91	1.16	90	0.8	7.8
rechargeable		1.0	6.1	0.82	1.12	90	0.8	5.4
		1.5	5.5	0.74	1.07	86	0.8	3.5
		2.0	5.2	0.70	1.00	82	0.8	2.5
120 mAh NiMH	3.5	0.7	7.1	2.00	1.01	105	0.8	9.1
rechargeable		1.0	5.7	1.60	0.95	92	0.8	5.6
		1.5	4.9	1.40	0.94	84	0.8	3.4
270 mAh NiMH	7.5	1.5	17.0	2.30	1.03	278	0.8	11.1
rechargeable		2.0	15.0	2.00	0.95	255	0.8	7.7
		2.5	13.0	1.70	0.93	234	0.8	5.6
		3.0	10.0	1.30	88	135	0.8	3.7

with each motor capable of up to 10 W of input power at the assumed efficiency of 50%. Finding an appropriate battery pack was difficult. The lithium-ion polymer (LiPo) rechargeable batteries favored at present were not available in the 1990s. At the time, the practical choices were lithium sulfur dioxide (Li/SO_2) primary (single use) cells, nickel-cadmium (NiCad) rechargeable cells, nickel-metal-hydride (NiMH) rechargeable cells, or lithium manganese-dioxide (Li/MnO_2) rechargeable cells. Table 4.1 compares some of the MAV-scale cells of these types that were available for MITE 2. These cells were tested at discharge rates applicable to powering an MAV. Note that the delivered capacity for all cells drops with increasing discharge rate, particularly for the lithium-based cells.

Roughly 50 g (1.75 oz) was allocated for the MITE 2B battery pack. From Table 4.1, note that packs with seven cells in series of the 110 mAh NiCads or the

270 mAh NiMH could be made at this weight. At a discharge rate of 2.0 amps, the NiCad pack would have an average voltage of $1.00 \times 7 = 7$ V, giving an average power of 14 W. This could keep the aircraft in level flight, but with little margin for climb or maneuver, and flight time would be less than 3 min. NiMH cells have a greater specific energy than NiCads, but also a higher internal resistance. The 270-mAh NiMH pack could be discharged at 3.0 A, but the voltage would drop to $0.88 \times 7 = 6.2$ V average, giving 18.6 W. Level flight power would be at 2.0 A and 6.65 V average, with flight times of 5–7 min.

The best overall battery for the MITE 2B estimated requirements was a three-cell series pack of 800-mAh Tadiran Li/MnO_2 rechargeable cells, weighing about 55 g (1.94 oz). Its peak power was $3.0 \text{ A} \times 2.40 \text{ V} \times 3 = 22$ W at 7.2 V average, and level flight power would be achieved at about $1.5 \text{ A} \times 2.56 \text{ V} \times 3 = 11.5$ W at 7.7 V average. Estimated flight time was 15–20 min. In practice, however, these cells were difficult to obtain. Much of the flight testing of MITE 2 was carried out with CR2 Li/SO_2 primary cells because they were readily available.

In early tests, three CR2s in series, weighing only 34 g (1.2 oz), flew MITE 2A (without payload) for 30 min. The CR2 cell, with a nominal voltage of 3 V and a rated capacity of 750 mAh, was available from several manufacturers and was generally used in cameras and other consumer electronics. As Table 4.1 shows, however, the CR2's capacity and voltage drop significantly as discharge current approaches 1.5 A. Four CR2s in series, weighing about 45 g (1.6 oz), were needed to provide adequate voltage for MITE 2B; the discharge characteristics were $4 \times 2.2 = 8.8$ V average for 25 min. at 1.0 A (8.8 W) and 8.4 V average for 10.3 min. at 1.5 A (13 W). This did not meet the initial estimate's requirements for climb performance, but operating the DC5 motors at 8–9 V instead of the nominal level of 6–7 V gave an increase in motor efficiency. Figure 4.10 shows the motor characteristics of the DC5 at 8.5 V. The maximum heating limit for the DC5 is 3 W, so that maximum continuous power is about 9 W input power at 65% efficiency, giving 5.9 W shaft power per motor. This is a definite improvement over the initial estimate of 50% motor efficiency.

Iterating back through Eq. (4.17) shows that level flight will be maintained with a battery output of 9.2 W, and the 13 W available from four CR2s in series should allow about a 5 deg climb. This is very poor performance and would probably have been unacceptable without one further mitigating factor. The cell voltage listed in Table 4.1 is an average taken over the course of the battery discharge. A fresh CR2 discharged at 1.5 A produced 2.2–2.3 V per cell for about the first 6 min., giving a total voltage of about 9 V and a power of 14 W for the four-cell pack. This gave sufficient power to climb when it was needed—at and immediately after launch—though performance was sluggish. After 6–7 min. of flight, the battery voltage began to drop, and climb became difficult or impossible. At about 9 min the battery voltage dropped below 7 V, and MITE 2B began a gradual powered descent.

Additionally, running the DC5 motors at 8–9 V instead of their designed level of 6–7 V shortened their operating life. Motor speed is proportional to voltage, so that running a motor on higher than rated voltage can push rpm past the motor's mechanical limits. This can cause the delicate rotor of a coreless motor to deform and come apart, as well as rapid bearing and brush wear. Overvolting a motor can be a short-term method of increasing power, trading motor life for additional performance. As with any component that is deliberately operated beyond its

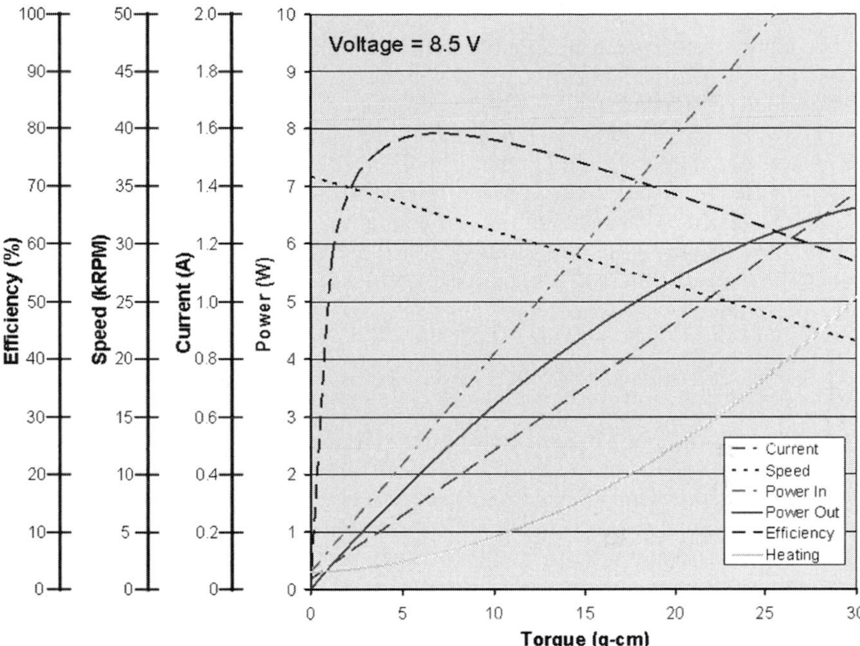

Fig. 4.10 Motor characteristics of the DC5-2.4 operated at 8.5 V.

design limits, the penalties in reliability, operating lifetime, and cost must be weighed against the system requirements.

Because the operating speed of the motors was increased in the transition from MITE 2A to MITE 2 B, the selection of propellers and gear reduction ratios for the DC5 motors should be reviewed to determine if they still match the airplane's performance. Using the techniques in Chapter 2 with a wing loading of 22.6 N/m^2 for MITE 2B, the stall speed V_S is estimated as

$$V_S \approx \sqrt{\frac{W_0}{S}(\mathrm{N/m^2})} = 4.8 \text{ m/s} \quad (16 \text{ ft/s}) \tag{4.18}$$

Propeller pitch speed should be at least twice the stall speed, and up to about three-times the stall speed for more aggressive performance. Propeller pitch speed for MITE 2 should therefore be 9.6–14 m/s (32–48 ft/s). Selection of commercially available model airplane propellers was somewhat limited because the MITEs required matching left and right hand props. Such propellers were available from K&P, including props of 15.2 cm diam × 16.8 cm pitch (6 × 6.6 in.) and 17.8 cm diam × 20.5 cm pitch (7 × 8.1 in.). From Chapter 2,

$$V_P(\text{m/s}) \approx 0.18 \times \text{rpm (in thousands)} \times \text{prop pitch (cm)} \tag{4.19}$$

The estimated minimum rotation speed ranges were 3170 rpm for the 15.2-cm prop and 2600 rpm for the 17.8-cm prop. For greater propeller efficiency, the larger propeller was preferred because of its lower disk loading. At 8.5 V, the

DC5 reached its heating limit of 3 W at about 25,000 rpm (see Fig. 4.10). The 8:1 reduction gear system already installed in MITE 2A gave a propeller rotation speed of 3130 rpm. This corresponds to a pitch speed of 11.5 m/s (38 ft/s), or about 2.4 times the estimated stall speed—a good match to the aircraft.

The twin DC5-2.4 motors, 8:1 gear reduction systems, and K&P 17.8 × 20.5 cm propellers of MITE 2A were retained for MITE 2B. The propulsion battery pack was increased from three CR2 Li/SO_2 primary cells in series to four CR2s in series. The extra weight of the payload was carried, but the performance, endurance, and motor life of MITE 2B were reduced considerably compared with MITE 2A. When the 800-mAh Li/MnO_2 rechargeable cells at last became available, MITE 2B's performance was improved.

4.4.4 Structure

As a flying wing, the basic airframe of the MITE was its wing structure. The forward fuselage pod and vertical stabilizers were added toward the end of construction. The airframe of MITE 2 and subsequent MITEs was built up from balsa with some carbon/epoxy reinforcement, in the manner of lightweight model airplanes (see Fig. 4.11). The wing ribs and main spar were cut from 1.6-mm (1/16-in.) balsa sheet by a computer-controlled laser cutting machine, a technique that yields high precision and part-to-part reproducibility. "Egg crate" assembly slots were included to reference the ribs to their proper locations on the spar, greatly easing construction. Initial airframe construction was done on a flat surface, to which parts were pinned to hold them in their proper positions for gluing. Cyano-acrylic (CA) adhesive and aliphatic resin were the main glues used. The flat-bottomed airfoil ribs resting on the flat building board ensured that the airframe would be built without unwanted twists or warps. Balsa leading and trailing edges were glued across the front and rear ends of the ribs, parallel to the spar. The ribs and spar were deliberately cut 0.8 mm (1/32 in.) thinner than the final airfoil shape on both their upper and lower surfaces. From the spar to the leading edge, the upper and lower surfaces were covered with 0.8-mm balsa sheet to form a "D tube," and balsa cap strips, 6.4 mm (1/4 in.) wide and 0.8 mm thick, were added over the edges of the ribs from the spar to the trailing edge. This increased strength and stiffness over the wing and gave a good foundation for the covering material. The leading edge was sanded to its final shape to complete the basic airframe. MITE 2 differed from other MITEs in that its wing tips were raked at the leading edge.

The propulsion motors, motor speed control, 72-MHz-band radio control receiver, and their associated wiring were added next. Two servomechanisms were installed behind the spar for the left and right elevons. MITE 2B and a later MITE 4 were equipped with a 2.6-GHz-band micro video transmitter installed between two wing ribs, with a small dipole antenna protruding above and below the wing near one wing tip. The radio control antenna was routed to the opposite wing tip to minimize interference between the two rf systems. Achieving rf isolation between electronic systems in close proximity is a problem that is often overlooked until the last minute, after assembly is complete. This can be a fatal error. Even though MITE 2's rf systems operated on vastly different frequencies, they were kept as far apart as possible. Frequency harmonics or raw rf power from a transmitter can overwhelm any other signals in the immediate vicinity, so that it is best to keep

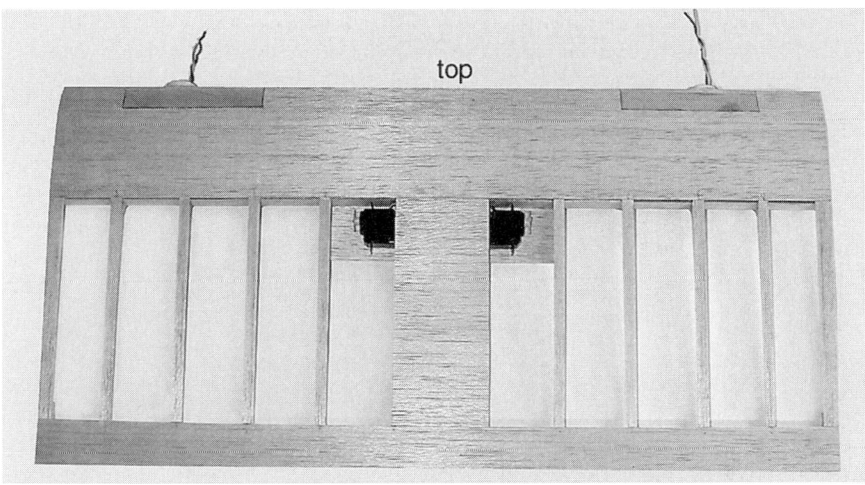

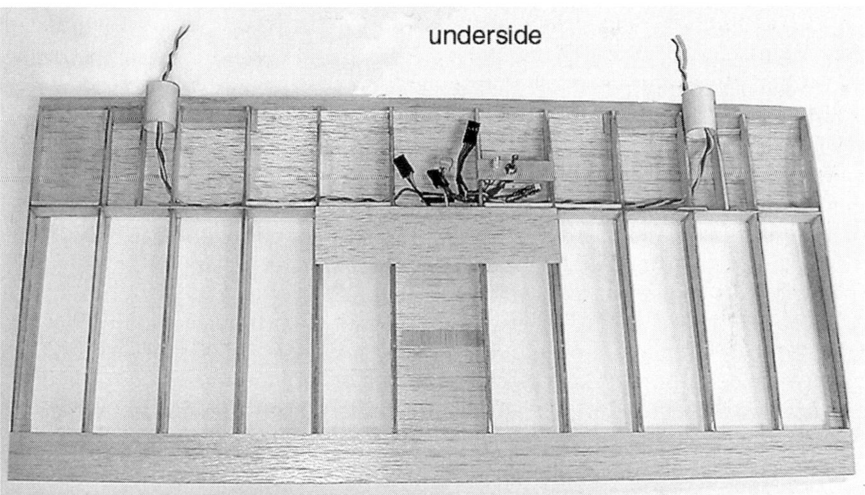

Fig. 4.11 Airframe structure for MITE 4, before covering.

receiving antennas well clear of transmitting antennas. Within the limited dimensions of an MAV, this needs particular care. Any arrangement of electronics and antennas should be thoroughly tested *before* they are sealed inside a wing or fuselage. Typical weight distributions of the MITE 2 vehicles are shown in Table 4.2.

Originally it had been planned to power the video payload from the propulsion battery, but this proved impractical. The heavy load of the motors caused the battery voltage to drop below the video system's operating voltage, and the motor commutation and the high-speed switching in the electronic speed control (ESC) created voltage spikes that interfered with the video when the two systems shared a common battery. These voltage transients caused sparking at the motor brushes that created rf noise. A 100-μF capacitor installed across the motor terminals

Table 4.2 Component weight breakdown of MITE 2A and MITE 2B

	Weight, g	
Component	No payload MITE 2A	With payload MITE 2B
RC command receiver	5	21
Control servos	6	6
Motor speed control	5	6
Motors	25	25
Props	7	7
Battery power supply	34	45
Cable harness, etc.	6	8
Airframe structure	41	41
Subtotal	129	159
Video system payload	0	52
Gross weight	129	211

reduced the noise, but this was not sufficient filtering for the sensitive video system. With more filtering and a solid state dc-to-dc converter, the payload could have been made to run on the propulsion battery, as was done on the Dragon Eye small UAV described in Section 4.7. For MITE, however, it was simpler to use a separate, small battery dedicated to the payload.

The MITE wing structure was covered with a nonwoven polyester tissue material that is marketed for model airplanes under the name Polyspan. The finished weight of Polyspan is similar to conventional model airplane tissue, but it is stronger, more puncture resistant, and impervious to moisture. It was attached to the balsa framework dry with clear nitrate dope, as smoothly as reasonably practical. Once attached, the material was heated with a heat gun to shrink it, removing the last wrinkles and tightening it overall. Polyspan is somewhat porous, so two to three coats of thinned nitrate dope were applied to make the covering airtight. The elevons, two 2.5-cm (1-in.)-wide plastic foam sheet strips running the full length of the trailing edge, were then added, hinged to the upper surface with plastic tape. A short fuselage pod was added below the wing center section, projecting ahead of the leading edge. This contained the propulsion battery compartment and, in some MITEs, a micro video camera. Twin vertical stabilizers of thin plastic foam sheet were attached near or on the wing tips.

With slight variations, this was the structure used on all of the MITEs. All of the units built were engineering test and demonstration aircraft, individually constructed by hand. Had the design gone into large scale production, it is likely that a different structure would have been used to reduce the costs, especially if the production MAVs were to be expendable. In this case, a more cost-effective method would be to machine a metal mold and manufacture MITE airframes from polystyrene or polypropylene foam. The foam airframes could then be populated with motors, batteries, flight-control systems, and payloads.

4.5 Flight Experimentation and Design Evolution

Although the climb and maneuver performance of MITE 2B was limited, it was an inherently stable flyer and a relatively smooth camera platform, flying at

Fig. 4.12 MITE 3, 25-cm wing span.

airspeeds of 4–8 m/s. A competent pilot could fly the vehicle via the onboard video system, and valuable experience was gleaned from numerous flights with this technique, as will be discussed later. It was obvious, however, that MITE 2B was inadequate for anything beyond limited endurance flights in calm wind conditions. Work continued to design a better MAV.

The fact that laser-cut balsa ribs and spars made for MITE 2 were available in large numbers led to a number of experiments with MITEs of different sizes. By removing or adding rib bays, MAVs with different wing areas were easily built from the same basic components. MITE 2 through MITE 5 were constructed in this manner. They all shared the basic MITE configuration and used the same airfoil and chord (25 cm), while the spans were varied. MITE 2 has already been discussed in detail; the other MITE variations are surveyed next.

4.5.1 MITE 3

MITE 3 (see Fig. 4.12) had a 25-cm (10-in.) span and was the smallest MITE to fly. It incorporated more powerful control system servos and ESCs than MITE 2. Gross weight without payload was 129 g (4.6 oz). The MITE 3 and subsequent MITEs also employed Li/MnO_2 secondary cells for propulsion batteries because of their superior energy and power density at relevant power levels. These batteries provided up to 9 W to each motor. DC5 motors were used on MITE 3, with 5:1 gear reduction and K&P 15.2 × 16.8 cm propellers. The improvement in specific power of the Li/MnO_2 batteries over MITE 2's CR2 cells, combined with lower weight and size, gave MITE 3 adequate performance while retaining MITE 2's inherent stability. MITE 3 did not carry a payload, however. It was used as a test aircraft only, including an investigation of the Zimmerman effect [4].

As mentioned earlier, the Zimmerman effect uses the helical slipstream from the twin, outboard mounted propellers to counter, to a certain extent, the wing-tip vortices. This effectively increases the wing aspect ratio, reducing induced drag. Because MITE 3 had the lowest aspect ratio (1.0) of all of the MITEs, and its propeller hubs were actually mounted at its wing tips, it was the best candidate for studying the Zimmerman effect. After flying in its normal configuration, with the propellers opposing the tip vortices, the left and right propellers were switched, and the direction of rotation of the motors was reversed. In this configuration the propeller slipstreams *reinforced* the tip vortices for an anti-Zimmerman effect. MITE 3 could fly in this configuration, but its performance was definitely diminished. In another test, the motors and propellers were reconfigured to turn in the same direction. Again, MITE 3 could fly without the propeller contra rotation, but lateral stability was degraded, and performance was reduced, though not to the extent that it had been in the anti-Zimmerman case. From these experiments NRL found that the Zimmerman effect and the contrarotating propellers were definitely advantageous to the MITE configuration. Quantifying the effect was difficult because flight instrumentation small and light enough to be flown on MITE 3 did not exist. The observations of the RC pilot on the ground, who flew the MAV in its different propeller configuration, were the only means of gathering data in these cases.

4.5.2 MITE 4

Though other variations were tried later, MITE 4 (see Fig. 4.13) became the standard MAV for the NRL program, offering the best combination of performance, payload capability, and compactness. It was used in the final mission feasibility demonstrations and carried payloads of 85–115 g (3–4 oz) for test purposes. When flown without a payload, MITE 4 was capable of aerobatic performance. The wing span was 47 cm (18.5 in.), with a wing area of 1200 cm^2 (1.28-ft^2), as opposed to 929 cm^2 for MITE 2. Flying weight without payload was about 210–260 g (7.4–9.2 oz), depending on the propulsion battery pack used. MITE 4 used more powerful coreless motors, the Faulhaber/MicroMo DC 1524, with 6.3:1 gear reduction and K&P 17.8 cm $\times$ 20.5 cm propellers. The motors were mounted at a 5-deg downthrust angle to counteract the airplane's tendency to pitch up at high thrust levels. The DC1524 is capable of up to 20 W input power each at motor efficiencies of 50–60%. It is wound for a higher voltage than the DC5-2.4 and runs well on three-LiPo cells in series (about 11 V). Initial flights were made on Tadiran Li/MnO_2 cells, but LiPo batteries became available before the end of the MITE program and were used for the later demonstration flights. A 300-g (10-oz) gross weight MITE 4 made a flight of 20 min while carrying both a video camera and transmitter and an electronic warfare payload. Both of these payloads had their own, separate battery packs and were operating throughout the flight. A photograph of a MITE 4 in flight is shown in Fig. 4.14.

4.5.3 MITE 5

The largest of the MITEs, MITE 5, had a 61-cm (24-in.) span, a wing area of 1550 cm^2 (240 in^2), and a flying weight of 270 g (9.5 oz) without payload

Fig. 4.13 MITE 4, 47-cm wing span.

Fig. 4.14 MITE 4 in flight.

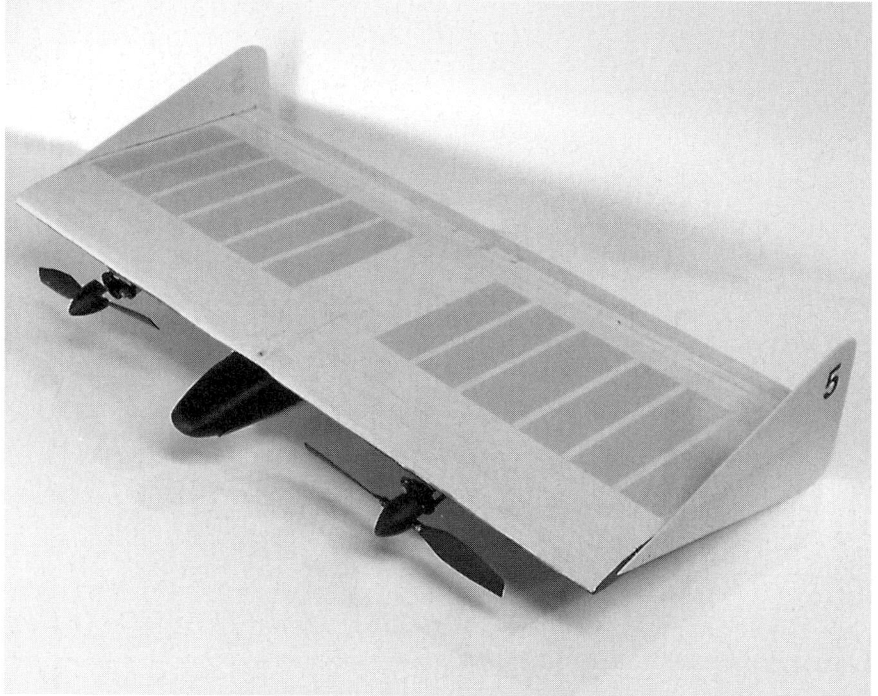

Fig. 4.15 MITE 5, 61-cm wing span.

(see Fig. 4.15). MITE 5 was essentially a payload carrier, used for the heaviest experimental payloads that were being developed in conjunction with the MITE, and for early testing of breadboard-level payloads. To reduce costs and increase durability, MITE 5 used Maxx Products, Inc., EPU 3 iron-cored brushed motors with 3.5:1 gear reduction. These were roughly equivalent to the MITE 4's DC1524 in power, though less efficient. MITE 5 performed well, but it also was a departure from the core concept of the MAV. Obviously, a larger aircraft will carry a larger payload; the point of an MAV is for all of the systems, including payloads, to be miniaturized to the greatest practical extent.

4.5.4 MITE 6

In the latter part of the MITE program, much of the effort shifted to the development of simple, low-cost navigation and guidance systems that could be integrated into an MAV. To save time and effort, these experimental systems were flight tested and proven as hand-built prototypes that could be miniaturized *after* the design was finalized. A difficulty arose because the prototypes were too bulky to fit easily inside the thin wing of a MITE 4, which by this time had been recognized as the optimum MITE. The MITE 4 fuselage pod was unsuitable as an electronics bay because its interior was already taken up by the flight batteries. MITE 6 (see Fig. 4.16) was identical to MITE 4 except that it used a thicker airfoil for greater

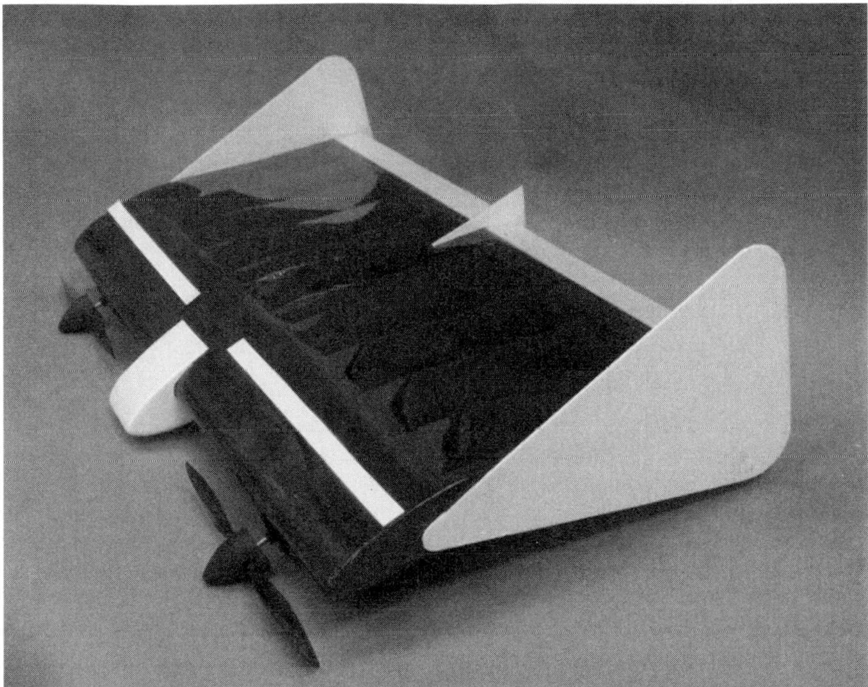

Fig. 4.16 MITE 6, with a boundary-layer trip at the 15% chord position.

internal volume. The airfoil chosen was a modified Liebeck 2573A of 11.6% thickness that was being used on the Dragon Eye small UAV, then under development at NRL. CFD studies warned that MITE 6, which had a chord Reynolds number of 8×10^4–9×10^4 compared with 4.7×10^5 for the Dragon Eye, would be extremely sensitive in pitch control. This prediction was accurate; MITE 6 was significantly less stable in pitch than MITE 4. An aerodynamic turbulator strip on the wing upper surface at the 15% chord position reduced this problem to an extent, but overall performance did not match that of MITE 4. It required higher power settings, indicating increased drag, and handling was somewhat erratic. Nevertheless, MITE 6 was a valuable platform for testing experimental guidance systems, as will be seen.

4.6 Stability and Flight Control

The thin-winged MITEs, MITE 2 through MITE 5, proved to have good stability in all three axes when flown by radio control. The 5-deg downthrust angle of the propellers on MITE 4 allowed the airplane to fly level at all power settings without changes in pitch trim. Pitch trim was adjusted immediately after launch on the first flight of a new MITE 4 and did not need to be changed thereafter. Once trimmed in pitch and roll, a MITE 4 that was banked into a turn continued to hold attitude and turn rate with the controls neutralized, hands off. Pilots reported that the airplane had no noticeable vices in handling. Flight tests were conducted with the center of gravity located at 15–20% chord position, with performance slightly better at 20%. Generally, MITEs were flown with the center of gravity at the 20% chord position, corresponding to a static margin of 5%. The center-of-gravity position was easily changed by moving the propulsion battery pack.

The MITE's blown wing probably contributed to another performance advantage that was observed: a reduced stall speed. The empirical stall speed estimation technique used in Sec. 4.4.3 was developed for model airplanes of conventional monoplane configuration. For MITE 2, this estimate was about 4.6 m/s, yet the airplane was able to maintain flight at speeds below 4 m/s. The energized flow from the propellers over the upper surface of the wing, combined with the high-angle-of-attack capabilities of the low-aspect-ratio wing, probably acted to delay flow separation.

Some MAVs have a noticeable yaw oscillation in flight. This is thought to be caused by intermittent vortex shedding at the wing tips and can be a severe limitation if the MAV's mission is to carry a video camera. The oscillation is transferred to the video image, hindering observation. None of the MITEs suffered from this problem, which seems more likely to occur in single-propeller MAVs. It is probable that the MITE's propeller wash, which covers most of the wing's surface including the wing tips, alleviates this effect. This effect has not been fully explained and was not investigated under the MITE program, as it seemed unnecessary to investigate a problem that did not occur.

Several techniques for navigation and guidance control for MAVs were researched during the MITE program. The main efforts concentrated on optically based systems: 1) optic flow for collision avoidance [11] and 2) monocular motion stereo to determine vehicle orientation with respect to features on the ground in six degrees of freedom and to integrate the changing position to determine ground track for navigation [12]. Experimental systems using both of these techniques were flown on surrogate test aircraft rather than on actual MITEs, but this work is beyond the scope of this case study. Two other navigation techniques were tested on MITE vehicles, however, and are discussed next.

4.6.1 Radio Control and Onboard Video Camera

Operating an MAV by conventional radio control, with the pilot keeping the aircraft in sight at all times, is useful for flight testing, but not for most missions envisioned for MAVs. At some fairly short distance from the pilot, perhaps 50–100 m depending on the size of the aircraft, the MAV can be seen only as a dot, making it difficult or impossible for the pilot to judge the aircraft's attitude and direction of flight. An RC pilot can only control what he can see. The simplest method for remotely piloting an aircraft out of sight of the operator is to transfer the operator's vision to the aircraft by means of a video camera in its nose. The RC pilot watches the video imagery on a screen or "virtual reality" video display goggles. He has the view that a pilot in the MAV would have. Because over-the-hill video reconnaissance is a prime mission for MAVs, a video camera and transmitter are likely to be all or part of the payload; flying "through the camera" adds a remote guidance capability for no additional weight. For this application, the camera should be angled so that the horizon appears at about 2/3 of the display

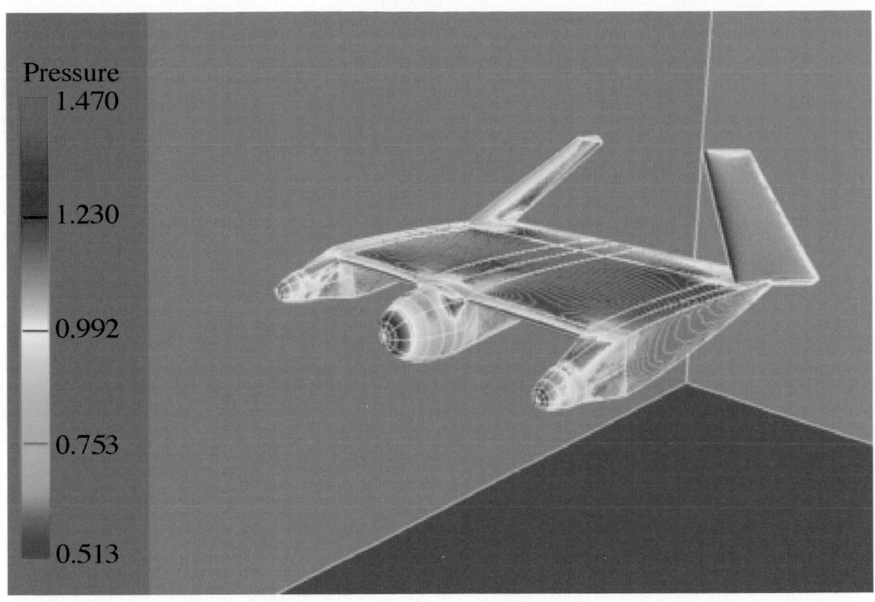

Fig. 4.7. Computational-fluid-dynamics simulation of the pressure distribution on the 15-cm MITE concept.

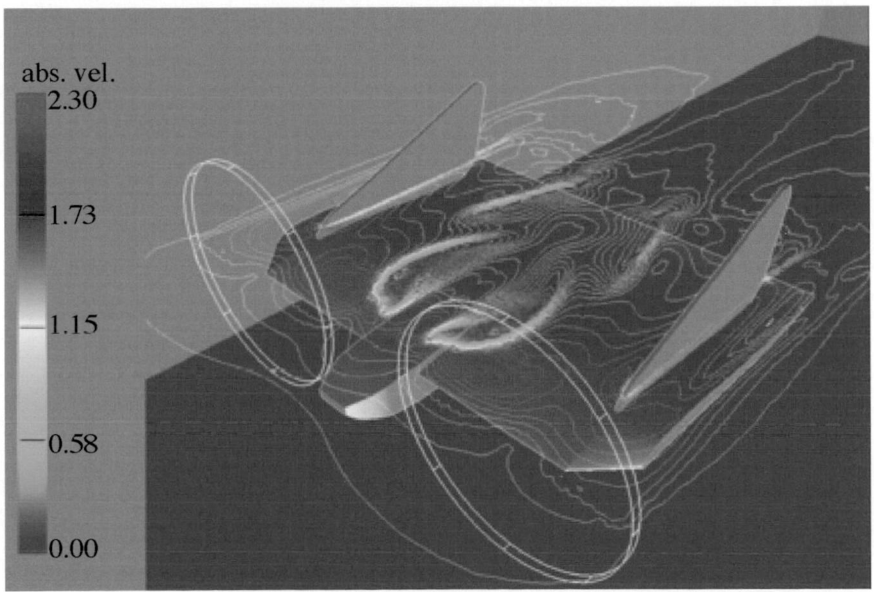

Fig. 4.8. Computational-fluid-dynamics simulation of the flowfield around MITE 2, with the propellers represented by actuator disks.

screen height when the MAV is flying level. This is a compromise between the pilot's need to see the horizon for orientation and the reconnaissance mission requirement of seeing objects on the ground; it is not optimum for either role.

The technique of flying through the camera was used on some MITE 2 and MITE 4 MAVs, but it is, at best, a stop gap method. It requires a great deal of practice on the part of the operating crew and is prone to numerous limitations and difficulties. Although the pilot has a view from the airplane, he has no physical sensation of its movement. For many people this quickly causes dizziness and nausea, a problem that is worsened by the fact that the pilot will probably have to wear a hood over his head. Most video screens and goggles cannot be seen well when used outdoors in sunlight. On a sunny day, this can be a problem even if the pilot is standing in the shade. The hood is needed to screen out stray light.

With only the view from the camera, navigation can be particularly difficult. All that the pilot has is the view ahead; unlike an onboard pilot he cannot turn his head to look in any other direction. Unless a compass is installed with a remote readout back to the ground-control station, he has no indication of the MAV's heading. To keep the horizon in view when the MAV changes its attitude, the video camera must have a wide field of view, so that what is seen is lacking in detail. It is therefore difficult to recognize landmarks unless they are large and prominent. As a result, it is easy for the pilot to lose track of the aircraft's location. If an assistant keeps the MAV in sight he can provide directions to the pilot, but this requires the MAV to be in sight. If lost, the pilot can try to set a course relative to the sun to return to the launch point or some known landmark, but he must know roughly where the airplane is to know what course to set. If this course requires having the sun in the field of view, the video image may be unusable.

Another difficulty is known as "video dropouts." The video transmitter in the MAV will be small, with a low transmitting power and equipped with an omni directional transmitting antenna. Inevitably, video transmission will be lost intermittently, usually for a few seconds at a time, leaving the pilot blind. This problem can be reduced by using a high-gain directional antenna for the ground station, held by an assistant who keeps it pointing in the direction of the MAV. This will improve both reception and transmission range. A commercial 2.4-GHz video transmitter with an rf power of 0.2 W can be received out to a range of about 2 km over a clear line of sight. Buildings, foliage, or other obstructions can cut this range significantly. The directional antenna can also give the antenna operator an indication of the MAV's direction relative to the ground control station, but at the cost of losing the video image while the antenna is moved in search of a signal. Additionally, the rf spectrum throughout the world is crowded, and other transmissions in the area can interfere with video or RC reception.

4.6.2 Autonomous Heading Hold

Ideally, MAVs should be capable of autonomous operation, navigating to their desired goals without the need for a human in the loop. This can greatly reduce the time required to train an operating crew because they do not have to learn to fly. For applications in which a large number of vehicles are to be flown simultaneously, autonomous operation is essential. For most reconnaissance missions this requires a sophisticated onboard guidance and navigation system capable of

Fig. 4.17 Autonomous heading and altitude hold MITE; the programmable calculator was used for operator interface on the ground.

flying to GPS waypoints. MITE, however, was largely intended as an expendable, one-mission vehicle that would use simple, low-cost autonomous guidance systems. For some potential MITE missions only the heading and altitude would need to be regulated, so that the aircraft would follow a preset course at a preset altitude. Any compensation for wind drift would have to be anticipated and corrected before flight, so precision navigation with this system would be impossible. One possible mission for such a system is the detection of chemical warfare agents. MITEs carrying chemical sensors could be launched on a range of courses, transmitting an alarm if chemical agents are detected. These MITEs would all be lost when their batteries were depleted. For this mission, however, in which the aircraft could become contaminated by chemical warfare agents, a disposable MITE might be preferable to a more expensive, multimission MAV.

A simple heading and altitude hold system was designed, built, and installed in a MITE 6 (see Fig. 4.17) by NRL flight controls engineer, Aaron Kahn [13]. Briefly, the system used three low-cost sensors for primary flight control: a magnetic compass, piezo rate gyro, and barometric pressure transducer. Flight-control processing was done on a simple microcontroller, interfaced through a programmable calculator. A commercially available flight stabilization system, the FMA Co-Pilot,* stabilized the aircraft in pitch and roll. This was essential, given the stability problems of the MITE 6. The Co-Pilot, a low-cost, lightweight device manufactured for model airplanes, senses the position of the horizon relative to the aircraft by averaging the infrared temperature difference between ground and sky. This stabilization system controlled the MITE's elevons to keep the aircraft level. The Co-Pilot's stabilization was sufficient that an inexpensive, two-axis magnetic compass could

*Data available online at http://www.fmadirect.com [February 2002].

be used to sense heading, instead of a larger, more expensive three-axis compass. Altitude and course corrections were implemented by controlling the twin propulsion motors, collectively for altitude change and differentially for heading change, so that the stabilization and guidance functions were separate. The rate gyro acted as a yaw damper. The MITE 6 with autonomous heading and altitude hold capability had a flying weight of 300 g (10 oz) and was demonstrated successfully. It was able to reach and maintain its preset heading and altitude, even when launched 180 deg off course. An RC system was included as a manual override so that this expendable system would not have to be expended during tests.

4.7 MITE Demonstrations and the Transition to the Dragon Eye UAV

A feasibility demonstration of the MITE as a delivery system for small electronic payloads was carried out in May 2002. Several flights were made by MITE 4s and the lone MITE 5 to a predetermined target. Operation of the payloads was confirmed while the MITEs were in the air and after landing on a rooftop and in an alley near the target. The payloads had separate battery packs and continued to operate as fixed sensors on the ground.

In 1999, NRL demonstrated the MITE 2 with its video camera payload at the Marine Corps Warfighting Laboratory in Quantico, Virginia. The airplane was flown by radio control, with the pilot watching the forward view from the camera. The USMC was impressed by the MITE's potential as a local reconnaissance tool that could be carried by one man on a battlefield. However, the performance that the Marines desired was beyond the capability of any of the MITEs. Additional USMC requirements were 1) fully autonomous operation so that operators would not need lengthy pilot training; 2) a payload capacity of 1 lb to allow carrying an infrared camera and communication links with a 10-km range; 3) sufficient speed and power to operate in 11 m/s (25 mph) winds; and 4) an airframe that could stand up to rough treatment in the field. These enhanced requirements required the development of a larger airplane, which was named Dragon Eye.

The first Dragon Eye experimental test aircraft was a scaled-up version of MITE 2, as shown in Fig. 4.18. This aircraft performed adequately, but it was felt that a vehicle with a more bird-like silhouette would be less conspicuous in the low altitude reconnaissance role. The airplane was therefore reconfigured to have a higher aspect ratio and a central, triangular vertical stabilizer, as shown in Fig. 4.19. The Dragon Eye system was refined in numerous tests and exercises conducted by NRL with the Marine Corps Warfighting Lab. NRL's function is research and development, not production, and so eventually the USMC awarded a contract to AeroVironment to adapt the Dragon Eye design to mass production and to produce the systems. Dragon Eye became operational with the USMC in 2003; it is the first electric-powered airplane, and the first backpackable reconnaissance airplane, to become an operational piece of military equipment.

4.8 Conclusion

The fundamental disciplines required to design an MAV such as the MITE are 1) low-Reynolds-number aerodynamics, 2) motor/battery systems, 3) stability and control, and 4) three-dimensional computational fluid dynamics. The student may be relieved to know, however, that all of these skills need not reside in one person. The MITE program was fortunate to have a team of skilled and dedicated engineers

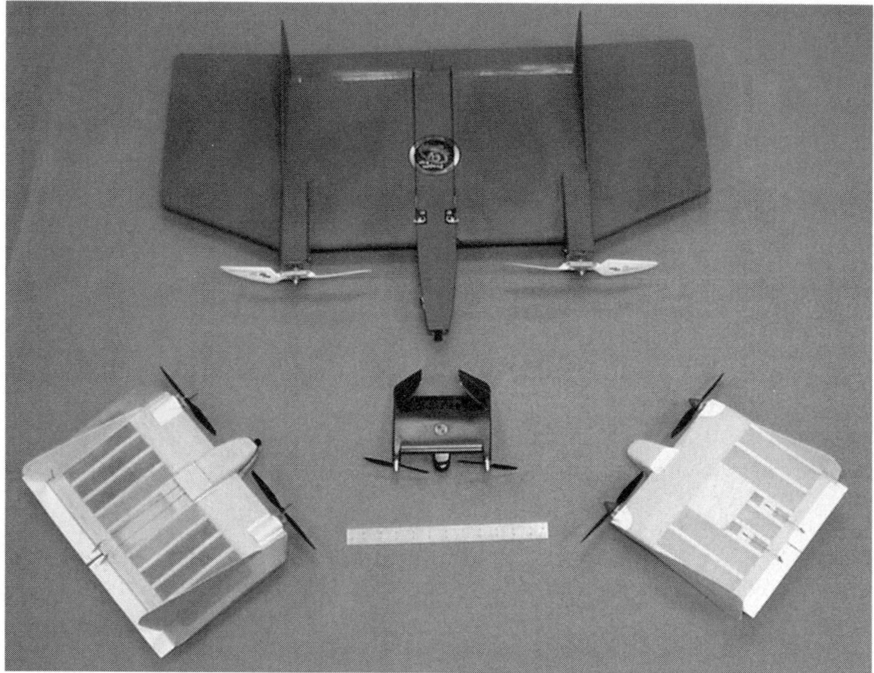

Fig. 4.18 First experimental Dragon Eye (top) with MITE 2B (left), MITE 3 (right), and a model of the original 15-cm MITE concept (center).

whose expertise covered all of these areas and overlapped in many. Additional talents are required to build and operate an MAV: 1) lightweight structures/composite materials, 2) electromechanical engineering, 3) microelectronics/control design and fabrication, 4) rf engineering, and 5) systems integration. It was in these areas that the flying model airplane experience of the NRL team members was invaluable. Creating MAVs and UAVs is a hugely multidisciplinary task; it is not simply an application of aeronautical engineering—it is flying robotics.

The key points of the MITE design were as follows: 1) a low-aspect-ratio, low-Reynolds-number wing that concentrated the wing area in a compact shape; 2) battery driven electric propulsion for reliable, near-silent operation; 3) most of the wing and control surface area blown by the propellers for enhanced lift and control; 4) dual, contrarotating propellers to counteract torque; 5) the Zimmerman effect, to reduce induced drag; and 6) simplifying flight control functions to create autonomous electronic guidance systems of minimal size and cost. Better MAVs can be built today than just five years ago. Since the MITE program ended in 2002, many of the component technologies have continued to improve, particularly batteries, small brushless electric motors, microelectronics, and micro sensors. All of the characteristics of the MITE just listed, however, remain valid, and the MITE configuration is still a good model for a simple, inexpensive micro air vehicle.

The experimental MITE series of MAVs was extremely successful. The payload delivery mission demonstration, which had been set as a goal for the program in

Fig. 4.19 Early NRL-built Dragon Eye used experimentally by the USMC.

its early days, was accomplished. A 300-g MITE 6, assembled from components with a total cost of under $500, flew autonomously, holding a preset heading and altitude. A MITE 4, also of 300 g, carried two payloads, a video camera/transmitter and an electronic warfare unit, for 20 min during a radio-controlled endurance demonstration flight. It is impossible to say if these were the *smallest* flight vehicles capable of performing valuable Navy missions, but they were indeed small and could perform valuable missions. In fact, the USMC required a larger vehicle, the Dragon Eye, which has a wing span of 114 cm and weighs about 2.7 kg. The Dragon Eye has greater capability than the MITE, but shares many of its key technologies: plank flying-wing configuration, battery energy source, inherent stability, and twin geared electric motors driving contrarotating propellers that blow over the wing. Dragon Eye is considered a small UAV rather than an MAV, yet is it small enough to be carried on the battlefield by one or two infantrymen. The demonstration of MITE 2 for the USMC in 1999 helped to finally solve a problem that has plagued foot soldiers since the beginnings of warfare. They no longer have to put a human scout at risk to find out what is waiting for them behind the next hill.

Acknowledgments

This work was sponsored by the U.S. Office of Naval Research. The author gratefully acknowledges the assistance of many of the people who made the MITE program a success: Donald Srull, Richard Foch, William Sandberg, Ravi Ramamurti, and Aaron Kahn. Additional thanks to Michael Baur, Eric Peddicord, and Michael Schuette for patiently helping an electrical engineer understand some of the ways of aerodynamics. Finally, many thanks go to Thomas Mueller for taking on the entire task and for bringing the rest of us along. An engineer could not ask for finer colleagues.

References

[1]Ailinger, Kevin, "Micro Air Vehicle (MAV) Development at NRL," Association of Unmanned Vehicle Systems International Conference 1997 (AUVSI '97), Arlington, VA, June 1997.

[2]Hoerner, S. F., *Fluid Dynamic Drag*, Hoerner Fluid Dynamics, Bakersfield, CA, 1965, Chap. 2.

[3]Kellogg, J. C., Bovais, C., Dahlburg, J., Foch, R., Gardner, J., Gordon, D., Hartley, R., Kamgar-Parsi, B., McFarlane, H., Pipitone, F., Ramamurti, R., Sciambi, A., Spears, W., Srull, D., and Sullivan, C., "The NRL Micro Tactical Expendable (MITE) Air Vehicle," *The Aeronautical Journal*, Vol. 106, No. 1062, Aug. 2002, pp. 431–441.

[4]Zimmerman, C. H., "*Airplane of Low Aspect Ratio*," U.S. Patent Number 2,431,293, 18 Nov. 1947.

[5]Drela, M., and Giles, M. B., "ISES: A Two-Dimensional Viscous Aerodynamic Design and Analysis Code," AIAA Paper 87-90424, Jan. 1987.

[6]Drela, M. "XFOIL: An Analysis and Design System for Low Reynolds Number Airfoils," *Low Reynolds Number Aerodynamics*, Lecture Notes in Engineering No. 54, edited by T. J. Mueller, Springer-Verlag, New York, 1989, pp. 1–12.

[7]Lohner, R., Yang, C., Cebral, J., Soto, O., Camelli, F., Baum, J. D., Luo, H., Mestreau, E., Sharov, D., Ramamurti, R., Sandberg, W. C., and Oh, C., "Advances in FEFLO," AIAA Paper 01-0592, Jan. 2001.

[8]Ramamurti, R., and Löhner, R., "Evaluation of an Incompressible Flow Solver Based on Simple Elements," *Advances in Finite Element Analysis in Fluid Dynamics*, FED Vol. 137, edited by M. N. Dhaubhadel et al., American Society of Mechanical Enginears, New York, 1992, pp. 33–42.

[9]Ramamurti, R., Löhner, R., and Sandberg, W. C., "Evaluation of Scalable 3-D Incompressible Finite Element Solver," AIAA Paper 94-0756, Jan. 1994.

[10]Ramamurti, R., Löhner, R., and Sandberg, W. C., "Simulation of the Dynamics of Micro Air Vehicles," AIAA Paper 2000-0898, Jan. 2000.

[11]Barrows, G. L., and Neely, C., "Mixed-Mode VLSI Optic Flow Sensors for in-Flight Control of a Micro Air Vehicle," *Proceedings of the SPIE*, Vol. 4109, Edited by Bains S. and Irakliotis, L. J., The International Society for Optical Engineering (SPIE), Bellingham, WA., 2000, pp. 52–63.

[12]Pipitone, F., Kamgar-Parsi, B., and Hartley, R., "Three Dimensional Computer Vision for Micro Air Vehicles," *Proceedings of the SPIE*, Vol. 4363, 2001, pp. 189–197.

[13]Kahn, A., and Kellogg, J. C., "A Low Cost, Minimum Complexity Altitude/Heading Hold Flight Control System," 21st Digital Avionics Systems Conference, AIAA and IEEE, Oct. 2002.

5
Flexible-Wing Micro Air Vehicles

Peter G. Ifju*
University of Florida, Gainesville, Florida
Roberto Albertani†
Research and Engineering Education Facility, Shalimar, Florida
and
Bret K. Stanford,‡ Daniel J. Claxton,‡ and Michael J. Sytsma‡
University of Florida, Gainesville, Florida

Nomenclature

C_D = drag coefficient
C_L = lift coefficient
$C_{L\max}$ = maximum lift coefficient
C_m = pitching moment coefficient
C_p = pressure coefficient
D = drag
D_{tool} = tool diameter
h_{scallop} = scallop height
L = lift
Re = Reynolds number
U = displacement in the X direction
V = displacement in the Y direction
W = displacement in the Z direction
x_{stepover} = stepover distance
ϵ_{xx} = normal strain in x direction
ϵ_{xy} = Shearing strain

Acronyms

AOA = angle of attack
AOI = angle of incidence
BR = batten-reinforced wing
CAM = computer-aided machining
CFD = computational fluid dynamics
CNC = computer numerical control

Copyright © 2006 by the American Institute of Aeronautics and Astronautics, Inc. All rights reserved.

*Professor of Mechanical and Aerospace Engineering.
†Assistant Visiting Professor.
‡Research Assistant, Mechanical and Aerospace Engineering.

DAQ = data-acquisition system
FSV = freestream velocity
GIU = graphical user interface
GPS = global positioning system
LAR = low aspect ratio
MAV = micro air vehicle
PIV = particle image velocimetry
PR = perimeter-reinforced wing
R = rigid wing
TS = test section
VIC = visual image coorelation
WT = wind tunnel

5.1 Introduction

Researchers at the University of Florida (UF) have developed a unique flexible-wing-based micro air vehicle. The University of Florida has produced numerous MAV designs that utilize a common design component, namely, a thin, undercambered, flexible wing made from a flexible skeleton and an extensible membrane material. The UF MAV departs from traditional MAV design by not restricting the design space to only geometric parameters that govern the aerodynamic behavior, but also exploring the compliance of the structure to achieve aerodynamic as well as control advantages. In this chapter we will expand on the motivation for developing the flexible-wing concept, design methodology, methods of fabricating flexible-wing-based MAVs, experimental characterization through wind tunnel as well as flight tests, modeling efforts, and applications.

Figure 5.1 illustrates a recent version of the UF flexible wing [1]. The culmination of nearly seven years of development is represented in the photograph of a 10-cm maximum dimension MAV that in 2005 was capable of flying for nearly 15 min. It is only one of a series of vehicles that have been developed over the years, which range in size from the 10-cm platform in Fig. 5.1 through a 30-cm fully autonomous vehicle equipped with inertial sensors, altimeter, airspeed indicator, global positioning system (GPS), and video cameras/transmitters. Figure 5.2 shows another version [2] with a 15-cm maximum dimension developed in 2004 equipped with a color video camera and capable of flying nearly 1 km from the ground station.

The development of the flexible wing utilizes a combination of biologically inspired design and the incorporation of modern composite materials. The wing is thin and undercambered, as are those of small birds and bats. In previous studies [3], it was shown that thin undercambered wings are more efficient than those with significant thickness. For birds and bats on the same scale as micro air vehicles, the wings have evolved towards the ideal thin undercambered shape as can be seen in Fig. 5.3. The University of Florida micro air vehicle is constructed with a carbon-fiber skeleton (analogous to the bone structure of the bat) and thin extensible membrane material (analogous to the skin of the bat wing). The overall aircraft configuration is a departure from the traditional flying-wing or lifting-body design. It has a distinct fuselage and wing, more similar to that of birds and bats.

The UF designs literally evolved from early versions whose development can be traced back to 1999. The initial inspiration for the flexible wing resulted from

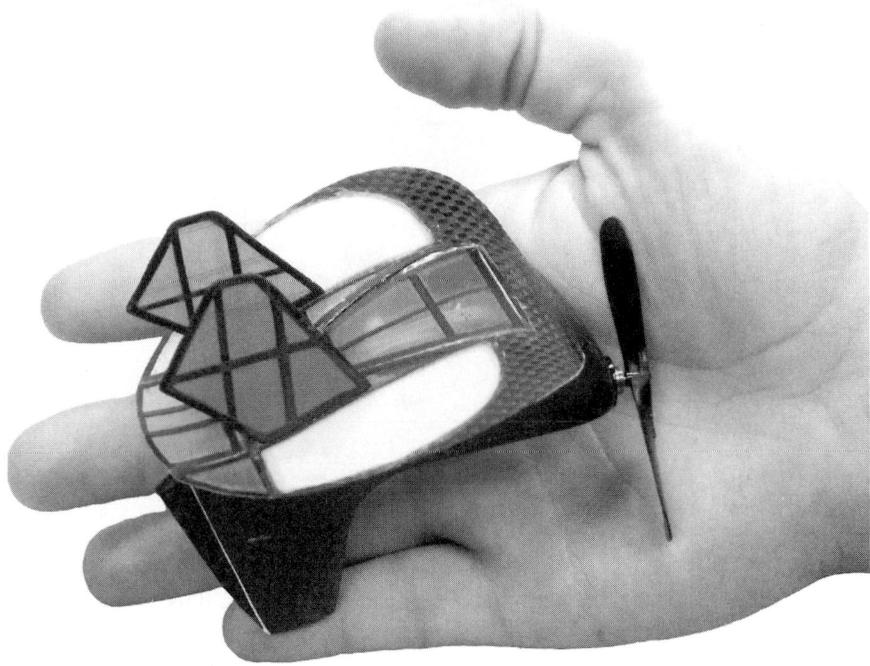

Fig. 5.1 One of the smaller University of Florida MAVs that competed in the 2005 International MAV competition in Seoul Korea.

a combination of watching early attempts at flying MAVs in combination with the author's knowledge and love for the sport of windsurfing. Early versions of MAVs developed at the University of Florida for the 1997 and 1998 International MAV competition proved to have poor controllability, especially in gusty wind conditions. The author's knowledge of the flexible sails used in windsurfing stimulated thoughts of developing a wing that would not be as affected by wind gusts as

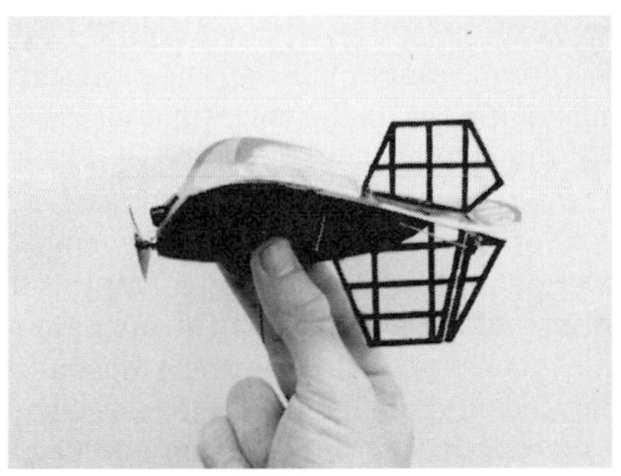

Fig. 5.2 MAV with 15-cm wing span equipped with video camera.

Fig. 5.3 Small bats and birds have thin undercambered wings. Bats have a flexible bone skeleton and an extensible skin membrane.

nominally rigid versions. From that point on, UF has explored the design space that is typically avoided in aircraft design, namely, compliance. Typical approaches to aircraft design obsess about geometric perfection of the wing airfoil shape and planform. From back of the envelope calculations, the UF MAV versions of 1997 and 1998 should have had sufficient efficiency to fulfill the design objectives, yet they were not controllable and required supreme pilot skill to operate. Our goal was to produce a vehicle that was easy to control, even when the weather did not cooperate.

The flexible nature of these wings can provide several nonobvious advantages over their conventional rigid counterparts [4]. The wings that we have fabricated with a carbon-fiber skeleton and extensible latex rubber skin have the ability to adapt to the airflow to provide smoother flight. This is accomplished via the passive mechanism of adaptive washout. In sailing vessels, adaptive washout is produced through twist of the sail. This greatly extends the wind range of the sail and produces more constant thrust (lift), even in gusty wind conditions. In the wings that we have designed, the shape changes as a function of the airspeed and the angle of attack (AOA). The adaptive washout is produced through extension of the membrane and twisting of the framework, resulting in angle of attack changes along the length of the wing in response to air speed and overall angle of attack. For example, as the plane hits a head-on wind gust the airspeed suddenly increases. The increased airspeed causes a shape change in the wing that decreases the lifting efficiency, but because the airspeed in the gust is higher, the wing maintains nearly the same lift. Once the airspeed decreases, the wing recovers to the original configuration. If there is a decrease in the relative airspeed, the angle of attack increases, and the wing becomes more efficient, and near constant lift is restored. The net result

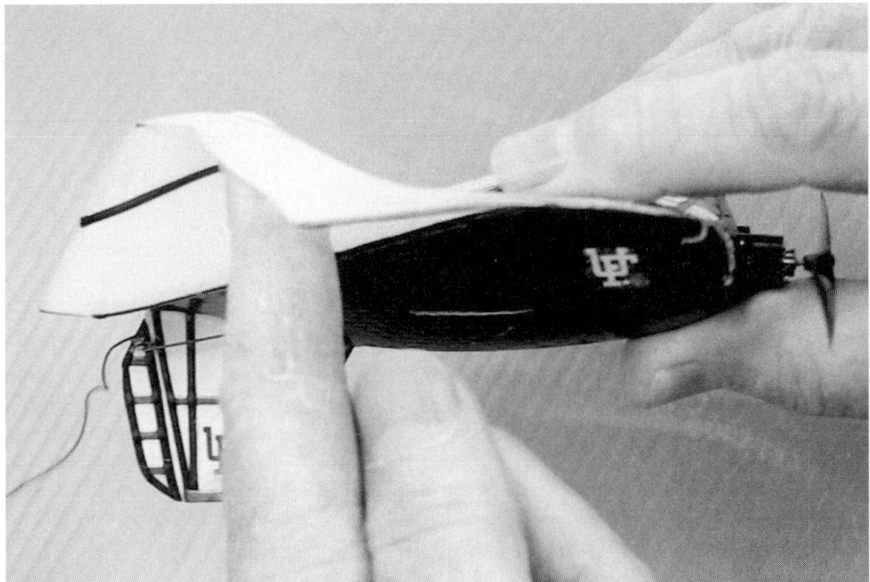

Fig. 5.4 Because of the light wing loading during flight, it was necessary to incorporate an extensible membrane to achieve adaptive washout.

is a wing that flies with exceptional smoothness, even in gusty wind conditions. The adaptive washout mechanism is subtle and must be "tuned" into the wings in order to work effectively. We have built hundreds of wing configurations and have been able to produce many wings with remarkably smooth flying characteristics. Figure 5.4 illustrates one of the flexible-wing configurations that is capable of developing adaptive washout.

For aircraft with very small inertia, as in the case of MAVs, changes in wing loading can immediately affect the flight path. As the aircraft becomes smaller and lighter, the need for suppressing the effects of wind gusts becomes more critical, especially if it is to be used as a camera platform. Additionally, as the airspeed of the vehicle decreases, wind gusts become a larger percentage of the mean airspeed of the vehicle. For example, our 15-cm (6-in.) aircraft flies between 8 and 18 m/s. On a typical day, the wind speed can vary by more than 5 m/s. For rigid wings, the lift can vary by 50% or more over the short period of time during the gust. To make matters more critical, gusts are not always head on. Because control of these aircraft is one of the most important hurdles, it is critical to suppress unwanted and sudden changes in direction, elevation, and orientation.

In nature, birds and bats display a similar form of adaptive washout. This passive mechanism can be observed on windy days by watching large soaring birds. The feathers at the wing tips flair to accommodate sudden changes in airspeed. To some extent, our design approach has been biologically inspired. We have observed both birds and bats and have designed our wings to have similar characteristics.

As mentioned earlier, the adaptive washout mechanism is subtle; therefore, the location and stiffness of the carbon-fiber skeletal members and thickness of the

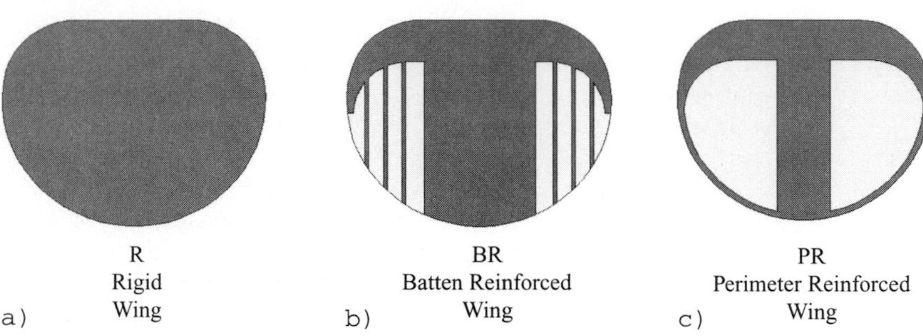

Fig. 5.5 Three configurations were studied extensively: a) the rigid wing (R wing), b) the batten-reinforced wing (BR wing), and c) the perimeter-reinforced wing (PR wing). The dark shaded area corresponds to carbon fiber while the light shaded areas correspond to regions covered with latex rubber.

latex membrane are critical. To define the design space for our flexible wing, we built numerous prototypes to learn how the geometry of the carbon-fiber skeleton affected flight performance. We also varied the relative stiffness of the different parts of the skeleton and were able to make observations in the field in order to qualitatively rank their performance. Using this relatively crude trial-and-error process, we were able to downselect the configurations that provided the best performance, then subsequently tested using more rigorous means via additional flight tests and wind-tunnel tests.

Two configurations stood out as having significantly improved characteristics compared to rigid versions of the same geometry. These two versions are shown in Figs. 5.1 and 5.2. A schematic drawing of the wing planforms is illustrated in Fig. 5.5. The first wing (Fig. 5.5a) represents the baseline design (nominally rigid wing), the middle (Fig. 5.5b) is composed of a relatively rigid leading edge and thin ribs called battens, and the third version (Fig. 5.5c) is composed of a relatively rigid leading edge and a flexible carbon-fiber perimeter. For the remainder of the chapter, we will refer to the three wings as rigid (R) wing, batten-reinforced (BR) wing, and perimeter-reinforced (PR) wing. The batten-reinforced wing has been extensively study by the authors [5–8] and was designed specifically to address gust alleviation. The perimeter-reinforced wing exhibits another distinct advantage over the BR and R wings in terms of increased stall resistance, enhanced lift, and increased static longitudinal stability. The BR wing was flown on literally hundreds of vehicles before we thoroughly studied the design in the wind tunnel, whereas the PR wing was first invented by UF students and tested in the wind tunnel before a vehicle was actually flown. Subsequent flight tests validated the wind-tunnel findings.

5.2 Description of the MAV

The wing is of course the most important component of the MAV; however, the vehicle efficiency and flight worthiness results from the entire configuration. Figures 5.1 and 5.2 illustrate that the vehicle is composed of a wing, a blended fuselage, a vertical stabilizer under the wing, and one or two vertical fins above the wing. The vehicles are designed to fit into a sphere that describes the maximum dimension; hence, the vehicles have a low aspect ratio (the trailing edge is actually circular) to maximize the available lifting surface. The leading edge is elliptical, allowing for the propeller to remain inside the maximum dimension sphere.

The wing airfoil currently being used has evolved over the years, but the basic concept is the same. It has a maximum camber of around 6% and recurve or reflex towards the trailing edge. The reflex is necessary to negate the otherwise negative pitching moment caused by the camber in the forward section. Note that there is no horizontal stabilizer; hence, the trailing edge of the wing must develop a downforce to right the aircraft during trimmed, level flight. The initial airfoil was modeled in two-dimensions by X-foils (a panel lattice code) coupled with a genetic optimization routine. A sixth-order polynomial was adopted to define the airfoil shape. The genetic optimization routine was used by first fixing the root-chord length at 15 cm (6 in.), the angle of attack at 6 deg, and the Reynolds number at 10^5 in X-foils and then performing the optimization with the objective function to maximize lift over drag, L/D. The optimization scheme was used to determine the coefficients of the polynomial in order to achieve the highest L/D. This airfoil has since been only slightly modified, mainly through trial and error. Initial designs only used a singly curved wing fabricated on a wing tool formed by bending a piece of sheet metal. This meant that the airfoil just mentioned only applied to the root-chord area because the planform is generally circular. More recent models utilize advanced tooling concepts (discussed in the next section) and allow for full three-dimensional wing shapes. The current preferred shape defines a reflexed portion near the root cord and blends into a nonrecurved section towards the wing tips. Additionally, polyhedral to improve global stability is used where the facets correspond to the fuselage and the two sides of the wings. The polyhedral improves overall stability. The wing is also composed of an elevator (discussed in the next paragraph) for pitch control and a hatch to access the internal components. All of these features are fabricated into the wing in a monolithic manner.

The control surfaces used on the MAV are an elevator located near the trailing edge of the wing and a rudder located near the trailing edge of the vertical stabilizer under the wing. The elevator produces pitch control by altering the overall pitching moment of the vehicle, whereas the rudder produces mostly roll with only minor yaw. This is because the control surface produces a side force under the center of mass thus producing a significant roll moment. In general the vehicle flies with nicely coordinated turns.

The fuselage blends with the wings and vertical stabilizer in order to create a more streamlined shape. The frontal profile is a rounded inverted triangle.

The propulsion system that has been adopted in recent years is a coreless electric motor powered by lithium polymer batteries. In the early years, small internal combustion engines were used. They suffered from numerous disadvantages including high noise levels, inconsistent performance, messy exhaust, poor form factor, lack of throttle adjustability, and production of more thrust than was needed for small MAVs. Typically, the propeller that is currently being used is an off-the-shelf propeller called the U-80 that is cropped to fit MAVs of differing maximum dimension.

Traditional rotary servos have proven to be more reliable than linear versions, or magnetic actuators; hence, we have predominantly utilized them. Small speed controllers and receivers round out the components. Additionally, payloads typically involve a color CMOS video camera and transmitter. The total vehicle weight varies with size; a 15-cm version with camera weighs around 55 g, and a 10-cm version, without video camera, weighs in at around 30 g.

5.3 MAV Fabrication Methods

UF's approach to MAV construction has consistently implemented composites in the airframe structure [9]. The advantages of composites in aerodynamic structures are manifold. Composites provide high strength, high stiffness, low weight, and easy formability over complex surfaces. Traditional aircraft materials such as foam, balsa wood, and monocoat are not appropriate materials for the flexible-wing construction. The wing geometry and structural layout require a skeleton of high strength and stiffness. Durability of the structure was of paramount importance because typical flight trimming requires a trial-and-error process in which the aircraft experience numerous crash landings. If the structure is compromised during that process, experimental acquisition of the trim condition would be nearly impossible. It can be said that the composite construction techniques developed at UF are the enabling technology that allows the exploration of flexible structures for MAVs.

The production of accurate and reproducible aircraft designs often necessitates some form of CAD software for development. There are several advantages of CAD, compared to more traditional development tools. In CAD the design can be precisely dimensioned to desired tolerances, symmetry is guaranteed, and design iteration can be as simple as modifying existing models. The major disadvantage of existing CAD software is that, even with the powerful tools available, a three-dimensional free-form surface such as a wing can be time consuming to construct. Additionally, there is an intermediate step between CAD design and a working physical model. Computer-aided machining (CAM) software is required to convert the geometry into a form that computer numerically controlled (CNC) milling machines can understand. With the implementation of CNC machining, most of the manufacturing process can be automated. Another method of automated production is called stereolithography, which involves building plastic parts in layers by tracing a laser beam on the surface of a vat of liquid photo-polymer. After many such layers are traced, a complete three-dimensional model is formed.

The University of Florida has developed an interactive wing generation program called MAVLab, which greatly simplifies rendering of complicated three-dimensional wing geometries. MAVLab CAD software attempts to eliminate the necessity of an expensive and complicated CAD/CAM program by creating a simple graphical user interface (GUI), which allows the user to design a wing based on a few basic constraints as shown in Fig. 5.6. Additionally, MAVLab can export this geometry to other common and useful formats such as those necessary for CNC machining and stereolithography. A simple methodology has been developed to produce MAV wings. This includes design by MAVLab, creating tool paths for milling, CNC milling, composite construction, and assembly.

5.3.1 Design Software Interface and Algorithms

MAVLab features a simple GUI designed to guide the user through the construction of a wing via simple geometric parameters. These parameters include wing span, root chord, dihedral angle, polyhedral ratio, sweep, and geometric twist. Figures 5.7 and 5.8 graphically depict these parameters.

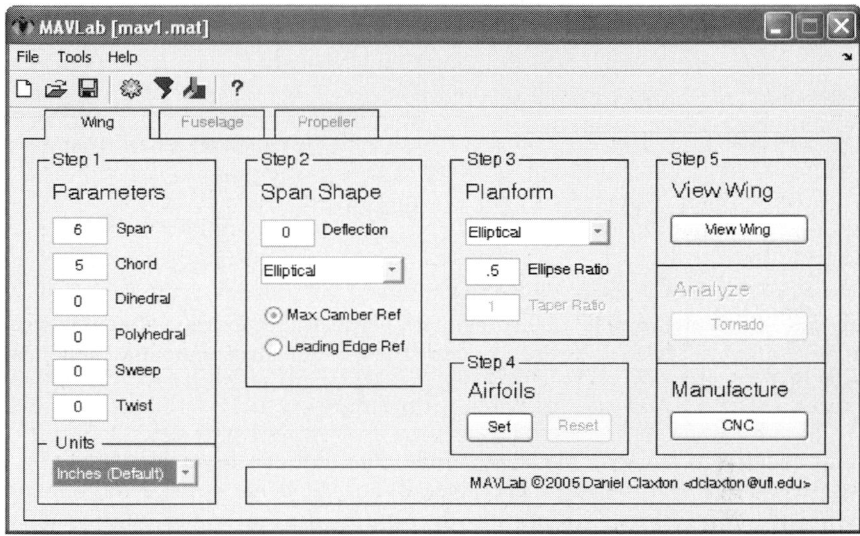

Fig. 5.6 MAVLab graphical user interface. MAVLab has the capability to create a wing geometry by modifying simple parameters such as wing span, chord, dihedral, polyhedral, sweep, twist, airfoil shape, airfoil distribution, and planform. Additional tools include aerodynamic analysis and CNC toolpath generation.

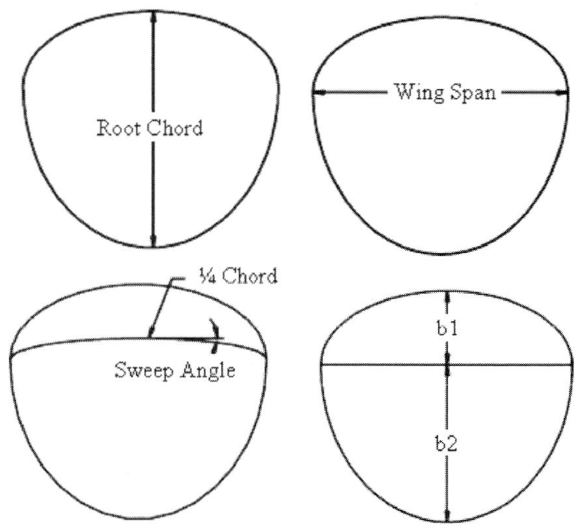

Fig. 5.7 Clockwise from top: root chord, wing span, ellipse ratio (b_1/b_2), and $\frac{1}{4}$-chord sweep.

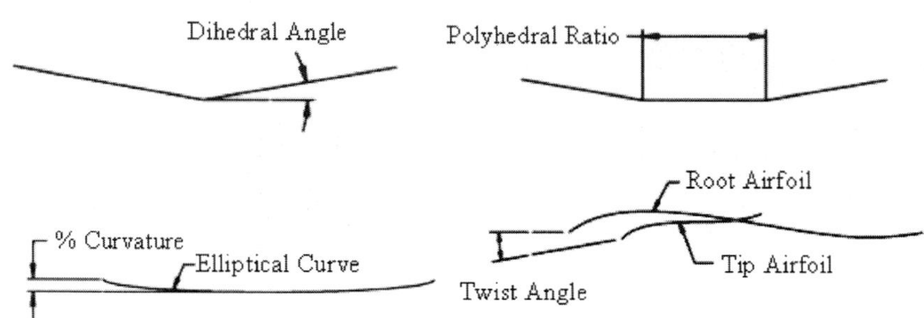

Fig. 5.8 Clockwise from top: dihedral angle, polyhedral ratio (ratio of nondihedral angle to dihedral angle), spanwise geometric twist, and spanwise curvature (elliptical example).

In addition to the basic parameters, the planform can be manipulated. This can be input as an arbitrary shape given in an ASCII file (similar to common airfoil coordinate files). Alternatively, the planform can be created by choosing a tapered or elliptical shape. The taper is modified by a taper ratio, and the elliptical planform is specified by an ellipse ratio. In the case of the ellipse, the planform is not necessarily a pure ellipse, rather a combination of two ellipses with identical dimensions for the semi-axis in the spanwise direction and independent dimension for the chordwise direction. The ratio of the two chordwise semi-axes is the ellipse ratio (Fig. 5.7).

Dihedral can be generalized by an arbitrary spanwise shape and can be loaded using a coordinate file or by selecting several preset geometries. The result is a smooth and continuous shape (Fig. 5.8). Some examples of span shape are elliptical, parabolic, and cubic distributions.

Airfoils can be created through an interactive drawing tool or by loading a coordinate file. Up to 21 unique airfoils can be specified along the semispan; intermediate locations are created by interpolation. Both singly curved and reflexed airfoils can be created this way.

Each input in MAVLab causes the model to update immediately, and a three-dimensional geometric display of the wing is regenerated. The user can interact with the model by rotating and changing the viewpoint with the mouse. An example of a wing is shown in Fig. 5.9.

It is often necessary to know the exact surface area, or average $\frac{1}{4}$-chord of a wing. These features can be challenging to determine from complicated planform geometry, without approximations. Routines in MAVLab have been developed to automatically return surface area (S), aspect ratio (AR), mean aerodynamic chord (MAC), and mean $\frac{1}{4}$-chord.

MAVLab offers several useful and common industry formats in order to facilitate exporting of the model. These formats include IGES (International Graphic Exchange Specification), STL, AutoCAD script (.scr), AVL (Athena Vortex Lattice, Mark Drela), and NC (CNC toolpath format).

One of the most useful aspects about MAVLab is its capability to generate G-code, the format that CNC machines need to control a toolpath. This process is as simple as clicking a single button. More advanced users can manipulate other

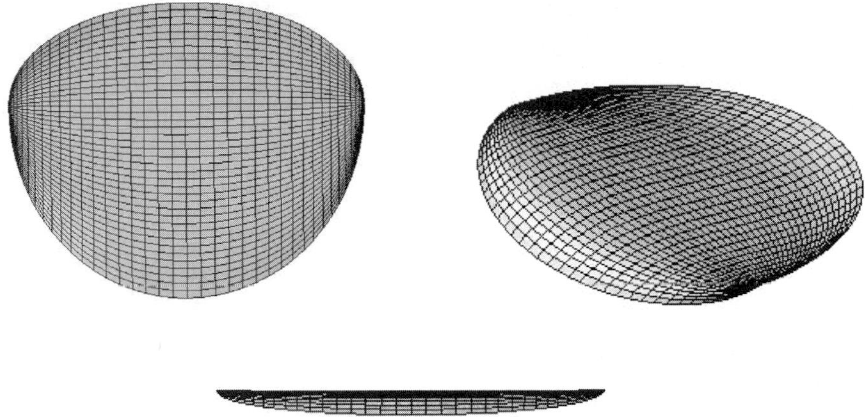

Fig. 5.9 MAVLab automatically updates the wing as the user inputs parameters. Rendering of wing from MAVLab, clockwise from top: top view, isometric view, and front view.

features such as milling machine tool diameter, scallop height, and direction of cut. A parameter specifies the diameter of the end-mill used to machine the part. For free-form surfaces, it is necessary to use a ball end mill, as shown in Figs. 5.10 and 5.11.

The scallop height is a physical parameter that directly affects the surface finish of the part. Mathematically, it is the maximum height of the ridges produced by the ball end mill, as shown in Fig. 5.12. The following equation relates the scallop height h_{scallop} to the step-over distance x_{stepover} and diameter of the tool D_{tool}.

$$h_{\text{scallop}} = \frac{1}{2}\left[D_{\text{tool}} - \sqrt{(D_{\text{tool}})^2 - (x_{\text{stepover}})^2}\,\right] \tag{5.1}$$

Step-over is the normal distance between adjacent paths of the tool.

Fig. 5.10 Wing tool milled on CNC machine.

Fig. 5.11 Ball end mill used to machine wing tools from high-density tooling board and aluminum. Typically, a $\frac{1}{2}$-in.-diam tool is used as opposed to a smaller tool, to reduce the machining time.

To create a toolpath, from the wing geometry, the surface must be finely meshed according to the desired scallop height and direction of cut. The surface normals at each of the grid points are calculated and scaled to the radius of the tool. Each normal is placed at its corresponding grid point. The order in which these points are arranged determines the toolpath.

Depending on the geometry, it might be necessary for the toolpath to follow various patterns. MAVLab's options for tool path pattern include zigzag (chordwise/spanwise) and spiral.

In addition to creating the toolpath, MAVLab also overlays the toolpath pattern over the wing surface and draws a box around the wing to signify the tool containment boundary. The length, width, and height of this box are also returned for reference.

Part of the engineering process of a wing is validating the design before it goes into production. This can be approximated with CFD. Though results cannot replace wind-tunnel and flight testing, CFD can be used to show trends in aircraft efficiency metrics such as lift/drag ratios. This, in turn, facilitates a quicker and more efficient iteration process. MAVLab offers several ways to implement CFD. It exports geometries in formats that can be converted to meshes for full Navier–Stokes CFD, as well as simpler vortex-lattice methods. It also has a plug-in feature that directly interfaces with a vortex-lattice solver called Tornado.

The basic idea behind MAVLab is to treat inputs as features. Each time a feature is changed, the model is updated and reassembled. Because this is done computationally, the wing surface is constructed from a finite grid of points (41 by 41). Cubic interpolation is used to create C^2 continuous surfaces. In other words, the second derivative of the surface is continuous.

The goal of UF's MAV production is to have the capability of constructing lightweight, accurate representations of the CAD design in a time-efficient manner. The production of precisely machined wing molds is the first step in this process.

The wing molds are made of high-density, high-temperature plastic tooling board, which offers three desirable characteristics: heat resistance, fast milling, and ease of surface finish. Alternatively, one can use a metal such as aluminum for tooling, increasing production time significantly.

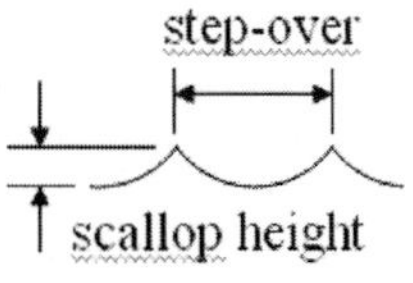

Fig. 5.12 Scallop height is the out-of-plane deviation of the surface created by the shape of the ball end mill. Stepover is the distance between passes of the tool.

a) 

b) 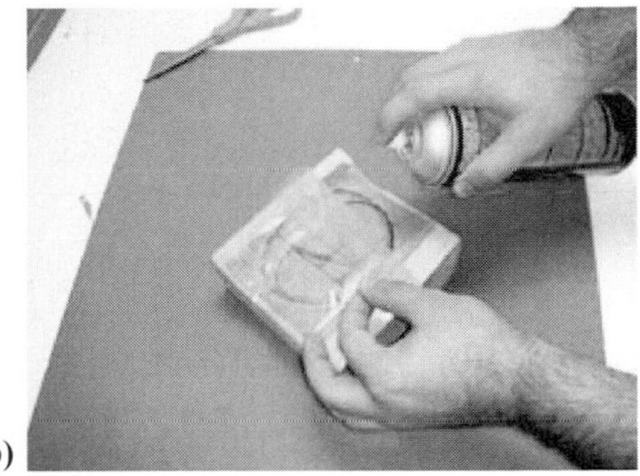

Fig. 5.13 Wing mold: a) drawing pattern and b) applying a release film.

The heat resistance is paramount because the tool must endure the heat of a cure cycle for every part produced. Milling allows for rapid, reproducible prototyping and reduced labor. After milling, the surface of a wing tool is not typically perfectly smooth and requires light sanding to remove the ridges left behind by the ball end mill (scalloping). Although the tooling board is dense, it quickly sands to a smooth finish.

To craft reproducible wings, lay-up patterns are required to show where the unidirectional carbon strips should be placed on the mold. One solution involves printing a pattern onto paper and then adhering it to the tool underneath the release film. This process can prove problematic for more complex curved wings because the two-dimensional paper does not conform well to the tool. To avoid this problem, MAVLab was modified to create CNC tool paths that can trace custom patterns onto the three-dimensional wing surface with a permanent marker. A completed wing tool is shown in Fig. 5.13.

To date, fuselage tools have been hand carved from tooling board; however, there is an ongoing effort to develop the capability to design and machine fuselage tools in the MAVLab framework. This will improve on tool accuracy (symmetry) and allow both negative and positive tools to be produced.

5.3.2 Composite Construction Methods

The primary composite utilized is carbon fiber and thermoset epoxy in a preimpregnated form, which demonstrates very high strength and flexibility after cure. The carbon-fiber cloth used in construction is a 0/90-deg plain weave, 176.3-g/m^2, 3000-fiber/yarn pre-impregnated fabric. In addition, pre-impregnated unidirectional carbon fiber is also used. Despite the high strength of the carbon fiber, 170-g/m^2 aramid Kevlar® cloth reinforcement is often used to prevent crack propagation in the 0/90-deg weave cloth on the wing leading edge.

Extreme care is taken in the wing lay-up procedure to take advantage of the precision allowed by the CNC wing tools. No viable automated process is available to lay up the wings, and so a new procedure was necessary for composite construction of the airframe. The following steps document the process.

1) Wing construction requires the application of Teflon® release film to the wing tool using spray glue. The tool is first lightly coated with spray glue, and

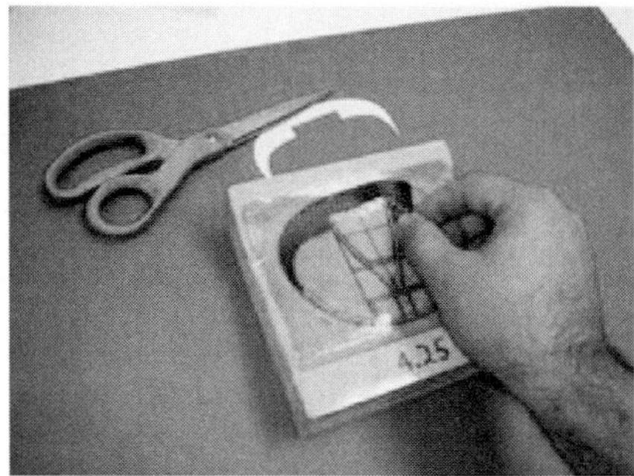

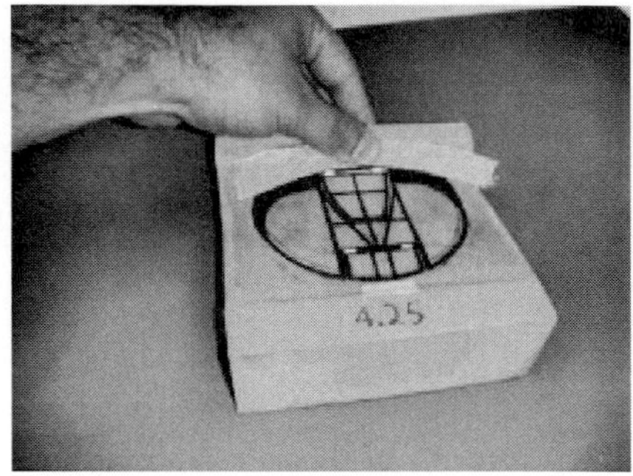

Fig. 5.14 Carbon-fiber weave and Kevlar® cloth are laid on the leading edge.

then Teflon® is delicately applied to the surface (Fig. 5.13). Care is necessary to prevent wrinkles from forming, as this will affect the final surface finish.

2) The leading 20% of the wing is constructed of a single layer of 0/90-deg cloth oriented at a 45-deg angle with respect to the chord line. A pattern is generally used to cut out a piece of carbon fiber to the proper shape. The resulting patch of carbon fiber is aligned with the pattern on the tool. The 0/90-deg cloth is then reinforced by a single additional layer of aramid cloth, as shown in Fig. 5.14.

3) Multiple 2-mm-wide strips of unidirectional carbon fiber are cut from a sheet of pre-impregnated material using a custom-built carbon strip-cutting tool. The portion of the wing that attaches to the fuselage is made by carefully aligning layers of unidirectional carbon-fiber strips with the pattern drawn on the tool (Fig. 5.15). The unidirectional carbon fiber is layered for strength and rigidity and is interwoven at intersections to create stronger mechanical joints. Hinges are created by interweaving Tyvek® material into the gaps between the control surface and frame.

4) Teflon® release film is applied to the fuselage tool. The complex curvature of the fuselage tool requires separate layers of 0/90-deg cloth to be applied to each side of the tool, and the layers are overlapped over the centerline. The fuselage is then placed on a plate that is shaped to the curvature of the wing (Fig. 5.16).

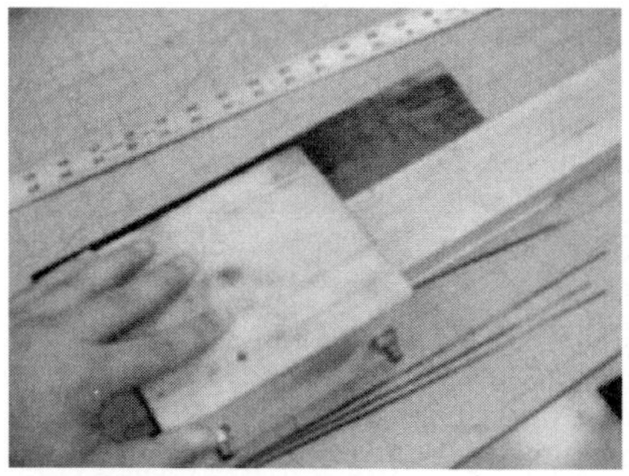

Fig. 5.15 Unidirectional carbon-fiber strips are cut and laid up according to the pattern drawn on the tool.

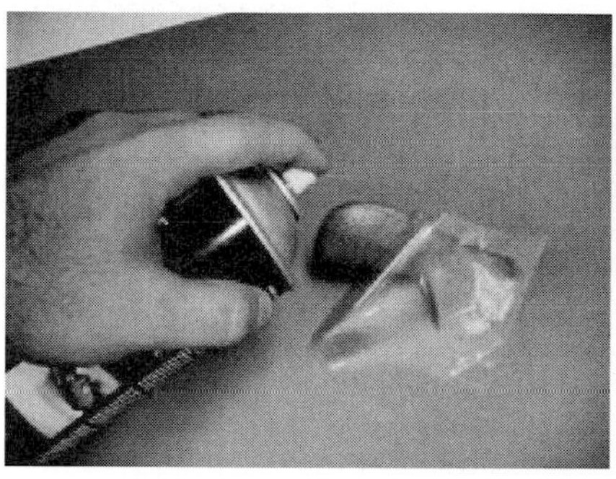

Fig. 5.16 Fuselage assembly. Release film is applied with spray glue, then carbon-fiber 0/90-deg weave is stretched over the free-form surface.

5) The wing and fuselage are covered with porous Teflon® release film and placed in a vacuum bag, as shown in Figs. 5.17 and 5.18. The bag is sealed, and a vacuum is drawn to a pressure of 0.01 atm absolute and placed in an oven at 130° C for 4 h.

6) Once removed from the cure cycle, the composite wing and fuselage are trimmed and joined using cyanoacrylate (CA) adhesive (Fig. 5.19).

7) The wing is then skinned with 0.1-mm latex rubber. The latex is applied by masking and then coating all contact surfaces of the wing with spray glue. Later the latex is permanently adhered with CA glue and trimmed.

8) Electric motor, ESC, receiver, and servos are then installed inside the fuselage, as shown in Fig. 5.20. The servos are mounted by adhering them to the carbon-fiber wall with two-sided tape and lashing them with aramid thread. The motor is installed with a 5-deg declined thrust line with respect to the wing.

9) Servos are connected to the control surfaces via 0.75-mm brass rods. Connections are secured with z-bends at both ends.

5.4 Wind-Tunnel Characterization of the Flexible Wing

A flexible wing, in comparison to the design of an equivalent rigid wing, introduces more independent variables in the design and in the experimental

Fig. 5.17 Peel ply is laid over the wing and then placed in a vacuum bag.

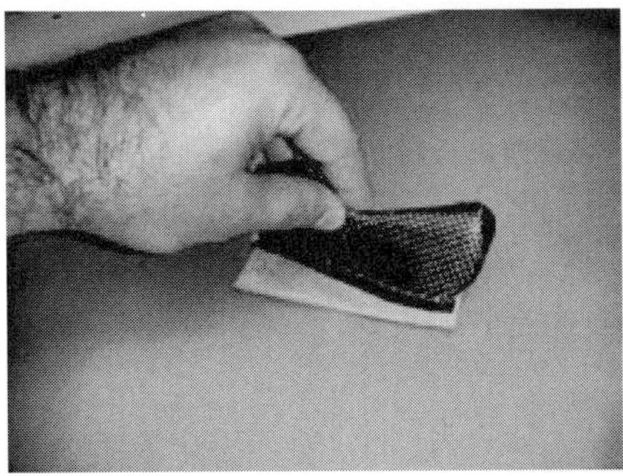
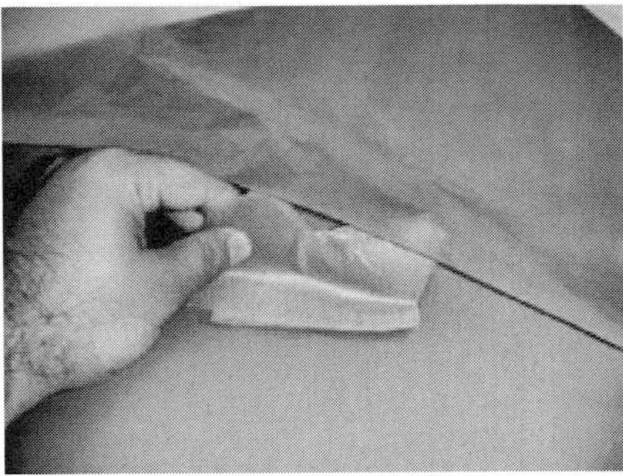

Fig. 5.18 Fuselage is placed on an airfoil-shaped tool, covered in peel ply and placed in a vacuum bag.

characterization process. In the case of the elastic wing, the AOA, the primary independent variable for standard experiments of rigid wings, assumes a multiple role because a change in the AOA at constant freestream velocity (FSV) will induce an elastic variation to the geometry of the wings. Furthermore, in the case of the flexible wing, the dynamic pressure influences the shape of the wing, introducing a substantial coupling between the aerodynamic characteristics and the Reynolds

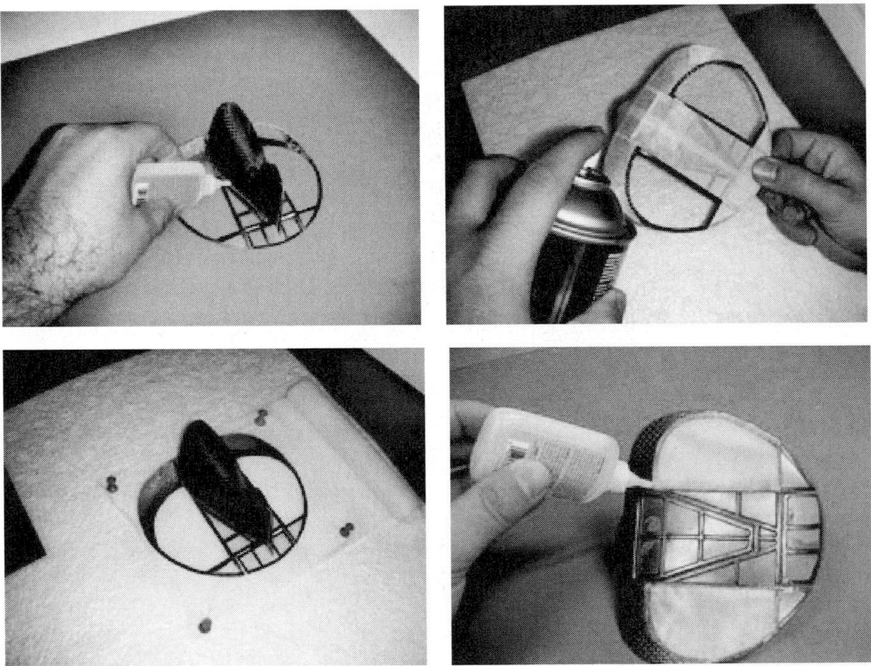

Fig. 5.19 Fuselage is glued to the underside of the wing, and latex is adhered to the top of the wing.

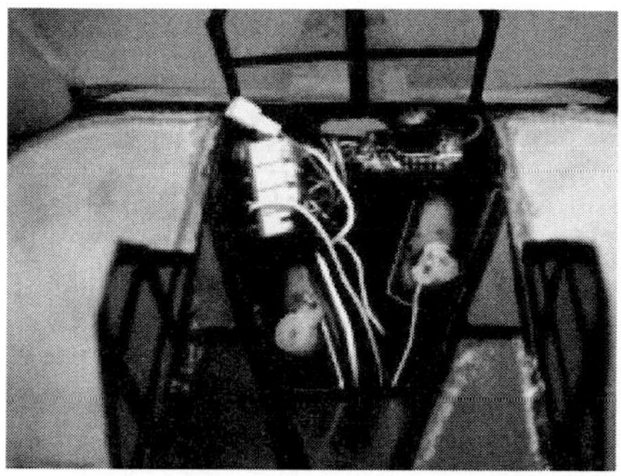

Fig. 5.20 Servos installed in fuselage with motor and radio. Brass connecting rods are used to mate servos with control surfaces.

number (RN), not exhibited with the rigid wing. Experimental results with the rigid wing in the wind tunnel demonstrated [10] and [11] that the RN varying during the tests from 6×10^4 to 1.5×10^5 did not appreciably affect the results.

A thorough survey of the available literature confirmed that an experimental study on a series of flexible low-aspect-ratio wings at low Reynolds number, with a parametric structural design, and performing aerodynamic measurements concurrent with out-of-plane static elastic deformation does not exist. Some landmark experiments on flexible wings, however, were obtained at the NASA Langley Research Center by wind-tunnel tests on an early generation of MAVs designed and fabricated at the University of Florida. Waszak et al. [12] tested vehicles with 152-mm (6-in.) wing span. The vehicles had a low mounted cruciform tail, and they were equipped with three different wing structures. The purpose of the research was to investigate the benefits of the flexible-wing concept, already postulated from results using numerical methods. The variation of structural compliance was obtained by varying the number of longitudinal stiffeners, or battens. Part of the investigation also consisted of measurements of the wing surface deformed by the aerodynamic load using a videogrammetry system.

The results of the MAV without the propeller proved that the flexible membrane wing allowed the vehicle to achieve higher angles of attack without stall and suggested the presence of benevolent characteristics induced by the wing flexibility, in terms of ability to adapt to atmospheric disturbances and to provide smoother flight. The latter is a very desirable attribute for an unmanned surveillance platform, but yet no systematic experimental evidence was offered to prove that statement.

Ettinger et al. [13] estimated the difference of the vehicle's lift curve vs airspeed between a flexible and a rigid wing. Jenkins et al. [6] recorded during flight tests the pilot controls input to two identical vehicles, one with rigid wing and the other with flexible wing giving comparative flight handling qualities. The elasticity of the wing was demonstrated to play an important but elusive role in the flight mechanics, dynamics, and energetics of a MAV with a flexible wing.

It was the goal of this work to perform a systematic characterization of the aerodynamics of low-aspect-ratio (LAR) wings at low Reynolds number including the effects of varying flexibility and to characterize the new perimeter-reinforced wing design. The experimental work utilizes the low-speed wind tunnel at the University of Florida using a six-component sting balance for the measurements of the aerodynamic coefficients, a visual-image-correlation (VIC) system for measurements of the full-field structural displacement and flow visualization using particle image velocimetry (PIV).

The deformation measurements in the wind tunnel represent an original and unique application of the VIC system. The aerodynamic results were compared and correlated with the flexibility of the structure, experimentally characterized for all wing designs. The data are based on wind-tunnel experiments of LAR wings with identical geometric features but with different structural design, thus various elastic properties. The main objective of the current wind-tunnel work is the experimental aerodynamic and geometric characterization of low-aspect-ratio wings at low Reynolds number, as affected by the following parameters: elastic characteristics of the structure, freestream dynamic pressure, and angle of attack.

5.4.1 Description of the Models

The wing planform and its associated airfoil shape were chosen from the flight line of the University of Florida MAV lab. The unloaded geometry was kept constant for all of the models tested. In this way the changes in any aerodynamic characteristics at the same dynamic pressure and AOA were only caused by the wing's flexibility and the consequent shape when loaded.

The first problem to be solved for a comprehensive experimental characterization of the wings was to establish the independent variables and their target values. The two main deformation shapes were identified as wing washout (geometric twist) and change of camber (aerodynamic twist). To obtain the two main shapes described earlier, it was decided to use a passive morphing technique by controlling the distribution and the elastic properties of the carbon fiber lay up in the wings; further control of the elastic deformation was obtained by the quantity of carbon fiber with respect to the latex skin.

Two main types of wing designs were used, the batten reinforced (BR) wing and perimeter reinforced (PR) wing, as depicted in Fig. 5.5. The rigid (R) wing was used as reference for aerodynamic comparison. After exploring a rather large design space associated with relative structural stiffness of the individual components of each wing, the best-performing examples of the BR and PR wings were compared to the R wing. For instance, four BR wings with variation in batten and leading-edge stiffness were tested. These variations were achieved through the number of plies utilized in individual battens, for instance. Four variations of the PR wing were constructed with differing degrees of membrane tension and perimeter stiffness. Again, for purposes of brevity, only the examples that showed significant promise in performance are presented in this text.

5.4.2 Experimental Apparatus

Because of the flexibility of the wings, their aerodynamic characteristics cannot be fully comprehended without a static aeroelastic characterization of the system performed in parallel to the aerodynamic measurements. Three main experimental arrangements were used in the wind tunnel for this research: a strain-gauge sting

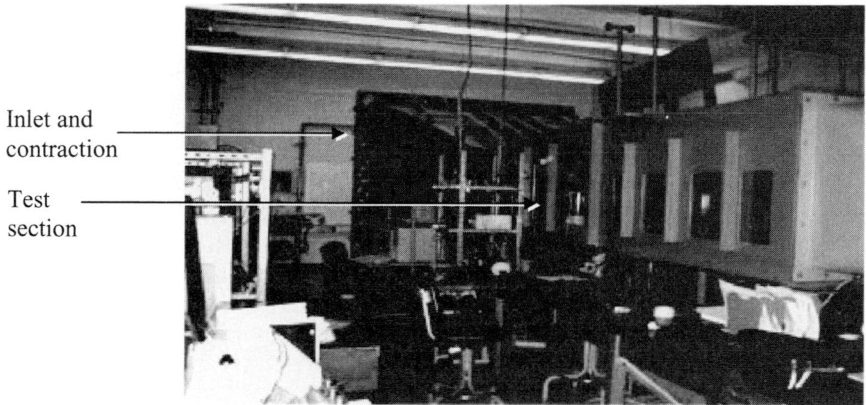

Fig. 5.21 Open-circuit wind tunnel used for the characterization of the flexible wing.

balance, for force measurements with the subsequent computation of aerodynamic coefficients; the VIC system, for the geometric and elastic structural deformation characterization; and a PIV system for flow visualization and computational-fluid-dynamics (CFD) model validation. The VIC and PIV work will be discussed in detail in subsequent sections. The following section describes the experimental techniques related to the measurement of the forces, with the test procedures and data analyses.

The wind-tunnel facility used during the tests was a horizontal, open-circuit, low-speed wind tunnel located in the Mechanical and Aerospace Engineering Department of the University of Florida, shown in Fig. 5.21. The wind tunnel, with an overall length of 10 m, has a square section entrance with a bell-mouth inlet type and several antiturbulence screens. The contraction section was designed to provide low turbulence levels in the test section, less than 0.1%. The test section was 914 × 914 mm with a length of 2 m. A downstream centrifugal-type fan regulated by a variable-frequency controller was operated remotely by a PC via the data-acquisition (DAQ) system used for the experiments running LabView software. The maximum velocity for test purposes was approximately 15 m/s, with a subsequent maximum Reynolds number of about 2×10^5.

To measure forces and moments, a six-component sting balance was procured from Aerolab and installed in the wind tunnel. The balance, with six channels each in a full Wheatstone-bridge configuration, is an internal five-force and one-moment sting balance. Two forces are normal to the balance's axis on a vertical plane (i.e., lift), two forces are normal to the balance's longitudinal axis on a plane perpendicular to the previous plane (i.e., side force), one force is aligned in the axial direction, and the only moment is around the longitudinal axis, typically the roll axis.

The forces generated by our aircraft utilize only 0.1 to 2.5% of the full-scale range of the balance; thus, the level of the electrical signals from the strain gauges is very low, in the range of a few microvolts, making the data processing particularly challenging.

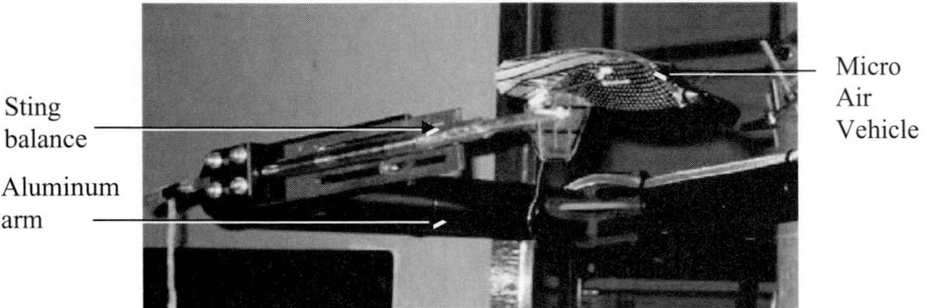

Fig. 5.22 Vehicle in the wind tunnel, mounted on the sting balance and on the aluminum pitching arm.

The balance, which holds the model under test, is connected to an aluminum custom designed arm, illustrated in Fig. 5.22. Calibration was achieved by loading the balance with certified weights, typically from 1 to 100 g, at precisely defined loading points and directions, using a special calibration rig. The voltage signals from the balance, on the order of a few microvolts, were recorded for each load and used to define the calibration matrix.

The DAQ used was a typical National Instruments (NI) modular system; the main element was a NI SCXI 1520 with an eight-channel programmable strain-gauge module, with full bridge configuration, excitation of 2.5 V and a gain of 1000 as typical settings. Other modules included in the system are the SCXI 1121 signal conditioner module, 1180 feed through with 1302 breakout and 1124 D/A module. A NI 6052 DAQ PAD Firewire provides A/D conversion.

A four-wire RTD thermocouple for airflow temperature measurement and the analog output from an absolute encoder for AOA measurement were connected with the SCXI 1121 module. A Heise differential pressure sensor, for dynamic pressure readings via a pitot tube, was connected to one of the PC's RS232.

A single-axis motion controller runs a servomotor, with the driver connected to the personal computer via the RS232 for firmware uploading. A 0-to 10-V analog signal was generated by the DAQ system and sent to the driver to command the pitching motion. The system was designed to perform tests at steady or unsteady AOA. Dynamic pitching tests can be performed to investigate a variety of phenomena, including vertical gust effects. The typical pitching velocity for an AOA sweep in steady tests was 1 deg/s; for unsteady tests the pitching frequencies can be up to 5 Hz with amplitudes up to 90 deg. The balance has a frequency response of several hundred Hertz. All components were assembled on a test rig stand next to the side of the wind tunnel.

5.4.3 Test Procedures

During a typical test, the sequence of the desired AOAs was from 4 to 36 deg. A set of zeros (signal with wind tunnel off) was acquired at the start and at the end of each run for further analysis and corrections as a result of signal drift, if needed. The software performed all data conversions in forces and displayed, in real time, the six coefficients (C_L, C_D, C_y, C_l, C_m, C_n) vs AOA. At the end of each run,

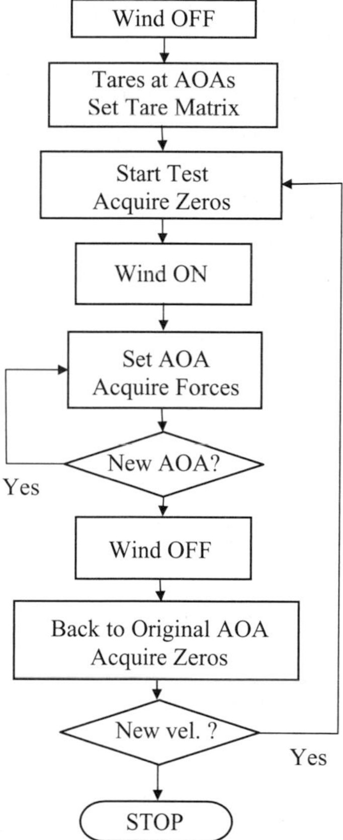

Fig. 5.23 Schematic of wind-tunnel test procedure.

the data were stored in a file for further analysis. The procedure is schematically illustrated in Fig. 5.23.

At each angle of attack, the output signals from the six channels of the sting balance were sampled 500 times at a frequency of 1000 Hz. The average and standard deviation were computed repeating this process 20 times for each test point, for a total of 10,000 samples in 10 s; a total average and standard deviation were calculated and stored for subsequent data analysis. The averages were used for the calculation of the loads, and the standard deviations were used to perform the uncertainty analysis. Additionally, the signals from the six channels of the balance were recorded with airflow off, at the beginning and at the end of each test sequence (i.e., angle of attack sweep), in order to account for any drift that might have occurred. The drift proved to be negligible.

Classical wind-tunnel corrections for blockage (streamline curvature, solid blockage, and wake blockage) were applied to the coefficients of lift, drag, and pitching moment. An additional correction was applied to the nominal values of the AOA. In the general case, any change in the model's position during wind-tunnel

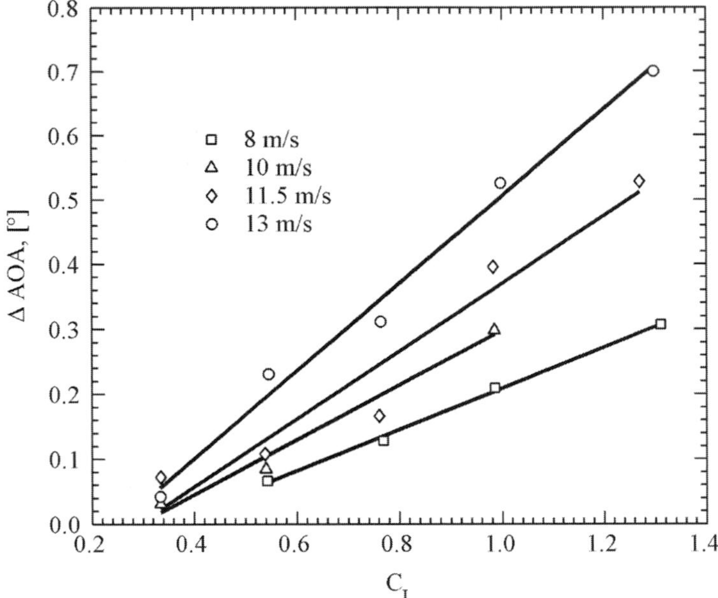

Fig. 5.24 Change of the AOA caused by rigid-body rotation vs C_L at four wind-tunnel freestream velocities for a typical PR wing.

tests can be considered as the summation of the flexibility of the holding structure of the sting balance and the model itself.

Specific tests were conducted with the three different types of wings, and the change in the AOA of the model as a rigid-body rotation was measured using the VIC system. In the linear region of the lift curve, the values of the change of AOA caused by system elastic deformation were plotted vs C_L at the available test flow velocities for the two types of flexible wings, PR and BR. An example of one plot is depicted in Fig. 5.24.

With the assumption that the wing rigid-body rotation was caused mainly by the lift component of the aerodynamic resultant, the corrections for the full range of AOA and velocities were estimated as a function of the lift coefficient using linear regression. Using linear equation expressions, the correction for the AOA for a particular type of wing was found as

$$\Delta\alpha = f(C_L, q) \tag{5.2}$$

The effective value of the AOA finally can be written as

$$\alpha_E = \alpha + \Delta\alpha \tag{5.3}$$

An error sensitivity analysis was performed for the three aerodynamic coefficients C_L, C_D, C_m, using the Kline–McClintock technique [14] for error propagation. The largest source of uncertainty was the uncertainty derived from the standard deviation of a given output signal. The quantization error, as a result of the use of 16-bit DAQ cards, was found negligible. Other minor sources of uncertainty were identified and accounted for.

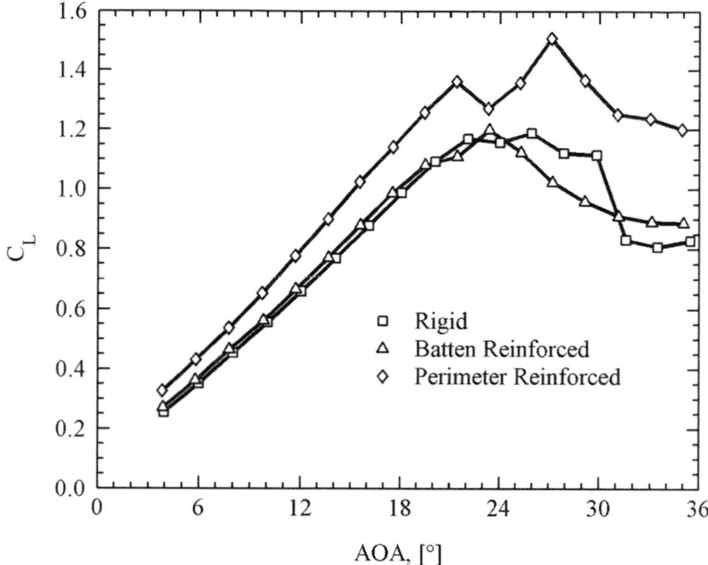

Fig. 5.25 C_L **vs** α **for the rigid, batten-reinforced, and perimeter-reinforced wings at** $V_\infty = 13.0$ **m/s.**

Plots of the aerodynamic coefficients C_L, C_D, and C_m with their respective error bounds about each data point were obtained. The error bound generally increased with the increase of the AOA, especially in the region surrounding the stall angle as well as with decreasing free-stream velocity. Typical values of C_L, C_D, and C_m at a Reynolds number of 11.5×10^4, at $C_L = 0.6$ (sufficiently away from the stall) were on the order of 3.32% for C_L, 10.55% for C_D, and 3.72% for C_m.

5.4.4 Sting Balance Results

This section represents the results from only three examples (a rigid wing, a batten-reinforced wing, and a perimeter-reinforced wing). The three illustrated here were taken from a systematic series of experiments on nine different wings (one rigid and eight flexible) with the fundamental aerodynamic data (C_L, C_D, and C_m). It is useful to remind the reader that the wings differ only in structural compliance, which was chosen to achieve various levels of flexibility. The presentation of the results can be organized in sections, with plots at different dynamic pressures for each structural design and plots at different flexibility at constant dynamic pressure.

A selection of the latter is presented in this work; each plot shows the aerodynamic coefficients of three different wings, with the distinctive structural design, at constant freestream dynamic pressure (FSV). The graphs can be used to compare how the aerodynamic characteristics of the wings change with the flexibility, while holding the dynamic pressure (FSV) constant. These results are also used to estimate the aerodynamic derivatives of the wings.

The results at a freestream velocity of 13 m/s are presented in Figs. 5.25–5.28 and are organized as follows: C_L vs α for rigid, BR, and PR wings; C_L vs C_D for

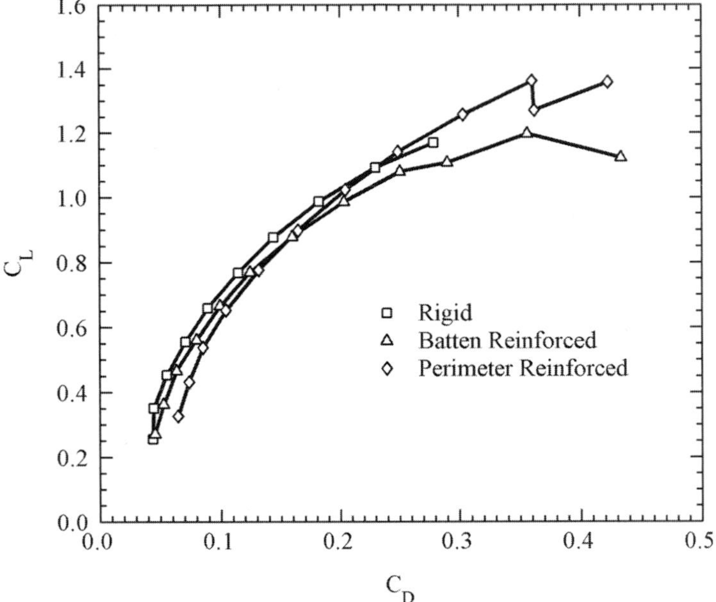

Fig. 5.26 C_L **vs** C_D **for the rigid, batten-reinforced, and perimeter-reinforced wings at** $V_\infty = 13.0$ **m/s.**

rigid, BR, and PR wings; L/D vs C_L for rigid, BR, and PR wings; and C_m vs C_L for rigid, BR, and PR wings.

All of the pitching-moment results reported in this work are measured about 25% of the root chord. The three flexible wings were selected for their relevance in the flying test program and because they were also tested in the wind tunnel on a complete vehicle with a powered propeller [1]; the rigid wing was used as reference. All of the other wings in the series were considered as well, and their results were used to compile a database with coefficients and aerodynamic derivatives. The tests were performed at Reynolds numbers between 79.4×10^3 and 12.9×10^4; the results presented correspond to a Reynolds number of 12.9×10^4.

Figure 5.25 presents the lift coefficient vs the AOA for the three wings. The perimeter-reinforced wing exhibited the highest maximum lift. Interestingly enough, for a Reynolds number of 12.9×10^4, the lift of the PR wing after stall is higher than the maximum lift of all other wings. The coupling between dynamic pressure and the wing-section shape for the flexible wings causes a different aerodynamic behavior at various dynamic pressures.

The rigid and batten-reinforced wings have similar values of lift at the different dynamic pressures, and the perimeter-reinforced wing shows a slightly higher lift at higher dynamic pressure. In general the PR wing exhibited higher drag (Figs. 5.26 and 5.27) with respect to the rigid and BR wing; thus, they have the lowest L/D ratio.

The lower L/D of the perimeter-reinforced wing is attributed to the less than optimal wing shape assumed by the latex in the deformed state, with a high camber shifted aft towards the trailing edge. The PR wings showed important advantages

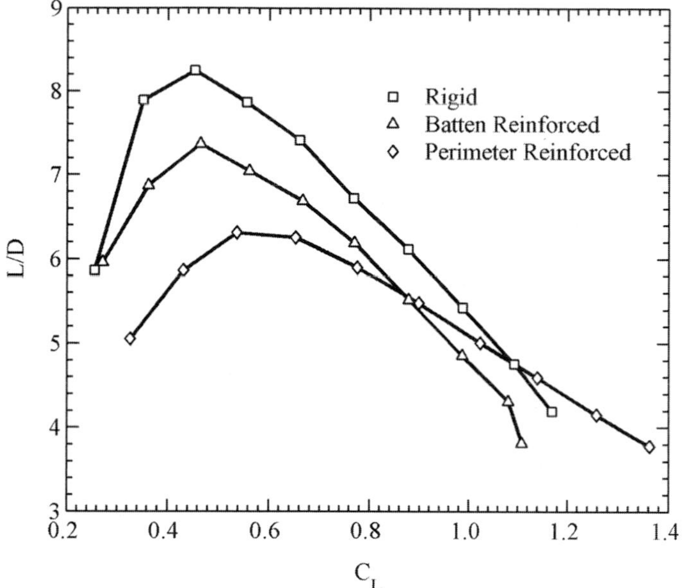

Fig. 5.27 L/D vs C_L for the rigid, batten-reinforced, and perimeter-reinforced wings at $V_\infty = 13.0$ m/s.

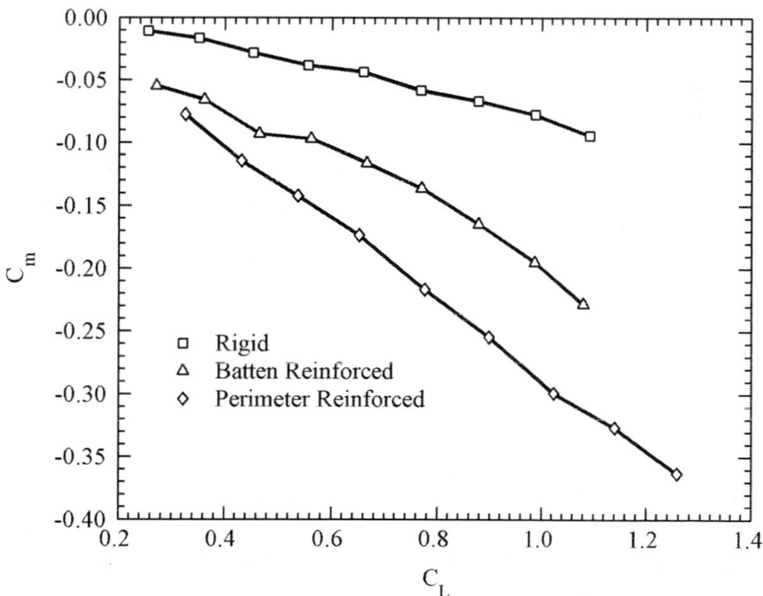

Fig. 5.28 C_m vs C_L for the rigid, batten-reinforced, and perimeter-reinforced wings at $V_\infty = 13.0$ m/s.

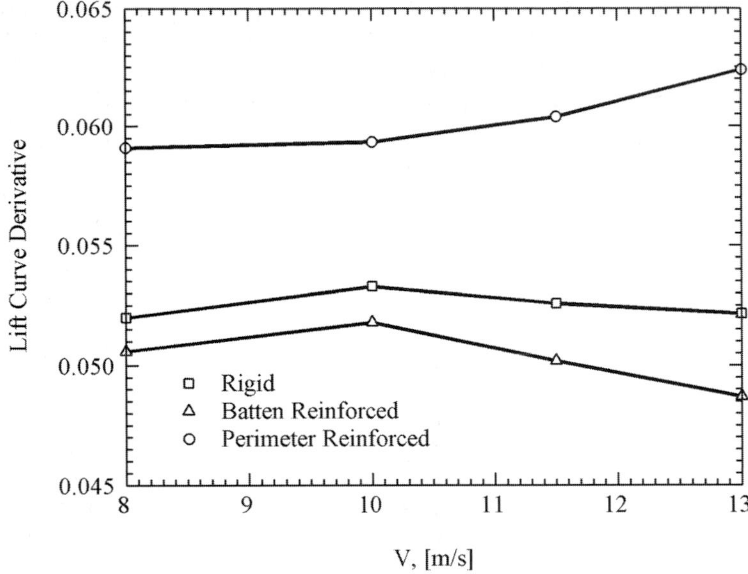

Fig. 5.29 Lift-curve derivative at the four FSVs tested in the wind tunnel.

in terms of static longitudinal stability, Fig. 5.28, offering the highest values of the $\mathrm{d}C_m/\mathrm{d}C_L$ derivative at any RN but especially at higher velocities. Furthermore the C_m over C_L curve of the PR wing remains linear for higher C_L in comparison to the other wings, improving the flight conditions at high AOA.

5.4.5 Combining Wing Aerodynamic and Deformation Results

A better general assessment of the aerodynamic characteristics of the flexible wings is obtained by considering global features such as aerodynamic derivatives $(C_L)_{\max}$, AOA at $(C_L)_{\max}$ and $(L/D)_{\max}$ for different FSVs. The results are reported in Figs. 5.29–5.33. Combining the observations of the aforementioned figures with the elastic deformation data, it will be possible to make some inferences on the aerodynamic features based upon the elastic deformation results.

Figure 5.29 shows the derivative $\mathrm{d}C_L/d\alpha$ vs the freestream velocity for the three wings. The value of the derivative is the highest for the perimeter-reinforced wing and the lowest for the batten-reinforced wing. The value has a slight tendency to increase with the FSV for the PR wings and to decrease with the FSV for the BR wings. This feature can be explained with the different deformation shapes (see Sec. 5.4.7) of the two groups of wings, featuring a geometric twist that decreases the local angle of incidence (AOI) in the BR wing case, decreasing the local lift. In the PR wing a relevant increase in local camber was observed, increasing the local lift.

The load factor caused by a vertical gust is proportional to the value of the $\mathrm{d}C_L/d\alpha$ derivative [15]; thus, for two geometrically identical wings subjected to the same gust, the wing with the lower derivative will experience a smaller load factor and smaller vertical acceleration; therefore, the BR flexible wings are characterized by more favorable gust sensitivity conditions with respect to the rigid and PR wings.

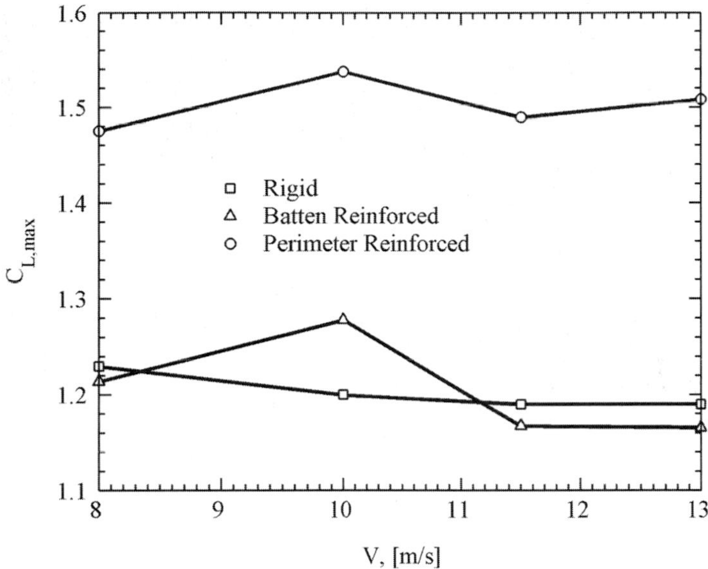

Fig. 5.30 Maximum lift coefficient at the four FSVs tested in the wind tunnel.

The perimeter-reinforced wing, as intended by its design, presents the highest value of the maximum lift coefficient, as displayed in Fig. 5.30, whereas the batten-reinforced wing shows lower values, similar to the rigid case. Indeed an inspection of the PR wing profile, shown in Sec. 5.4.7 for a FSV of 13 m/s, the maximum velocity tested, shows a measurably higher camber.

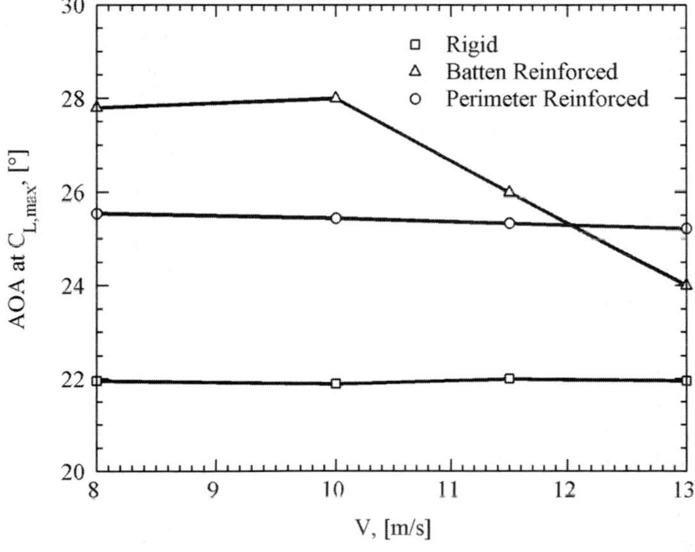

Fig. 5.31 AOA at maximum lift at the four FSVs tested in the wind tunnel.

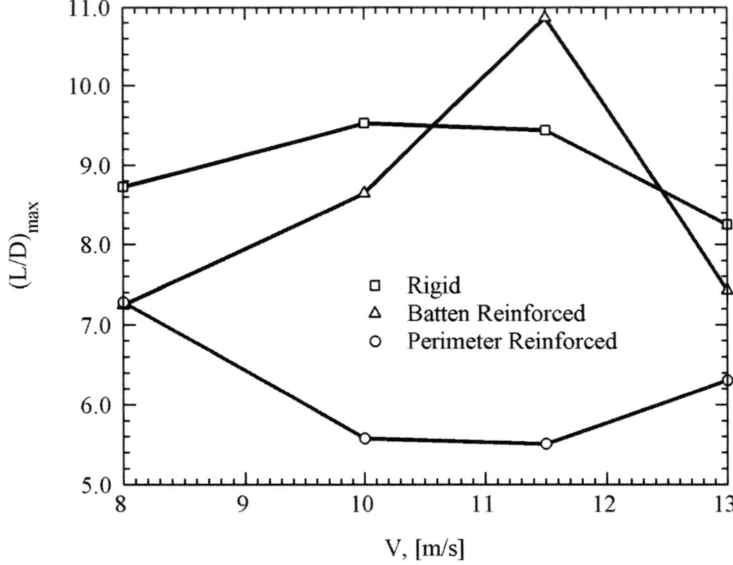

Fig. 5.32 Maximum lift-over-drag ratio at the four FSVs tested in the wind tunnel.

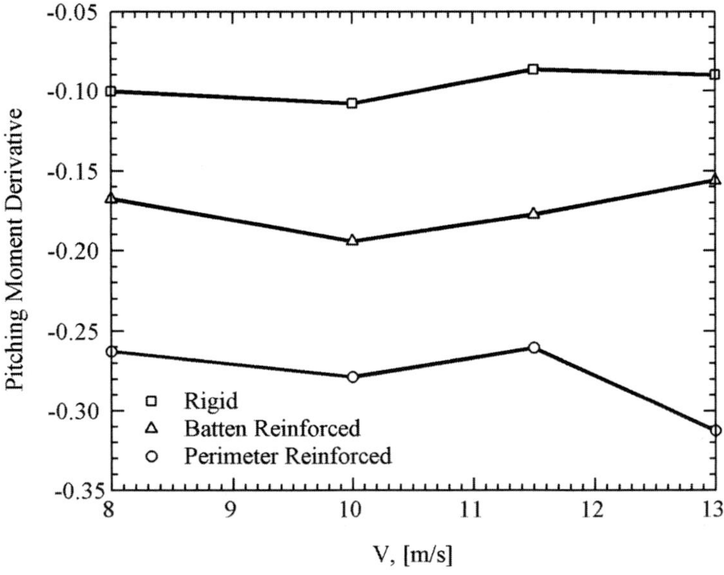

Fig. 5.33 Pitching-moment derivative with respect to lift at the four FSVs tested in the wind tunnel.

The AOA for maximum lift, shown in Fig. 5.31, has the lowest value for the rigid wing, as expected (but still relatively high because of the low aspect ratio of the wing). The highest values were measured for the batten-reinforced wing, showing the effectiveness of the adaptive washout in terms of stall angle delay. The relevant nose-down twist of the BR wing was visualized in the deformation tests discussed in Sec. 5.4.7.

In general, the maximum lift-over-drag ratio, as shown in Fig. 5.32, exhibited an overall maximum for the rigid wing, while the PR wing featured the lowest values and the BR wing's values were intermediate between the PR wing and the rigid wing. The lowest values of the PR wings are explained by the less-then-perfect shape assumed by the wing section when loaded and deformed by the combination of increase of AOA and dynamic pressure. The BR wing does not deform to an optimum shape, but the presence of the battens, with their own shape, increases the performance of the wing. Furthermore, in the case of the flexible wings, an amount of strain energy is used for elastic deformation of the structure at the expense of the kinetic energy in the wake, creating an extra drag with respect to the rigid case.

The last feature of the flexible wings that was studied was the static longitudinal stability characteristic. Longitudinal stability is of paramount importance for MAV flight, thus was our primary target for design improvement for the next generation of MAV wings. Figure 5.33 shows the pitching-moment derivative with respect to lift for the three wings.

All of the flexible wings have generally a higher negative pitching moment with respect to the rigid wing; therefore, the flexible wings have a higher nose-down tendency. This should not be a surprise given that the reflex part is less effective in the flexible wing because of elastic structural deformation. The derivative values for the BR wing are approximately twice those of the rigid, whereas derivative values for the PR wing are up to three times the values for the rigid wing. The observed deformation of the latex membrane by the aerodynamic load, and the subsequent shift downstream of the position of the camber, explain the higher nose-down pitching moment observed in the PR wing. Furthermore, the PR wing generally displays the linear part of the pitching-moment curve extended to higher AOAs rendering the PR wing easier to control at higher AOAs.

One of the motivations of the project was to determine if the flexible wings could offer any advantage with respect to a rigid wing, such as lift improvement, better controllability, and reactions to gusty conditions. Conspicuous differences were detected in the elastic static deformation characteristics between the batten-reinforced wings and the perimeter-reinforced wings. The elastic geometric twist of the wings under static aerodynamic loads is insignificant for the PR wing and present for the BR wing. Accordingly, the BR wing is potentially less sensitive to gusts because of the lower values of the lift derivative. Additionally, it maintains the original shape with an increase of dynamic pressure, and thus is better suited for high-velocity flight conditions. The PR wing with the largest amount of latex exhibited higher lift and therefore can be considered more appropriate for slower and more agile vehicles.

5.4.6 Experimental Techniques to Measure Wing Deformations

The flight performance of flexible micro air vehicles has indicated several desirable properties directly attributed to the elastic qualities of the wing, one of which

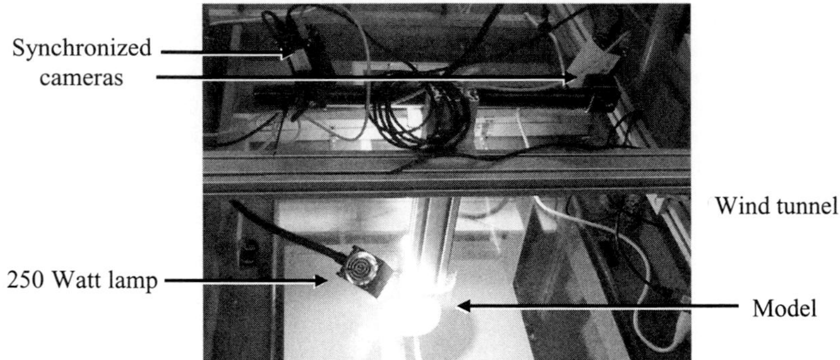

Fig. 5.34 Wind tunnel with VIC system mounted on top.

is adaptive washout. This phenomenon is a direct measure of the wing deformation that occurs as the load increases. The deformation automatically provides a measure of gust rejection (which can be observed in flight) and assists to delay the onset of stall as the angle of attack increases. With the aerodynamic characteristics of the wing under consideration so dependent upon its inherent flexibility, it is of great interest to experimentally quantify the shape change of a flexible wing subjected to various flight conditions. Deformations of particular importance are billowing (a change in the wing's camber), the aforementioned washout (a change in the spanwise angle of incidence), and bending (a change in the wing's dihedral). A well-documented wing deformation field lays the groundwork for investigations into numerical models (experimental validation of the complex fluid-structure interaction over a flexible MAV) and provides the link between wing elasticity and aerodynamic performance.

Any experimental method used to characterize the change in shape of a flexible wing must have the capability to determine both the original wing shape and the corresponding displacement field caused by pressure loading. A VIC system, developed and introduced by researchers from the University of South Carolina [16–18], was used for this purpose. The fundamentals of visual image correlation include determining the displacements of a specimen under load by tracking the deformation of a randomly distributed speckle pattern previously applied to the surface. The random pattern is digitally acquired by imaging sensors before and under load and mathematically processed by finding the region in the deformed image that maximizes the normalized cross-correlated score with respect to a small subset of the image taken while no load was applied (reference image). Recovering three-dimensional structure from two imaging sensors is called stereo-triangulation, an idea analogous to that of human vision.

To be able to capture the three-dimensional features of the models, synchronized twin cameras, each looking from a different viewing angle, were installed on the wind-tunnel ceiling, as shown in Fig. 5.34. Optical access into the test section is through a float (nonoptic quality) glass window. The effects of a glass interface have already been carefully investigated [19], and the cameras are calibrated through the window to ensure minimal distortion effects. Two continuous 250-W lamps illuminate the model, enabling the use of camera exposure times of 5 to 10 ms. Energy

Fig. 5.35 Speckled micro air vehicle model.

emitted by the lights, a potential hazard for the specimen, was not a concern because of the cooling effect of the wind-tunnel flow rate, typically from 8 to 13 m^3/s.

The preparation of the specimen is relatively straightforward, but requires attention to some important details, like the choice of background color on the target surface (to blend with the wing skin) and skin finish (diffuse) to minimize or eliminate noise levels that result during image processing. For MAV testing in particular, the top surface of the wing must be coated with a light-colored paint. The wing skin, consisting of a strip of thin, lightly colored, and partially transparent latex rubber is then speckled with a flat black paint. The fine speckle and the final treatment of dulling paint do not significantly change the mechanical properties of the latex skin. The latex is stretched out around a frame and then glued to the wing structure. After the excess latex skin has been trimmed away, the results can be seen in Fig. 5.35.

The following procedure was adopted to systematically quantify the deformation of a flexible wing subjected to all relevant flight conditions. For a given model angle of attack, a reference image is taken with zero wind velocity. The next step is to start the wind tunnel up to a desired dynamic pressure. When stable conditions are achieved, a second picture is taken of the deformed wing. Images are stored for future postprocessing. Then the wind is stopped, and the model is automatically moved to the next AOA. If so desired, aerodynamic loads can be measured at the same time. As the current digital image correlation setup is only suitable for static conditions, prestall flight conditions are maintained. Low-speed VIC is capable of extracting data from a vibrating wing, but only on a single snapshot basis.

5.4.7 Wing Deformation Results

Typical data results that can be obtained from the VIC system consist of the geometry of the wing surface in discrete X, Y, and Z coordinates (where the relevant coordinate system is a body axis, and thus the wing's angle of attack is not

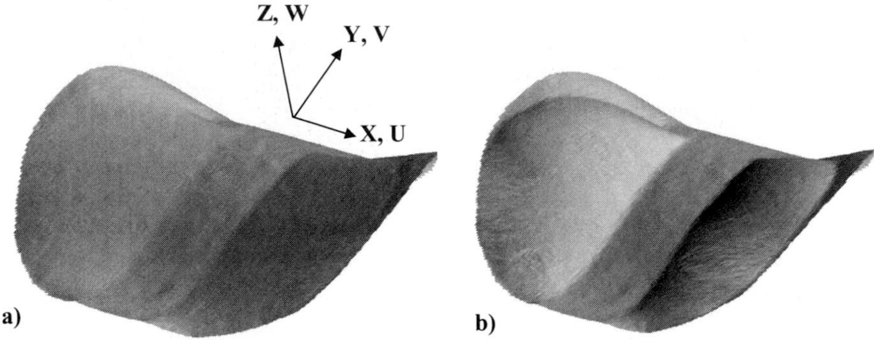

Fig. 5.36 Surfaces of a perimeter-reinforced wing: a) undeformed and b) loaded.

evident from the geometry), and the corresponding displacements along the wing (U, V, and W, using the aforementioned coordinate system). These displacements imply the use of two images: a reference wing configuration and a "deformed" wing configuration, just described. The VIC system calculates the wing geometry in both cases, and the difference between the two yields the flexible wing's displacement field. The acquired displacement field is composed of both the elastic deformation of the wing and rigid-body rotations/translations inherent within the wind-tunnel setup. As only the former is desired, care must be taken to filter out the latter [8]. A final post-processing option involves calculating the in-plane strains based upon the gradient of the displacement field.

Shape changes of a perimeter-reinforced wing can be seen in Fig. 5.36. As can be expected, wing deformation is an order of magnitude higher in the latex skin areas. The carbon-fiber skeleton is simply too stiff for the VIC system to read appreciable values of displacement. This would also imply an inability of the VIC to read the strain in the carbon-fiber skeleton. Sensitivity tests have shown that the maximum resolution of the system is approximately 0.01 mm in displacement and 500 microstrain.

Figure 5.37 shows the W-displacement contours of a batten-reinforced wing, found at 12-deg angle of attack, with a wind speed of 13 m/s. Several interesting conclusions can be drawn from this plot. The primary region of deformation is

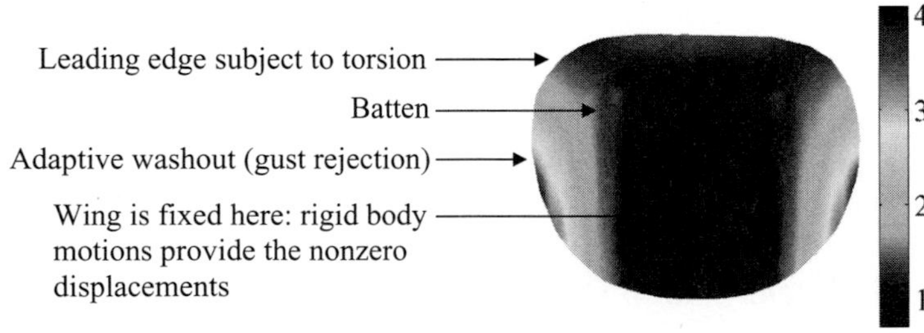

Fig. 5.37 W displacements (mm) of a batten-reinforced membrane wing at 12-deg angle of attack and 13-m/s wind speed.

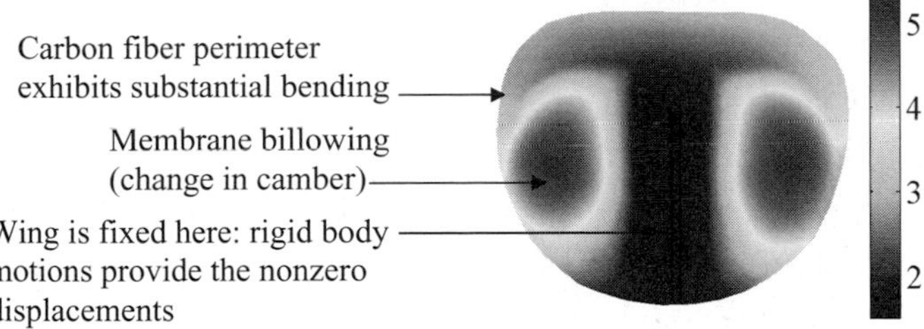

Fig. 5.38 W displacements (mm) of a perimeter-reinforced membrane wing at 22-deg angle of attack and 13-m/s wind speed.

found at the outward trailing edge of the wing, where the battens are forced to bend upwards as a result of wind loading. (This is the primary mechanism for adaptive washout). A second region of significance is the appreciable deformation of the corners of the leading edge. This implies that the pressure at this point is strong enough to subject the carbon-fiber leading edge to torsion. Also of note is that the displacement resolution of the VIC system is fine enough to pick out the slender battens that run through the membrane skinning of the wing. Lastly, that the lowest displacement value on the contour plot is nonzero (rather, 0.5 mm) means that the entire wing is subjected to a rigid-body translation/rotation that, as just mentioned, must be filtered out. This motion derives primarily from the small amount of play found in the model arm of the wind tunnel.

A similar plot can be composed for a perimeter-reinforced wing (Fig. 5.38), albeit with very dissimilar results. The largest difference between the two wings is found in the structural boundary conditions provided to the latex membrane skin. Where the batten-reinforced wing keeps the skin in a one-dimensional state of strain (perpendicular to the battens) and derives the majority of its flexibility from bending, the perimeter-reinforced wing is akin to stretching a membrane over a drumhead. This promotes a more two-dimensional strain state, and the primary mode of deformation is billowing (change in camber).

Figures 5.39 and 5.40 give the airfoil shapes for the batten-reinforced and perimeter-reinforced wings, respectively. The vertical dimension has been exaggerated, all at the same span section ($X = 50$ mm) for zero wind velocity (reference

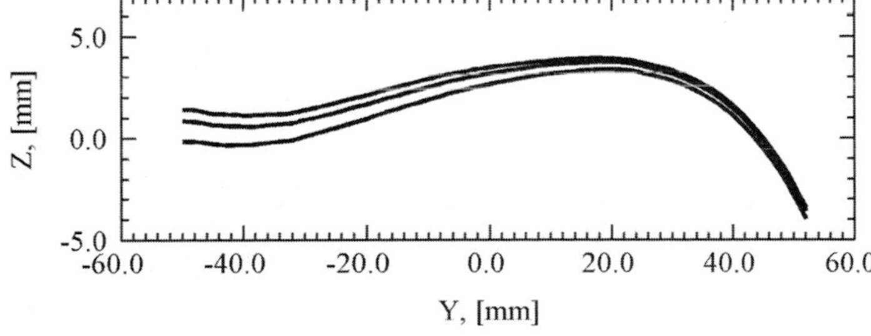

Fig. 5.39 Cross-sectional slices of a batten-reinforced wing at 12-deg angle of attack, for various wind speeds (0, 8, and 13 m/s).

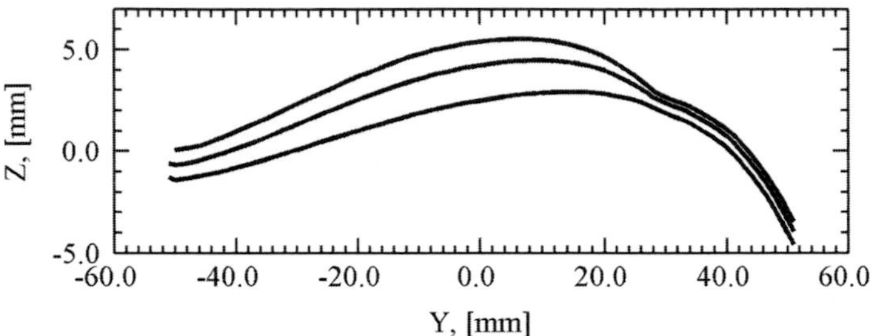

Fig. 5.40 Cross-sectional slices of a perimeter-reinforced wing at 12-deg angle of attack, for various wind speeds (0, 8, and 13 m/s).

configuration) and for two wind velocities (8 and 13 m/s). These plots were composed by slicing through VIC surface data. Each section has the leading edge on the right and the reflex trailing edge to the left. The change in airfoil shape with higher wind speeds is evident. For the batten-reinforced wing there is a notable rotation of each loaded section about the leading edge, which again implies adaptive washout. The airfoil shape (sectional shape) basically remains the same, with small changes in camber and camber location. Additionally, the rotation of the foil as a function of the freestream velocity increases with position along the span. This can be seen in Fig. 5.41, where the plot shows the angle of incidence along a line from the center of the wing (0 mm) to the wing tip. The washout is pronounced at the tip as can be seen by the diverging lines.

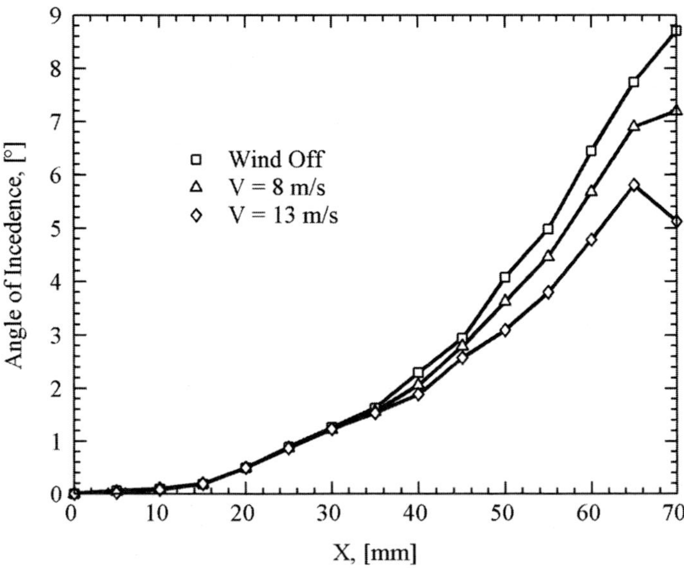

Fig. 5.41 Angle of incidence for the batten-reinforced wing. Data are collected using VIC at two different freestream velocities and at a constant AOA of 12 degs.

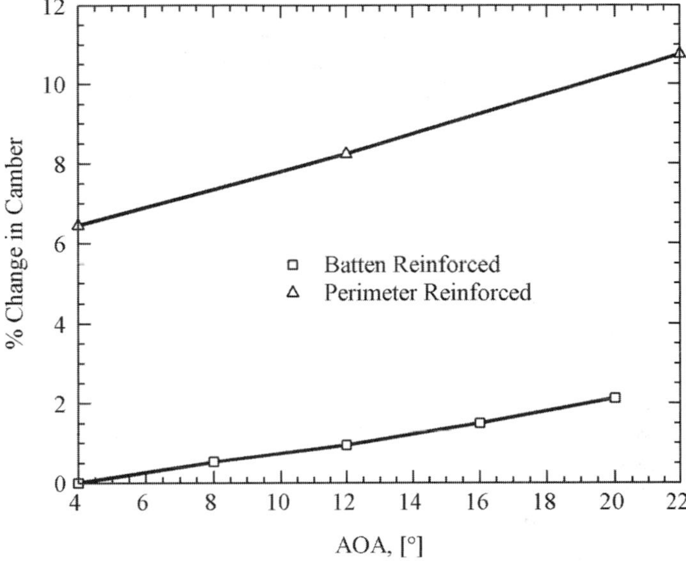

Fig. 5.42 Variation of the camber vs AOA for the batten-reinforced and perimeter-reinforced wings. The data, at the same wing section, are measured using the VIC. The freestream velocity was at the constant value of 13 m/s.

The perimeter-reinforced wing demonstrates negligible washout, but, on the other hand, demonstrates a substantial change in the shape of the wing section. The camber increases with freestream velocity as well as shifting aft with freestream velocity. The plots in Figs. 5.42 and 5.43 show the increase in camber as a function of angle of attack and the shift in camber as a function of angle of attack. The perimeter-reinforced wing is highly subject to wing-section distortion as compared to the batten-reinforced wing.

As just mentioned, VIC provides a postprocessing option of calculating the strains from a given displacement field. The system uses the small-strain approximation by calculating the gradient of the displacement field. Although there is no structural limit on a possible displacement configuration to ensure good accuracy via VIC (as long as the entire wing area of interest remains in the camera's view and focus range), the strain field must be small to ensure that VIC's calculations are valid. Turning one's attention to the accumulated strain in a flexible wing, noise is also an issue. As evidenced by Figs. 5.37–5.40, the displacement fields obtained from VIC have very low noise characteristics. But, as with any experimental data fed through a numerical system, the strain data can be very noisy. Resolution is a third factor: as stated, the lowest strain VIC appears capable of resolving is about 500 microstrain. Thus, the strain in the regions of carbon fiber registers only as noise, and low wind speeds and/or low angles of attack generally provide a strain field that falls beneath VIC's resolution.

Figure 5.44 gives the strain field (ε_{xx} is strain perpendicular to the flow, and ε_{xy} is shear strain) for a batten-reinforced wing at 12-deg angle of attack and 13-m/s

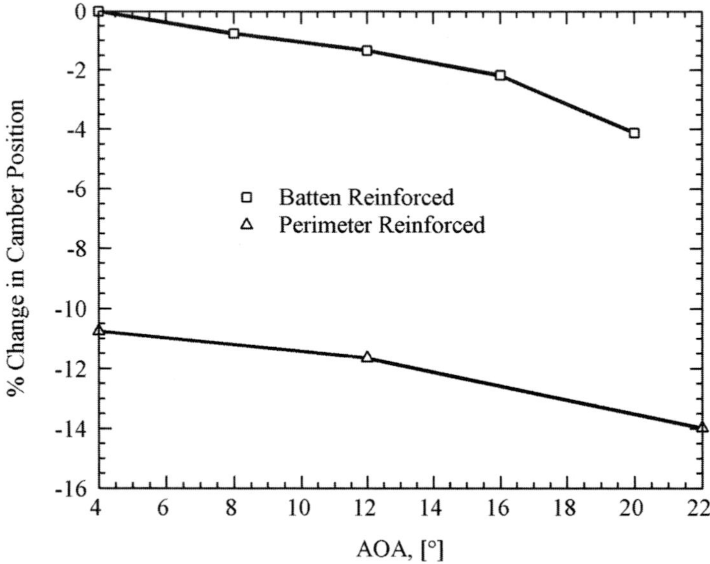

Fig. 5.43 Variation of the position of camber vs AOA for the batten-reinforced and perimeter-reinforced wings. The data, at the same wing section, are measured using the VIC. The freestream velocity was at the constant value of 13 m/s.

wind speed. A similar plot is given in Fig. 5.45, for a perimeter-reinforced wing at 22-deg angle of attack and 13-m/s wind speed. As expected, appreciable strain is found only in the membrane skin. Concerning Fig. 5.44, the strain in the battens should be near zero. Although the contours do slightly trace out the batten boundaries, the blurriness is caused by the moving average algorithm employed by VIC. Both normal strain fields (in the perimeter- and batten-reinforced wings) display areas of substantial negative strain, which would imply a compressive force. This idea is counterintuitive: the thin latex membrane used for MAV wings offers no

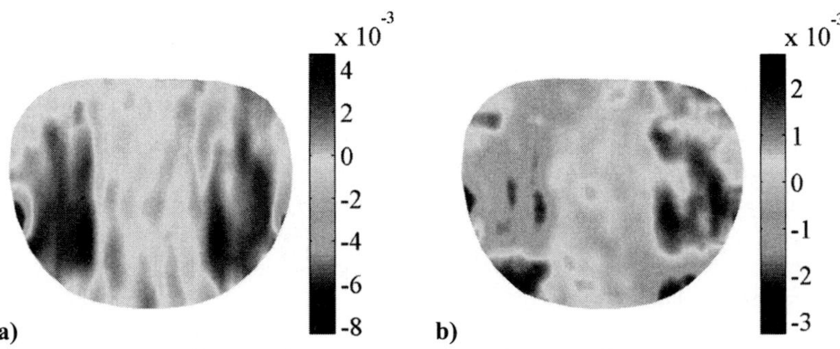

Fig. 5.44 Strain field of a batten-reinforced wing at 12-deg angle of attack and 13-m/s wind speed. a) strains perpendicular to the flow and b) shear strains.

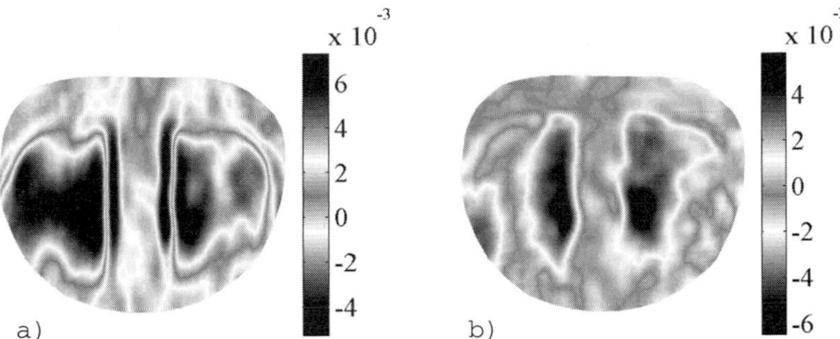

Fig. 5.45 Strain field of a perimeter-reinforced wing at 22-deg angle of attack and 13-m/s wind speed: a) strains perpendicular to the flow and b) shear strains.

resistance to compression; it will simply wrinkle. The key to the problem is latex prestrain. Any area where VIC calculates a negative strain is simply a relaxing of the prestrain imposed on the membrane before it is glued to the carbon-fiber skeleton. This prestrain can be measured if one uses the speckled sheet of unstretched latex as a reference image, rather than the undeformed MAV wing.

Other than the wind speed and model angle of attack, the elastic displacement field along the MAV wing is also presumed to be directly affected by the slipstream wash created by a propeller. Intuitively, the maximum displacement found within a flexible MAV wing will increase both with higher wind speeds and with higher angles of attack. In addition to this, the absence of any propeller will lead to a displacement field that is relatively symmetric (within the bounds of experimental uncertainties) about the centerline of the MAV. As can be seen in Fig. 5.46, the asymmetric effect of a spinning propeller mounted to the fuselage of a MAV plays a role in the displacement field of the elastic wing. Figure 5.46a is a perimeter-reinforced MAV with a freely spinning propeller. Figure 5.46b shows substantial asymmetric shape changes as the propeller is given power. As the angle of attack and/or wind speed of the model is increased, the propeller's effect on the wing displacement field diminishes.

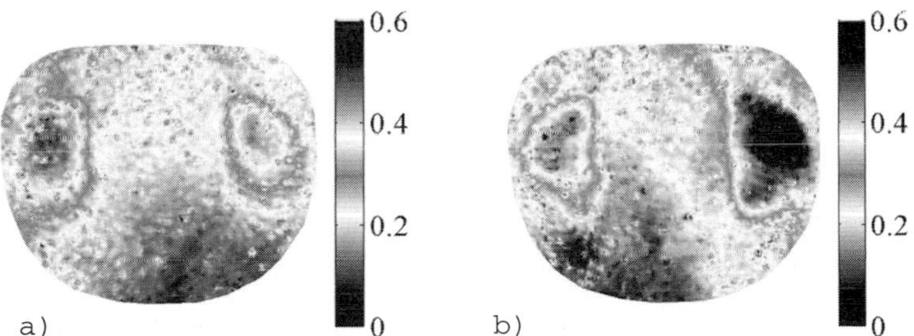

Fig. 5.46 W displacements (mm) of a perimeter-reinforced membrane wing at 4-deg angle of attack and 8-m/s wind speed, with a) a freely spinning propeller and b) a propeller powered by 7.4 V.

5.4.8 Flow Visualization

In addition to load testing and deformation studies in the wind tunnel, flow-visualization methods have been utilized to study the rigid and flexible wings. These studies have mainly been used in a qualitative sense in order to gain basic understanding of the flowfield including locating areas of recirculation and flow separation. Additionally, these results can confirm whether CFD results provide meaningful simulations.

MAVs operate at a very low speed and Reynolds number, which can be modeled as constant property, incompressible fluid without significant error. This property, in addition to their small size, allows MAVs to be tested full scale in small, low-speed wind tunnel (WT) facilities. Several low-speed flow diagnostic methods can be employed to determine velocity fields in such a condition, of which perhaps the most useful is PIV [20–23].

The PIV system operates by taking two consecutive photographs of an illuminated flowfield and then cross correlating the photos to create a two-dimensional velocity vector field. Particle image velocimetry enjoys many benefits over other flow diagnostic methods, although at the same time it suffers from many disadvantages. PIV is nonintrusive, but the process only senses the velocity of seed particles in the flow that might not follow the actual path of the fluid. PIV requires a simple setup of a special PIV camera, a laser system capable of discharging two consecutive bursts, and a synchronizer system. PIV is a full-field technique that instantly captures the two-dimensional velocities in the field of view of the camera. The time resolution of PIV, however, is poor compared to other methods because of technical limitations.

The University of Florida's Mechanical and Aerospace Engineering (MAE). Department owns a two-dimensional PIV system developed by TSI, Inc. The TSI system includes a dual-head 50-mJ ND-YAg laser, a 1600×1200 pixel gray-scale PIV camera and a synchronizer unit that synchronizes the camera exposures to the laser pulses. Both the synchronizer and the camera are connected to a dual-processor Pentium Xeon 2-Ghz system with 4 Gb of RAM. The PC runs Insight, the software required to access the synchronizer and camera as well as perform the vector validation. All parameters necessary for velocimetry are input to Insight and are passed on to the hardware.

The MAE department also operates a moderate-speed, low-turbulence subsonic recirculating wind tunnel (MSLTWT) with an optically windowed $60 \times 60 \times 240$ cm test section (TS). The TS has a pitching arm installed, which is connected to a servo actuator to control angle of attack of a model. The arm has a six-component sting balance, which has been calibrated to the small loads that MAVs create in flight. The wind tunnel also has a pitot static tube installed upstream of the test section and has a temperature probe inside the test section. A diagram of the experimental setup is depicted in Fig. 5.47.

Velocity fields of flow over MAV wings were gathered at various spanwise locations with the laser sheet oriented parallel to the incoming flow and the gravity vector. The models were created from a competition mold for a 15- cm reflexed wing with dihedral. A rigid wing, BR wing, and PR wing were tested at multiple angles of attack at several flight speeds. At each speed and angle of attack, the velocity field was found first over the suction side of the wing, and then the model was rotated rollwise 180 deg and then pitched into the same AOA, and the velocity

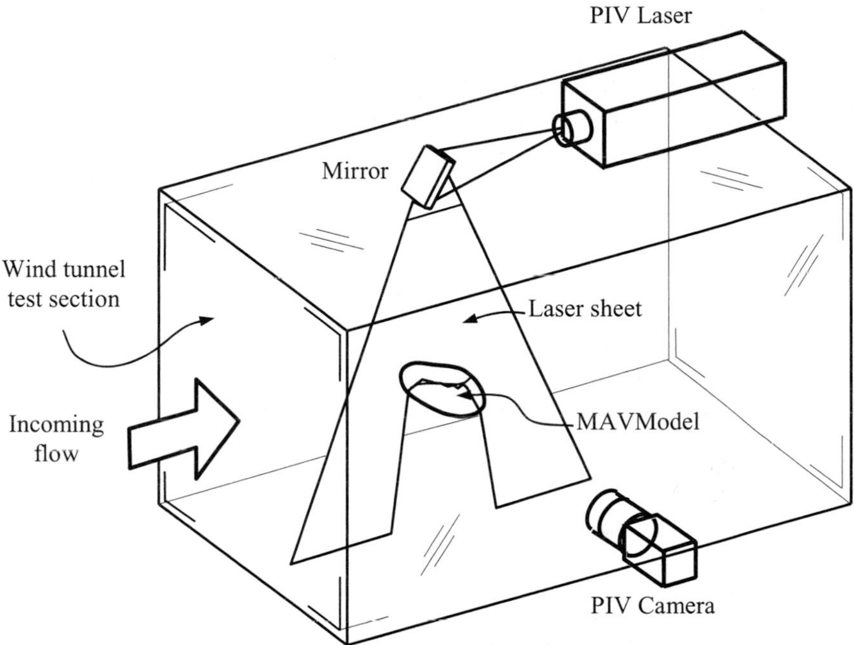

Fig. 5.47 Schematic of the PIV setup.

field was found for the pressure side of the wing. At speeds less than 20 m/s, it was determined that smoke from a theatrical haze machine would provide sufficient seeding. The haze machine creates fairly uniformly, sized particles by vaporizing propylene glycol over a filament and then blowing the seed into the WT downstream of the test section. The fan driving the WT mixes the seed to a homogenous distribution. The seed has good endurance in the WT, with one application lasting 10 min of continuous run time.

The laser heads were set up on a spanwise traverse attached to an optical table on top of the TS. A mirror was utilized to aim the laser sheet through the ceiling window down through the TS. The laser could be traversed so that the sheet could cross the span of the wing with much accuracy. The camera was oriented normal to the laser-sheet plane outside the TS and was positioned so as to optimize seed visibility while keeping the entire profile of the wing in the frame. The computer and synchronizer were placed so that the operator could avoid looking at the bright laser light and also so that the light from the computer screen could not be intercepted by the PIV camera.

Three wings were tested using PIV: rigid, batten-reinforced, and perimeter-reinforced. As would be expected, the flowfields over the centerline of the wings were very nearly identical. This is because the wing is very rigid in that area and the deflection caused by air loading is very small. Additionally, three-dimensional flow effects, if present, should be quite small along the centerline because it is an axis of flow symmetry. Poor velocity correlations were obtained around the spanwise location of the wing tips. This poor correlation is most likely caused

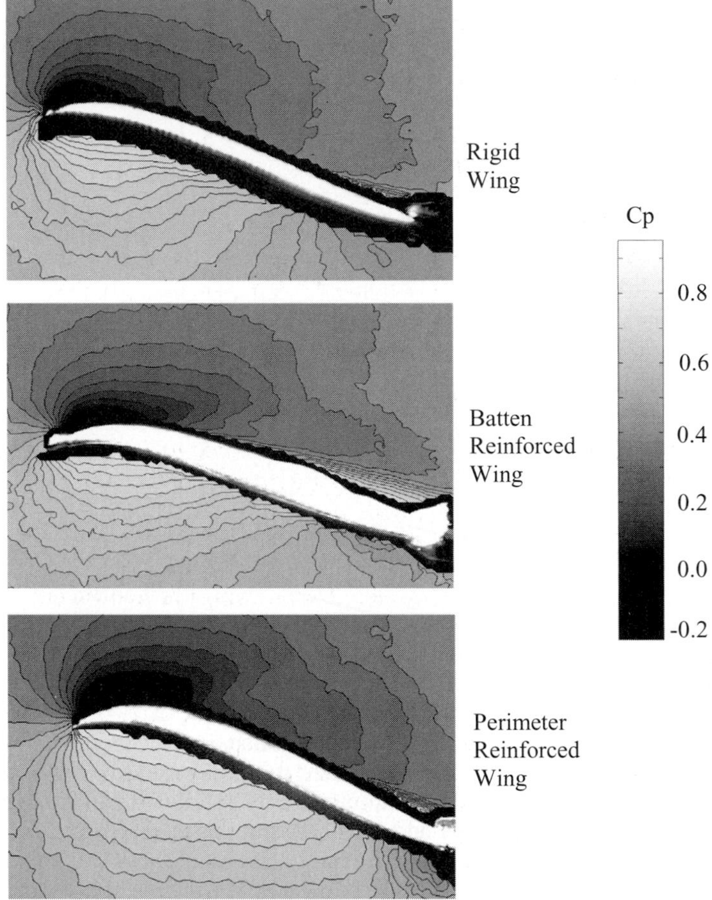

Fig. 5.48 Pressure coefficient plots C_p of flow around 6-in. MAV wing, with span location 40 mm from center, where $U_\infty = 10$ m/s and AOA = 15 deg.

by the large out-of-plane component of air velocity because of the strong wing-tip vortices of the low-aspect-ratio MAV wing. The highest-quality outboard PIV correlations were obtained around 40 mm away from the center span of the wing. In this region the effect of the third component of velocity was diminished while still showing the flow effect of the flexible wings. Typical PIV data were utilized to create Fig. 5.48, which displays the pressure coefficient C_p, neglecting the x component, distribution around MAV wings at a span location of 40 mm.

Each experiment captured 150 image pairs with a *dt* of 15 μs. The region of interrogation was sized at 32 × 32 pixels based upon the seeding capabilities of the haze machine. After the PIV images were captured, the vectors were calculated in Insight, and the resulting vector fields were ensemble averaged to create a mean vector field. The mean vector field of the suction and pressure side of the wing were combined in MATLAB®, and the velocity vector data were utilized to calculate a field of velocity magnitude, pressure coefficient C_p, vorticity, and global turbulence. Of most interest were plots of C_p, which indicated the normalized pressure distribution around the wings. Poor vector calculations were present around overexposed areas of the wings leading to the inability to resolve boundary-layer information.

For the images depicted in Fig. 5.48, the wings were placed in the wind tunnel at an angle of attack of 15 deg and a flight speed of 10 m/s. This flight condition was chosen because it is common for a typical 15-cm MAV to obtain this AOA during flight transients. This also represented a flight condition in which the PR wing operates more efficiently than either the BR or rigid wings. The C_p plots prepared indeed indicated the difference between the wings. The pressure coefficient was calculated using the formula neglecting the x velocity component:

$$C_p = 1 - \left(\frac{V\text{mag}}{V_\infty}\right)^2$$

The PR wing C_p plot indicates that the low-pressure region on the top of the wing extends over a longer region of the wing and has an overall lower pressure. The difference in the low-pressure region alone can account for the enhanced lifting properties of the PR wing. Furthermore, it seems that the high-pressure region on the pressure side of the PR wing extends further aft than it does on the other wings. Load tests on the wings indicated that the wings have different pitching-moment properties, and this can be visualized by the pressure distributions. The BR wing has what appears to be a C_p gradient on the aft portion of the suction side. This gradient is most likely caused by the turbulent boundary layer in that region decreasing the mean velocity field, thereby decreasing the calculated value of C_p.

Streamlines were calculated in MATLAB® based upon velocimetry data and superimposed over images of the seeded flowfield to illustrate the flow characteristics (Fig. 5.49). A difference in the streamline plots between the PR, BR, and rigid wing was noted, which might affect lift. On the pressure side of the aft portion of the BR and rigid wings, the streamlines can be observed to come closer together indicating a flow acceleration. This flow acceleration is likely caused by the pronounced reflex in that area of the wing. However, the PR wing does not exhibit so strong an acceleration effect indicating a lessened effect of reflex. This makes sense as the PR wing has no physical supports, which enforce the trailing-edge shape. This effect is important because it affects the pitching moment of the PR wing and perhaps increases its lift over the other types of wings. The streamlines that end in space are artifacts of the PIV system's inability to closely resolve the boundary along the wing.

PIV can add valuable insight into the pressure distribution and the flowfield for many applications. It is possible to obtain high resolution and high accuracy velocity field data for a number of applications. PIV has been applied to MAV research to help better understand the fluid dynamic causes of wing loading and the resulting causes of pitching moment.

5.5 Modeling of Flexible Wings

This section represents work to date on the development of models of flexible wings to simulate their behavior. Accurate solid and fluid models are required in

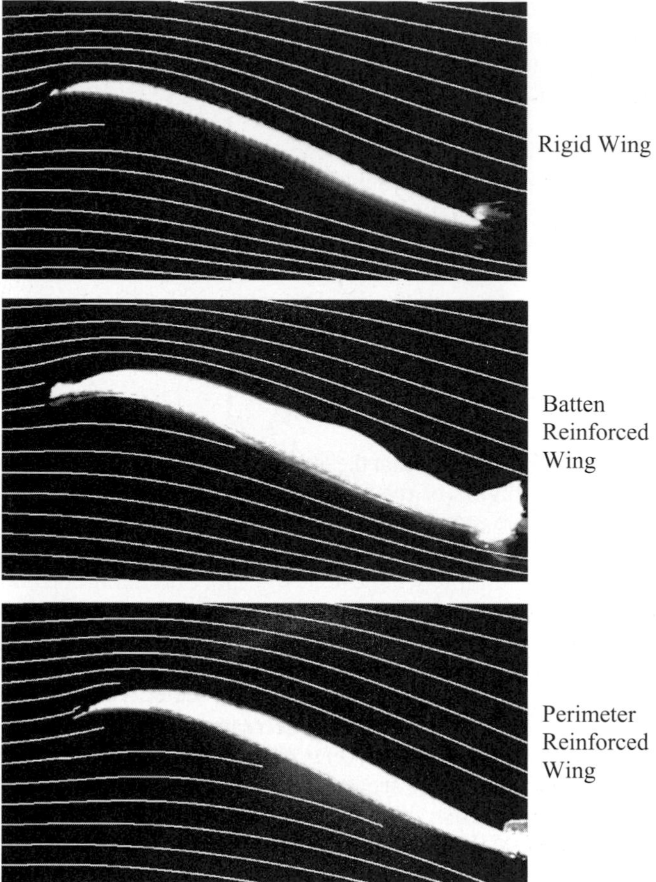

Fig. 5.49 Streamline plots superimposed over PIV flow image.

order to simulate fluid-structure interaction. Only initial attempts to interact the fluid and structural models are presented in this text.

5.5.1 Structural Modeling of Flexible Wings

An accurate numerical model (via finite elements) of the flexible MAV wing can provide vital aeroelastic information concerning the complex fluid-structure interaction over the wing. A static model can be easily verified using the visual image correlation techniques discussed earlier. A dynamic model can be verified using either high-speed correlation techniques, or ground vibration testing. These models can be combined and then coupled with an unsteady fluid solver for a complete numerical simulation. A static structural model for the batten-reinforced and perimeter-reinforced wing will be considered here.

Nonlinear structural modeling of flexible winged micro air vehicles has been studied by Lian et al. [24], using a slightly homogenized wing. (Regions of carbon

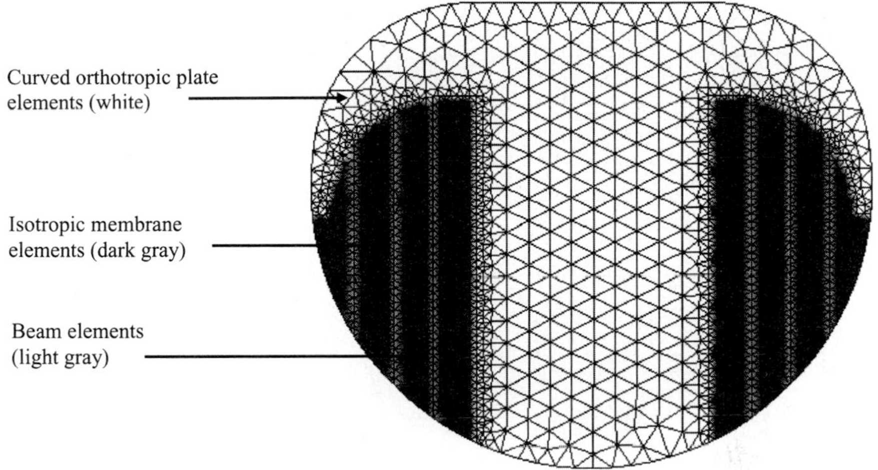

Fig. 5.50 Batten-reinforced wing discretization.

fiber are modeled as rigid, and battens are modeled as heavy membranes.) The composite nature of the wing will be fully realized here, albeit with a linear model. A linear membrane wing model is justified as follows. Using VIC to measure the deformation of a wing under steady level flight in pre stall conditions (wind tunnel testing), the average out-of-plane wing displacement is on the order of 5 mm. This value is very small compared to the length scale of the wing (150 mm). In-plane wing displacements are an order of magnitude smaller than out of plane (~0.5 mm) and can thus be labeled as minor second-order phenomena. Strain fields are also low (~0.004 or 0.4% strain). Small strain approximations are usually violated at about 10% strain. Potential pitfalls to a linear model (large deflections might introduce a geometric nonlinearity or hyperelastic membrane rubber effects) are considered minimal because of the relatively small deformation field.

The discretization of the wing can be seen in Fig. 5.50. Three basic elements were used to capture the physics of the wing's elasticity. Triangular curved plate elements can accurately model the bending and twisting modes of a carbon-fiber laminate. Orthotropic properties and different stacking sequences are easily included. The battens of the MAV (three are equally spaced on each side of the wing) are made of slender strips of flexible unidirectional carbon fiber and are modeled by a two-node beam element. Transverse shear deformation, torsion, and in-plane displacement are all neglected. The final, and most important, aspect of flexible-wing modeling, is the membrane model. Prior investigations have opted for a hyper-elastic Mooney–Rivlin model [24], popular because of its relative simplicity and good accuracy through large stretch ratios. Here we opt for a linear model where the out-of-plane displacements are governed by Poisson's equation. In-plane displacements are effectively ignored, and the tension in the latex is assumed constant (equal at all times to the pretension). Experimental VIC results indicate that these are all reasonable approximations.

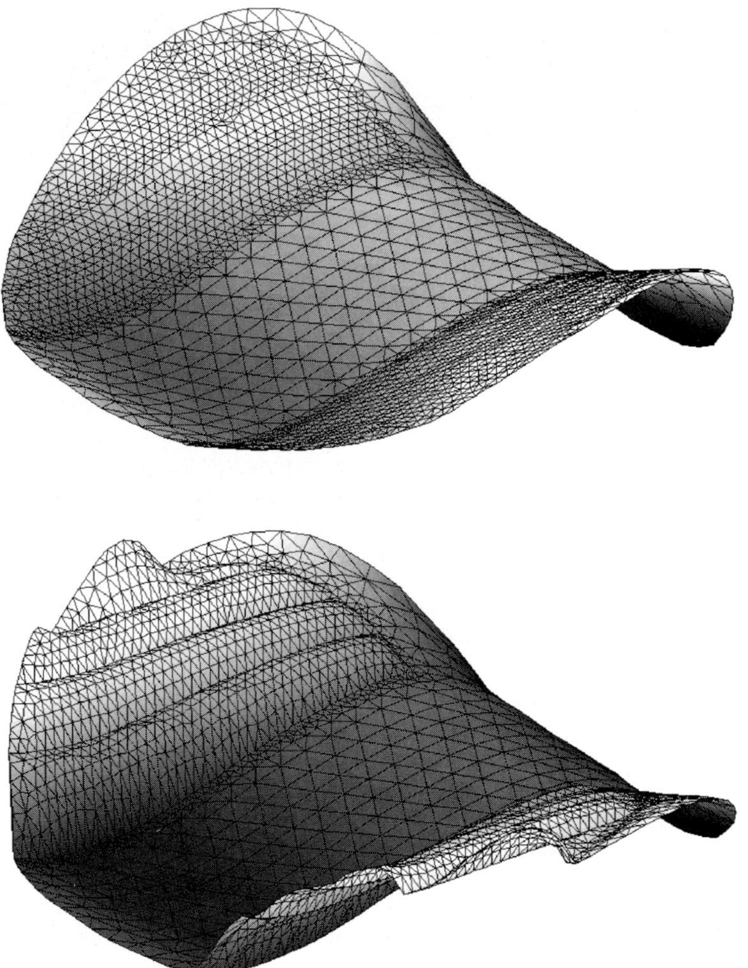

Fig. 5.51 Batten-reinforced wing before and after an applied unit pressure load.

As an initial inspection of how the model performs, the interaction of the three elements as a result of a uniform pressure load can be seen in Fig. 5.51. Similar to the experimental VIC results, substantial displacement occurs primarily in the membrane wing (rather than areas of carbon fiber), with adaptive washout indicated at the trailing edge. Also note the billowing of the membrane skin in between each batten. The salient features of the physical wing appear to be captured in the structural model. Although the loading is not representative of that which the wing experiences during flight, the interaction between the three different elements is by first approximation captured.

Visual image correlation presents an ideal method for full-field finite element model validation. Each element just described can be validated separately by performing a series of simple loading cases in the structures laboratory, using VIC

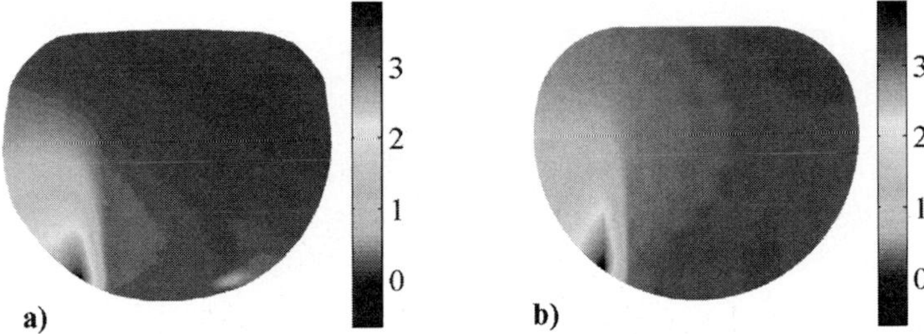

Fig. 5.52 Out-of-plane displacements (mm) of a batten-reinforced wing subjected to a 7-g dead weight at its outer left batten and clamped at the trailing edge: a) experimental and b) numerical.

to measure the corresponding displacement fields, and then replicating the setup with finite elements. Relevant material parameters (the flexural rigidity stiffness matrix of the composite carbon fiber, flexural stiffness of the battens, and the tension force resultant of the latex membrane) are tweaked until good agreement is reached between the experiment and the numerical model. Such a process results in a very accurate static MAV model. Figure 5.52 shows the results from such a test. A 7-g dead weight was hung from the outer left batten of a wing clamped at the trailing edge. The contour plots for experimental and numerical out-of-plane displacements throughout the wing match very well.

A structural model of the perimeter-reinforced wing was developed in the same manner as that for the batten-reinforced wing described earlier. It was validated by applying a transverse point load to the left membrane wing. This is a straightforward numerical simulation. From an experimental standpoint, this was accomplished by attaching the tip of a thin string to the membrane wing, off of which a series of small weights were hung. VIC was then implemented to measure the displacement field of the wing as a result of the dead weight. The results from one of these tests (in terms of out-of-plane displacement, or W) can be seen in Fig. 5.53, with suitable correspondence between numerical and experimental data.

5.5.2 Computational-Fluid-Dynamics Studies

To help understand the aerodynamics of our MAVs, CFD studies were performed [25]. CFD was particularly helpful in providing insight into the three dimensionality of the flow. Although the true nature of transition between laminar and turbulent flow is still unsolved, CFD can nevertheless provide trends that can be used to feed-back in our design process. In this section we will only briefly address the methodology and present in more detail results and key findings. In keeping with parallel tests performed in the wind tunnel,neither the fuselage nor the propeller were taken into account.

Three-dimensional, incompressible, viscous, Navier–Stokes equations written in curvilinear coordinates were solved for steady laminar and turbulent flow over a rigid wing of 5-inch. span. The scope of the CFD work included studies of various turbulence models, grid refinement, stability, and convergence. The wing geometry used for computations is a faithful CAD representation of the wing designs seen in

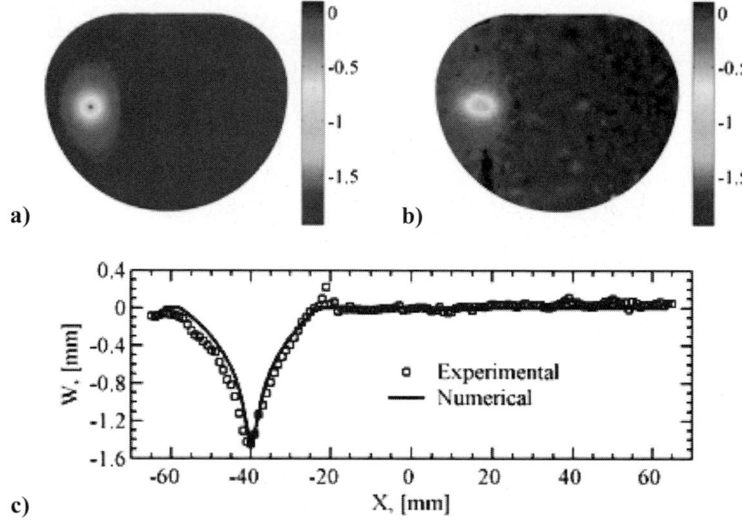

Fig. 5.53 Deformation caused by a 3-g weight hung from the center of the membrane wing: a) numerical out-of-plane displacements, b) experimental out-of-plane displacements, and c) displacement data at a spanwise section (bottom).

Figs. 5.1 and 5.2 (the same CAD file was used to mill the actual wing mold, thus minimizing error along this front), but the wing areas composed of latex membrane are simply modeled as rigid.

The computational domain can be seen in Fig. 5.54, with the MAV wing enclosed within. The dimensions of the computational domain are given in terms of the root chord length, c, and are all placed far enough from the MAV so as not to affect the

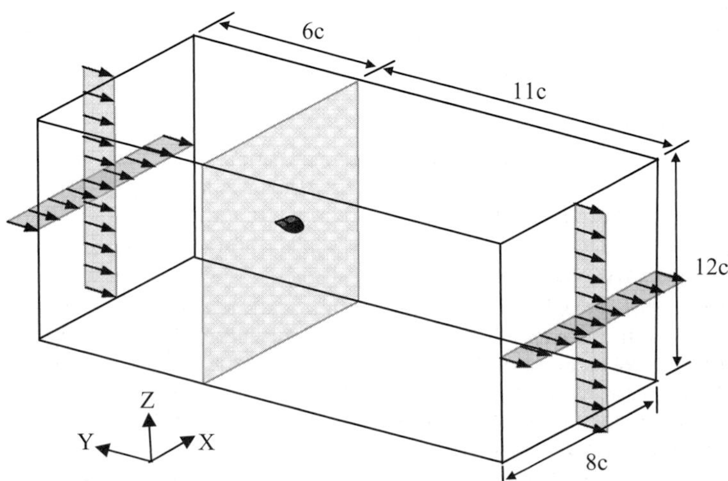

Fig. 5.54 Computational domain and flow conditions. All dimensions are in terms of *c*, the root-chord length.

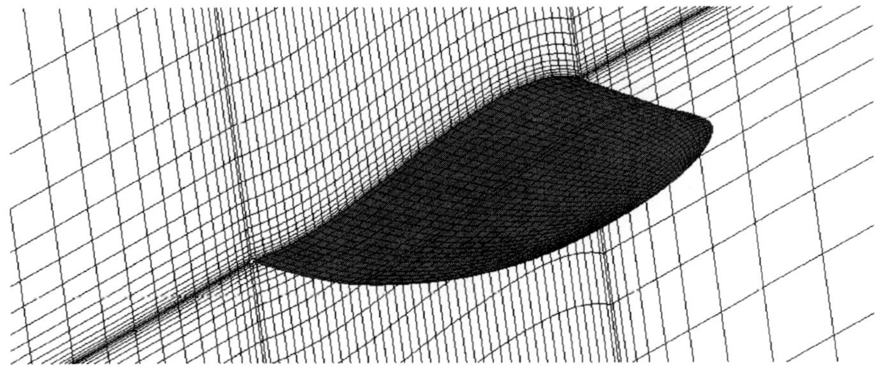

Fig. 5.55 Detail of the mesh used for the fluid solver near the wing surface.

flow. Inlet and outlet boundaries are clearly marked by the flow vectors. The shown configuration is that for 0-deg angle of attack simulations. For any other AOA, the lower and upper surfaces will also see a mass flux. (This is a more straightforward approach than remeshing for each new AOA.) The side surfaces are modeled as slip walls. (These walls impart no friction into the flow and thus no boundary layer forms.) Many of these boundary conditions represent a divergence from the actual flow conditions in the UF wind tunnel, but these differences are not thought to play a significant role. Trial runs were performed using a fluid density of 1.225 kg/m^3, constant viscosity of 1.79×10^{-5} kg s/m, and inlet velocities of 10 s and 13 m/s corresponding to Reynolds numbers of 7.1×10^4 and 92.5×10^3, respectively.

In the interest of simplicity at little expense to accuracy, symmetry is exploited by modeling only half of the computational domain. (The plane of symmetry is also modeled as a slip wall). Different views of the grid (approximately 210,000 nodes fill the entire domain, with 42×31 nodes on the wing surface) can be seen in Fig. 5.55. A well-developed pressure-based solver is employed, and further particulars concerning the computational modeling of flow over a rigid 5-inch micro air vehicle wing can be found in Stanford et al. [25].

In Fig. 5.56 the pressure contours on the wing surface are plotted for an angle of attack of 6 and 15 deg for freestream velocities of 10 and 13 m/s. The low-pressure area on the upper wing surface towards the wing tip, associated with the tip vortex core, can be visualized for both angles of attack. The large pressure drop on the wing tip at 15-deg angle of attack indicates a strong wing-tip vortex.

On the lower wing surface at 6-deg angle of attack, for both freestream velocities, a negative pressure gradient is present towards the leading edge indicating a possible separation area especially on the central wing panel. Towards the trailing edge the negative curvature of the airfoil accelerates the flow on the lower wing surface, indicated by the pressure drop as shown in the figure. On the upper wing surface the negative pressure gradient towards the trailing edge indicates yet another possible separation area.

At 15-deg angle of attack, the pressure gradient on the lower wing surface is smooth, accelerating the flow from the leading edge towards the trailing edge and preventing separation. Additionally, the rotational motion generated by a strong

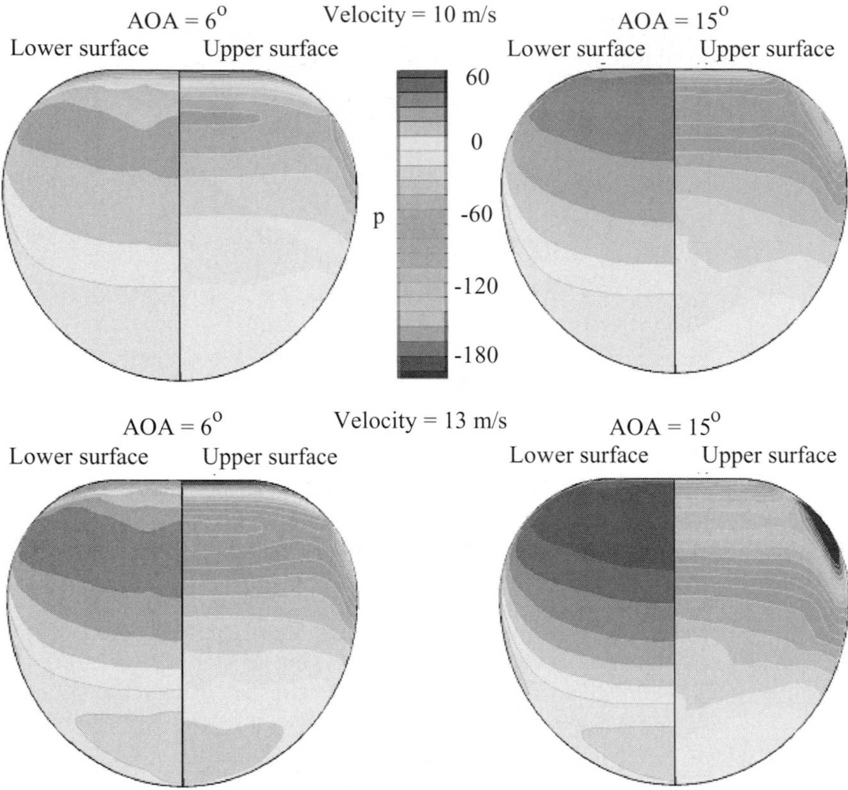

Fig. 5.56 Pressure contours on the upper and lower wing surfaces for angles of attack of 6 and 15 deg at inlet velocities of 10 and 13 m/s.

tip vortex accelerates the flow, and, consequently, the pressure drops towards the wing tips. The same behavior is observed for both Reynolds numbers, with larger pressure magnitudes corresponding to the higher Reynolds number.

On the upper wing surface and 15-deg angle of attack, the airfoil reflex helps create an adverse pressure gradient towards the trailing edge as can be observed. The increase in pressure is most visible near the root because the recurve of the wing is greater there as well as most distant from the wing-tip vortex. In Fig. 5.56 one can notice that for the lower Reynolds number the area of increased pressure towards the trailing edge is slightly larger than that corresponding to a higher Reynolds number, so that a larger recirculation bubble is expected.

As noticed from the pressure contours, the negative curvature toward the trailing edge generates an adverse pressure gradient that leads to flow recirculation. Also near the leading edge at small AOA, the pressure gradient indicates another recirculation area. This can be clearly observed in Fig. 5.57, where the streamlines are plotted for 3- and 15-deg AOA in a vertical plane at the middle of the central wing panel (root).

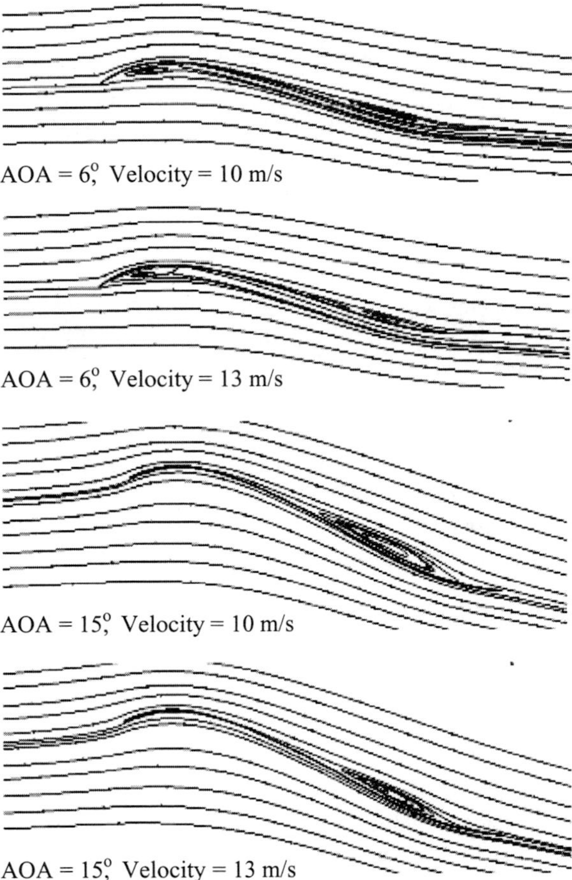

Fig. 5.57 Streamlines in a plane parallel to the middle of the wing (root) for angles of attack of 6 and 15 deg at inlet velocities of 10 and 13 m/s.

At smaller angles of attack (6 deg), the main separation bubble can be seen toward the leading edge on the lower wing surface. Another smaller recirculation area can be observed toward the trailing edge on the upper wing surface. Generally the separation area is contained within the central wing panel. As the Reynolds number increases, the recirculation zones tend to be smaller because the viscous effects are slightly reduced.

As the AOA increases, the leading-edge separation bubble on the lower wing surface disappears while the separation bubble on the upper wing surface near the trailing edge increases as a result of the adverse pressure gradient generated by the negative curvature of the airfoil. This is clearly seen where the streamlines are plotted for angles of attack of 9 and 15 deg. Again the recirculation zone is larger in the case of the smaller Reynolds-number flow.

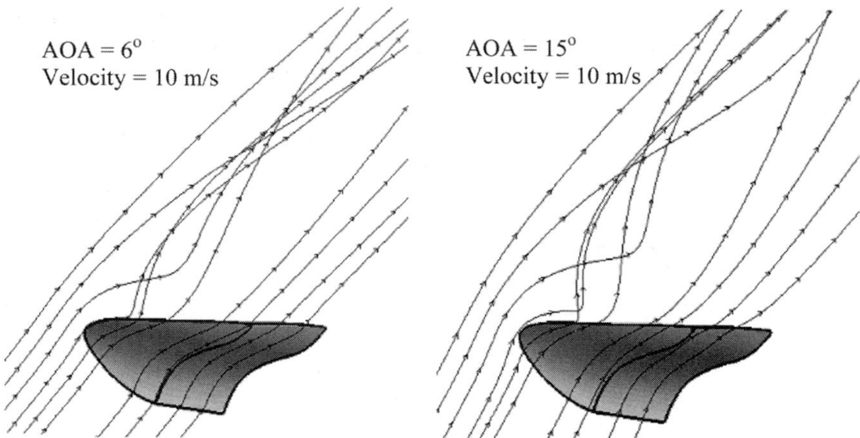

Fig. 5.58 Streamlines behind the wing showing the wing-tip vortex for angles of attack of 6 and 15 deg at inlet velocities of 10 m/s.

In Fig. 5.58 the streamlines behind the wing are plotted for angles of attack of 6 and 15 deg. The intensity of the wing-tip vortex is increasing with the increase of the AOA. Also the separation on the upper wing surface close to the root can be noticed from the streamlines, because they do not follow the wing profile.

Figure 5.59 shows more clearly the link between the pressure gradients and flow behavior. The pressure contours on the upper wing surface are plotted along with

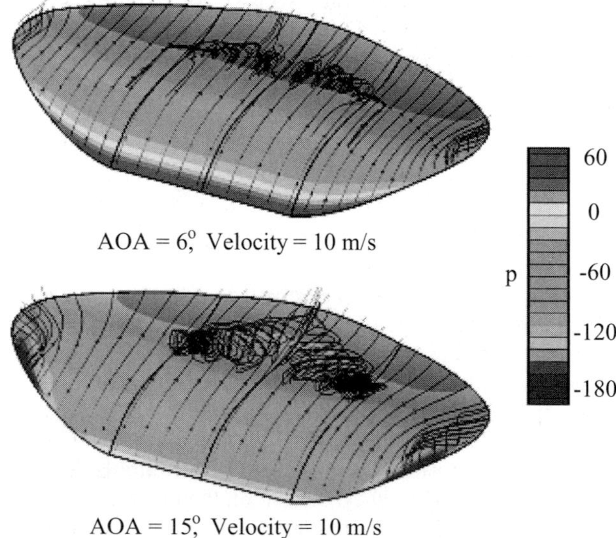

Fig. 5.59 Streamlines and pressure contours on the upper surface of the wing for angles of attack of 6 and 15 deg at inlet velocity of 10 m/s.

the streamlines for 6 deg and 15 deg angles of attack at 10 m/s. At 6-deg AOA and lower Reynolds number, the separation zone on the upper wing surface is small and occurs mostly on the central wing panel. As the Reynolds number increases, the recirculation zone on the wing upper surface is reduced as a result of diminished viscous effects. The wing-tip vortex initiation area can be observed in Fig. 5.59 as well as its affected zone that is confined near the wing tip.

At 15-deg AOA one can see the increase of the separation zone on the upper wing surface. For higher Reynolds number the recirculation bubble is smaller than in the case of lower-Reynolds-number flow. For both Reynolds numbers the wing-tip vortex is stronger as indicated by the low-pressure zone near the tip. The area affected by the vortical flow is increased to almost 50% of the half-wing span from the wing tip. The tip vortex initiation zone is closer to the root compared with the flow over the wing at 6-deg AOA. The stronger tip vortex drives the flow towards the wing tip as the vortex core low-pressure zone increases. This illustration clearly shows how three-dimensional the flow is on our low-aspect-ratio-wing. The recirculation volume is confined mainly to the central portion of the wing, and the wing flow remains attached.

5.5.3 Fluid/Structure Interaction Modeling

The deformations of the flexible wing are generated by aerodynamic loads. The wing shape is thus altered, and the aerodynamics change. To model this fluidstructure interaction, an iterative process can be adopted, where the loads are generated for the initial geometry, deformation of the wing is then modeled, and the CFD is updated for the new geometry. This process is repeated until the solution converges to a stable geometry. The process can be computationally expensive, and we are still developing the technique. As a first approximation, we have taken the aerodynamic loads from the CFD discussed earlier and imposed them on our structural model. For illustrative purposes we will demonstrate the result on a perimeter reinforced wing.

The out-of-plane displacements of a flexible membrane MAV wing are given in Fig. 5.60 (12-deg AOA at 13 m/s). The deformation fields (numerical and experimental) match very well. Most of the displacement measured in the regions of carbon fiber is simply noise, rather than genuine displacement. The numerical model provides smooth shapes, whereas noise is evident in the experimental data. Nevertheless, the agreement is quite good.

Further data, documenting the displacement field across a sweep of angles of attack, are given in Fig. 5.61. In terms of the deformed shape of the billowed membrane, the numerical models predict that the trailing edge of the wing actually deforms downward for low angles of attack. This is a result of the negative pressure resultant predicted by the CFD at the trailing edge of the wing. This is a phenomenon that is never witnessed through the experimental investigation (VIC) and suggests that the pressure distribution over the trailing edge as predicted by CFD is strongly dependent upon the fluid-structure interaction (which is ignored here). Further deviations in membrane shape occur within the maximum camber. The position of the maximum camber as calculated through the numerical model is well predicted at high angles of attack, but placed too far forward at lower angles.

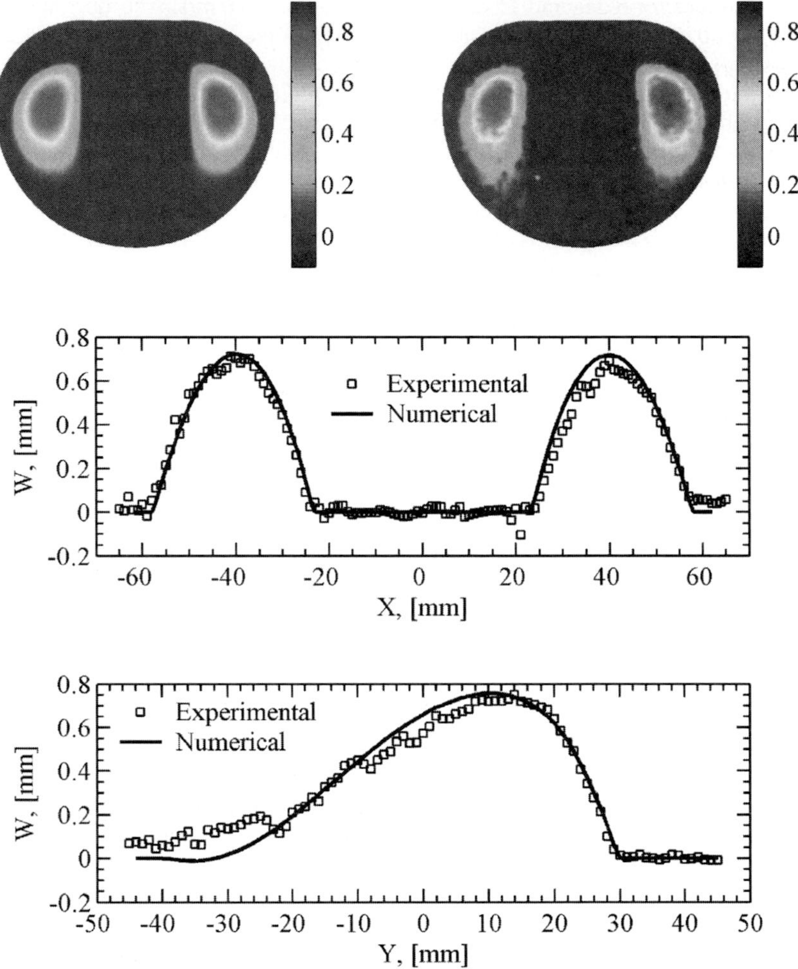

Fig. 5.60 Out-of-plane displacement contours and data comparing CFD/structural model with experimental VIC data for a PR wing at 12-deg AOA and 13 m/s. The displacement plots are taken at a spanwise section (middle plot) and a chordwise section (lower plot).

5.6 Flight Testing

The ultimate validation of a MAV design is its flight worthiness. The basic aerodynamic efficiency can be assessed in the wind tunnel; however, its ability to fly in real-world settings, such as gusty conditions, can only be assessed through flight tests. The research effort to develop the flexible-wing-based micro air vehicle relied extensively on flight tests to determine the relative merits of various wing designs. For the most part our early conclusions were dominated by direct pilot feedback, which typically included qualitative language. Remarkably rich

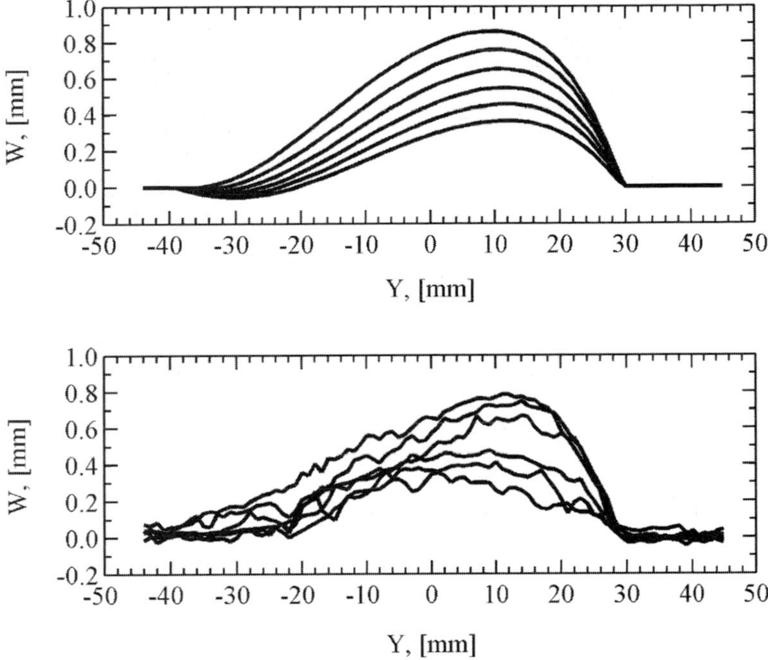

Fig. 5.61 Fluid-structure model vs experiment for the PR wing at various angles of attack taken at a chordwise section.

data on flight quality can be derived from such tests, and thus this is an irreplaceable component in the design process. This method is very effective for getting planes to fly and for trimming them out, however, it is not very scientific and lacks quantitative measure. Our challenge was to develop a quantitative method to assess flight quality. A method was developed to record pilot input on the two control axes, namely, roll and pitch. From experience, we noted that when experienced pilots fly RC planes over a set course the frequency of feedback is higher for poorly behaved aircraft. The converse is true for well-behaved or easy-to-fly aircraft. For unstable aircraft, high-frequency corrections to the flight path are required to maintain the course, making the pilot quite busy. By recording the input and plotting the results in the frequency domain, we could characterize how well an aircraft flew. High-frequency input was direct evidence of poor flight quality.

On a common fuselage, we tested four configurations to verify this method of assessing flight quality [6]. A batten-reinforced flexible wing was tested against a nominally rigid wing fabricated from a continuous sheet of carbon fiber with an undercambered configuration of the same shape as the flexible wing. The flexible wing was tested in three conditions; a forward c.g. on a calm day, a forward c.g. on a windy day, and an aft c.g. on a calm day. Typically an aircraft with a forward c.g. will be stable, and one with an aft c.g. will be erratic. The baseline, rigid wing was set to maximize its stability by setting the c.g. forward.

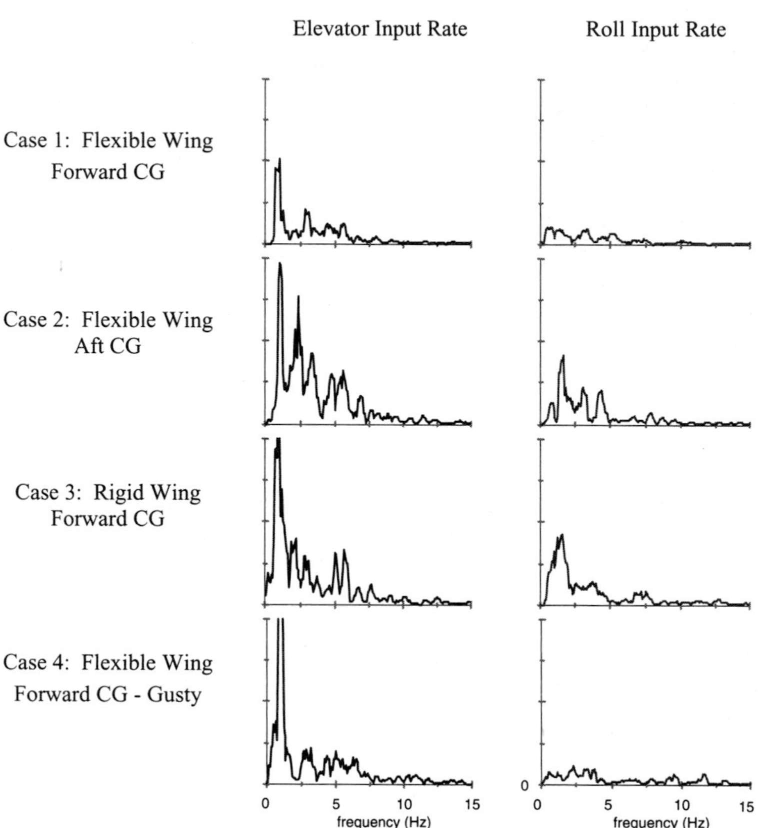

Fig. 5.62 Frequency of pilot input illustrates that high-frequency input, associated with unstable aircraft, is minimized by the flexible wing.

Figure 5.62 shows the spectral decomposition of the pilot input. It can be seen that the high-frequency pilot input was far less for the forward c.g. flexible wing on the calm and windy day than that for the other two configurations. This was the case for both the roll and pitch commands. This method confirms in a quantitative sense the qualities that our test pilots have been reporting about the flexible-wing concept. The flexible wing is easier to fly and more forgiving in windy conditions than nominally rigid versions.

Another measure of the flight worthiness of our vehicle designs is assessed every year at the International Micro Air Vehicle competition, which was discussed earlier in the book. The University of Florida won the competition from 1999–2006 with the flexible-wing design. From 1999–2004 UF utilized the batten-reinforced wing, and in 2005 UF unveiled the perimeter-reinforced wing in Seoul, Republic of Korea.

5.7 Conclusion

By only modifying the compliance of a MAV wing, one can achieve vastly different aerodynamic characteristics, as can be seen in the wind-tunnel data. Pilots also report significant differences as a function of wing compliance. Many of the flexible structures have resulted in vehicles that have improved characteristics over rigid-wing versions. These flight characteristics, although difficult to quantify in the traditional manner (wind-tunnel sting balance measurements), are documented in terms of deformation measurements and flow visualization as well as pilot feedback. The physics of how these wings perform has been documented, and modeling efforts are underway to predict how future versions of the compliant wing behave. It is our long-term goal to have a rigorous effort to optimize the compliance for a specified benefit, whether for improvement in control or for enhanced efficiency in specific flight regimes. By improvements in our models and validation through advanced experimental methods, the likelihood of success is strengthened. The true proof of these designs will be determined by the eventual use in the real-world environment.

Acknowledgments

The author would like to acknowledge the generous support of NASA Langley Research Center, especially Marty Waszak and David Raney, the Air Force Office of Scientific Research, the Air Force Research Lab at Eglin Air Force Base, and the National Science Foundation. Additionally, gratitude is extended to the many students who contributed to the research including Roberto Albertani, Bret Stanford, Daniel Claxton, Mike Sytsma, Scott Ettinger, Dragos Viieru, Mujahid Abdulrahim, and Kyu-Ho Lee. I would also like to thank colleagues Paul Hubner, David Jenkins, Wei Shyy, Rick Lind, Andy Kurdila, and Mike Nechyba. A loving thanks to my wife Diana Caro and my two daughters Marcela and Daniela.

References

[1]Albertani, R., Stanford, B., Hubner, J. P., and Ifju, P., "Characterization of Flexible Wing MAVs: Aeroelastic and Propulsion Effects on Flying Qualities," *AIAA*, 2005.

[2]Albertani, R., Stanford, B., Hubner, J. P., Lind, R., and Ifju, P., "Experimental Analysis of Deformation for Flexible-Wing Micro Air Vehicles," *AIAA Paper 2231, SDM*, Austin, TX, April 2005.

[3]Jenkins, D. A., Shyy, W., Sloxsan, J., Klevebring, F., and Nilsson, M., "Airfoil Performance at Low Reynolds Numbers for Micro Air Vehicle Applications," University of Bristol, Paper 29, *Thirteenth Bristol International RPV/UAV Conference*, March 30–April 1, 1998.

[4]Ifju, P. G., "Flexible-Wing-Based Micro Air Vehicles," *Compliant Structures in Nature and Engineering*, edited by C. H. Jenkins, Wit Press, 2005, Chapter 8, pp. 171–192.

[5]Ifju, P. G., Jenkins, D. A., Ettinger, S., Lian, Y., Shyy, W., and Waszak, M. R., "Flexible-Wing-Based Micro Air Vehicles," AIAA Paper 2002-0705, Jan. 2002.

[6]Jenkins, D. A., Ifju, P. G., Abdulrahim, M., and Olipra, S., "Assessment of the Controllability of Micro Air Vehicles," 16th Bristol International RPV/UAV Conference, Paper 30, *Micro Air Vehicle Conference,* April 2001.

[7]Albertani, R., Stanford, B., Hubner, J. P., and Ifju, P., "Aerodynamic Characterization and Deformation Measurements of a Flexible Wing Micro Air Vehicle," *SEM Annual Conference*, Society of Experimental Mechanics, Paper 302, June 2005.

[8]Ifju, P. G., Jenkins, D. A., Ettinger, S., Lian, Y., Shyy, W., and Waszak, M. R., "Flexible-Wing-Based Micro Air Vehicles," *CEAS Aerospace Aerodynamics Conference*, London, 2003.

[9]Ifju, P. G., Ettinger, S., Jenkins, D. A., and Martinez, L., "Composite Materials for Micro Air Vehicles" *SAMPE Journal*, Vol. 37, No. 4, July/Aug. 2001, pp. 7–12.

[10]Torres, G. E., "Aerodynamics of Low Aspect Ratio Wings at Low Reynolds Numbers with Applications to Micro Air Vehicle Design," Ph.D. Dissertation, Dep. of Aerospace and Mechanical Engineering, Univ. of Notre Dame, IN, April 2002.

[11]Albertani, R., "Experimental Aerodynamic and Elastic Deformation Characterization of Low Aspect Ratio Flexible Fixed Wings Applied to Micro Aerial Vehicles," Ph.D. Dissertation, Dep. of Mechanical and Aerospace Engineering, Univ. of Florida, Gainesville, Dec. 2005.

[12]Waszak, M. R., Jenkins, L. N., and Ifju, P. G., "Stability and Control Properties of an Aeroelastic Fixed Wing Micro Aerial Vehicle," AIAA Paper 2001-4005, Aug. 6–9, 2001.

[13]Ettinger, S. M., Nechyba, M. C., Ifju, P. G., and Waszak, M., "Vision-Guided Flight Stability and Control for Micro Air Vehicles," *Proceedings of the IEEE International Conference on Intelligent Robots and Systems*, Vol. 3, 2002, pp. 2134–2140.

[14]Kline, S. J., and McClintock, F. A., "Describing Uncertainties in Single Sample Experiments," *Mechanical Engineering*, Vol. 75, No. 1, Jan. 1953, pp. 3–8.

[15]Shufflebarger, C. C., "Tests of a Gust-Alleviating Wing in the Gust Tunnel," *NACA Technical Notes* No. 802, Langley Memorial Aeronautical Lab., Washington, DC, April 1941.

[16]Sutton, M. A., Cheng, M., Peters, W. H., Chao, Y. J., and McNeill, S. R., "Application of an Optimized Digital Correlation Method to Planar Deformation Analysis," *Image and Vision Computing*, Vol. 4, No. 3, 1986, pp. 143–151.

[17]Sutton, M. A., Turner, J. L., Bruck, H. A., and Chae, T. A., "Full Field Representation of the Discretely Sampled Surface Deformations for Displacement and Strain Analysis," *Experimental Mechanics*, Vol. 31, No. 2, June 1991, pp. 168–177.

[18]Lichtenberger, R., and Schreier, H., "Non-Contacting Measurement Technology for Component Safety Assessment," Limess Messtechnik u. Software GmbH, Pforzheim, Germany, 2002.

[19]Sutton, M. A., and McFadden, C., "Development of a Methodology for Non-Contacting Strain Measurements in Fluid Environments Using Computer Vision," *Optics and Lasers in Engineering*, Vol. 32, No. 4, Aug. 2000, pp. 367–377.

[20]Westerweel, J., "Fundamentals of Digital Particle Image Velocimetry," *Measurement Science Technology*, Vol. 8, Issue 12, 1997, pp. 1379–1392.

[21]Melling, A., "Tracer Particles and Seeding for Particle Image Velocimetry," *Measurement Science and Technology*, Vol. 8, Issue 12, 1997, pp. 1406–1416.

[22]Huang, H., Dabiri, D., and Gharib, M., "On Errors of Digital Particle Image Velocimetry," *Measurement Science Technology,* Vol. 8, Issue 12, 1997, pp. 1427–1440.

[23]Raffel, M., Willert, C. E., and Kompenhans, J., *Particle Image Velocimetry, A Practical Guide*, Springer-Verlag, Berlin, 1998, pp. 374–381.

[24]Lian, Y., Shyy, W., and Ifju, P., "Membrane Wing Model for Micro Air Vehicles," *AIAA Journal*, Vol. 41, No. 12, Dec. 2003, pp. 2492–2494.

[25]Stanford, B., Viieru, D., Albertani, R., Shyy, W., and Ifju, P., "A Numerical and Experimental Investigation of Flexible Micro Air Vehicle Wing Deformation," AIAA-2006-440, AIAA Aerospace Sciences Meeting and Exhibit, Reno, NV, Jan. 9–12, 2006.

6
Development of Micro Air Vehicles with in-Flight Adaptive Wing

Motoyuki Aki*
University of Arizona, Tucson, Arizona
Martin Waszak†
NASA Langley Research Center, Hampton, Virginia
and
Sergey Shkarayev‡
University of Arizona, Tucson, Arizona

Nomenclature

A	=	aspect ratio (b^2/S)
b	=	wing span
C_D	=	drag coefficient $= D/(0.5\rho V^2 S)$
C_L	=	lift coefficient $= L/(0.5\rho V^2 S)$
$C_{L\max}$	=	maximum lift coefficient
C_{L_α}	=	lift-curve slope, 1/deg
C_M	=	pitching-moment coefficient about quarter-chord point of the root chord,
	=	$M/(0.5\rho V^2 S\,\bar{c})$
c_0	=	root chord measured along the longitudinal axis of the wing
$\bar{c}$	=	mean aerodynamic chord measured along the longitudinal axis
D	=	drag force
d	=	position of the maximum reflex
h_i	=	height of the maximum inverse camber
h_z	=	camber height at z cross section in spanwise direction
h_0	=	camber height at the root of the wing
L	=	lift force
M	=	pitching moment about quarter-chord point of the root chord
m	=	mass
Re	=	mean-aerodynamic-chord Reynolds number
S	=	wing planform area
T_P	=	thrust force

Copyright © 2006 by Motoyki Aki, Martin Waszak, and Sergey Shkarayev.
Published by the American Institute of Aeronautics and Astronautics, Inc. with permission.

*Graduate Research Assistant.

†Senior Scientist.

‡Associate Professor.

T_S = throttle setting
t = thickness of the wing
V = freestream velocity
V_T = wind-tunnel speed
W = weight
W/S = wing loading
α = angle of attack, deg
ρ = air density

6.1 Introduction

The 1990s saw the beginning of micro air vehicles (MAVs) being designed and successfully flown. Because of their small size, MAVs are often considered for surveillance applications in urban environments. It follows that the aircraft should travel to the target as quickly as possible, loiter at slow speeds, and then return to the point of origin quickly and land safely.

One method of slowing an aircraft for loitering is to increase the angle of attack and increase the power to the motor to avoid losing altitude. Because MAVs are power limited, such a maneuver can result in a near-stall situation. Landing is another problem with MAVs. Noting the absence of landing gears on most MAVs, the landing impact is absorbed by the fuselage, motor with propeller or wing, damaging the structure. Therefore, the ability to decrease landing speed is important for effective flight performance. To achieve the desirable performance during the flight, MAV can be coordinated and controlled by deformation of wing structure. This control technique involves a rapid change of wing geometry: sweep angle of the leading edge, wing area, camber of the airfoil, wing span, etc. These changes of geometry redistribute an aerodynamic pressure spanwise and chordwise, creating a desirable combination of aerodynamic forces and moments. This innovative design will allow adjustments in the wing geometry, making it adaptive to specific flight conditions.

Many insects and birds have developed flight mechanisms that allow them to demonstrate sophisticated maneuvering capabilities at low speed, including vertical takeoff and landing, hovering, steady flight, as well as forward, backward, and sideward accelerated motions. They are capable of maintaining an effective airfoil over a significant range of wing shape changes. The size and performance requirements for MAVs have initiated an interest in adaptive-wing shaping. Because downsizing the craft makes the structure flexible, an adaptive-wing concept is naturally suited for these vehicles. Mimicking the best natural flyers can be a helpful experience, however, MAV design and optimization still builds upon the fundamentals of aircraft design.

Improvement of aeromechanical properties of MAVs is a challenging problem because of the imposed limitations on the size. These limitations force the designer to use a low-aspect-ratio wing. This not only maximizes the wing area and thus reduces the loading for a given weight (although it is a primary reason), but it also increases the chord Reynolds number *Re*. The lifting capacity poses the primary challenge for an MAV. The reduction in aspect ratio also changes the mean lift coefficient generated by a wing reducing it in a significant way. Even today with the abundance of knowledge regarding MAVs, more research is needed, particularly in the area of low-aspect-ratio wing optimization for the different phases of MAV flight (i.e., ingress/egress, loiter, and landing).

In the wind-tunnel testing conducted by Schmitz [1], model wings of rectangular planform, with aspect ratio of 5 were tested in the range from $Re = 2 \times 10^4$ to 1.7×10^5. A 12% maximum thickness wing, 2.9% thickness flat plate (0% camber), and cambered (5.8% camber) plate were investigated. It was discovered that at Reynolds numbers around 4.2×10^4, the maximum lift coefficient of a thin, cambered-plate wing ($C_{L\,\max} = 1.05$) was nearly twice that obtained for a traditional 12% thickness wing. Results of this study demonstrated the superiority of the thin, cambered wing for use in the low-Reynolds-number regime (low Reynolds number is assumed to be less than 2×10^5). These data, although limited to one cambered-plate wing, paved the way for further research into the benefits of cambered-plate wings at low Reynolds numbers. Laitone [2] studied rectangular wings with aspect ratios of 5 for $Re = 2 \times 10^4$ to 7×10^4. Results of these studies supported the conclusions [1] that a 5% circular arc, 1% thick wing produced the highest lift-to-drag ratio L/D and highest $C_{L\,\max}$.

In recent years, there has been a concerted effort to try to determine the optimum airfoil configuration and planform shape for these low-Reynolds-number flight vehicles. Selig et al. [3] developed an inverse computational approach for designing airfoils, providing the designer an effective tool for achieving desired aerodynamic performance. Using this approach, several series of low-Reynolds-number airfoils were created and validated through wind-tunnel testing.

Pelletier and Mueller [4] conducted a series of low-Reynolds-number tests of thin flat-plate and cambered-plate wings of low aspect ratios. Test models were of rectangular planform. Both flat plates as well as various 4%, circular-arc cambered-plate wings were tested, all of approximately 2% thickness. This work showed that the cambered-plate wings outperformed the flat plates. Torres and Mueller [5] studied the effect of wing planform and aspect ratio on the lift and drag of flat plates of thickness-to-chord ratio 1.96% and of aspect ratios 0.5, 1.0, and 2.0 that were tested for $Re = 7 \times 10^4$ to 1.4×10^5.

The wing loading of MAVs needs to be low in order to provide adequate controllability. The design constraint on the maximum dimension of MAVs can dictate the selection of a circular planform, giving a maximum wing area for a given maximum dimension. For the same reason, many MAVs utilize a flying-wing configuration. The biggest factor driving the design of the flying-wing aircraft is pitch stability and control. Large negative, nose-down pitching-moment coefficients were found for cambered, low-aspect-ratio wings at the low Reynolds number [4]. This moment can be compensated for by introducing a wing with inverse camber (reflex).

Effects of camber in the S-shape airfoil on the aerodynamic coefficients of circular-wing models were studied with the help of wind-tunnel testing [6]. Four micro-air-vehicle wind-tunnel models were built with 3, 6, 9, and 12% camber. A very significant feature of these designed wings was that the ratio of inverse camber to regular camber was kept constant and equal to $\frac{1}{3}$.

These models were tested in the low-speed wind tunnel at velocities corresponding to mean aerodynamic chord Reynolds numbers of 5×10^4, 7.5×10^4, and 1×10^5. Positive, nose-up pitching-moment coefficients were found with all cambers at the lowest Reynolds number. The test results showed that the 3% camber wing gives the best lift-to-drag ratio of the four cambers and theoretically would be the optimal choice for high-speed, efficient flight. It is theorized that the 6 and 9% camber wings will give the best low-speed performance because of their high lift-to-drag ratios and mild pitching moments near their stall angles of attack.

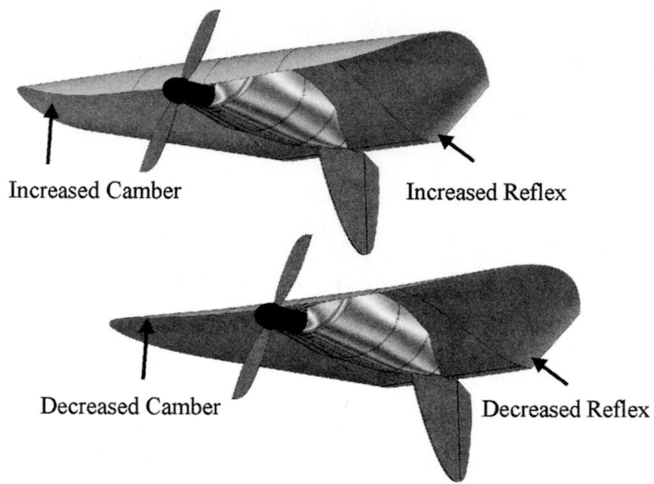

Fig. 6.1 Concept of the variable-camber wing.

If a thin and flexible MAV wing can be built, the camber of the wing can be modified in flight in a way that allows the lift coefficient to be varied without substantially changing the angle of attack. This type of wing could be classified as a thin, variable-camber wing. From the preceding discussion, fortunately, it also appears that a cambered-plate wing gives the best aerodynamic characteristics for low-Reynolds-number applications; the thin, cambered wing gives better performance characteristics as Reynolds number decreases. By applying wing-warping technology in the control of MAVs, particularly through utilization of smart materials, traditional control surfaces eventually can be eliminated, reducing weight and improving aerodynamics. In addition, the flight envelope of MAVs will be increased because of an in-flight, variable camber. Thin, cambered wings are, by their very nature, deformable and hence are ideally suited for MAV designs.

An adaptive wing is not a very new concept. The Wright brothers utilized control of a deformable wing by a symmetric warping of the trailing-edge geometry for pitch and an asymmetric warping for roll. The concept of in-flight adaptive wing or even morphing of the airplane found significant attention in applications to larger airplanes.

Past work at the University of Arizona has resulted in some interesting findings regarding the benefits of micro air vehicles with adaptive wings [7] and [8]. The concept of an adaptive wing micro air vehicle was introduced, and a flying test model was developed that was capable of camber change from 3 to 9%. Test flights were conducted with this vehicle demonstrating the flight speed variation by inducing a camber change.

In this chapter, a design case study is presented for the University of Arizona MAV with adaptive wing (UA-AW). The concept of the UA-AW with a variable camber wing is illustrated in Fig. 6.1. The wing design features a change in its cross-sectional geometry through a synchronous increase/decrease of camber and reflex.

The present study focuses on the following topics: 1) conceptual design of variable-camber wing MAVs; 2) experimental investigation of effects of Reynolds number, camber, and planform shape of the wing on the aerodynamics of MAVs; and 3) designing, construction, and testing of MAVs.

6.2 Conceptual Design

6.2.1 Mission Requirements and the Concept of Adaptive-Wing MAV

MAV design starts with the formulation and analysis of potential customer needs and priorities. For the current design, the purpose of the micro air vehicle is surveillance applications in an urban scenario, where the target of interest is 150–500 m from the launching site. A schematic of a mission profile typical of a short-distance surveillance mission is shown in Fig. 6.2. The vehicle should be capable of unassisted hand launch from a stationary position. After reaching an altitude of about 100 m, the vehicle flies to the target with a speed of 8–10 m/s. The MAV will loiter at a speed of 4–5 m/s for about 8 min while making a video of the target. Then it comes back at speed of 8–10 m/s. Thus, the desirable flight endurance is at least 10 min and a speed in the range of 4–10 m/s.

A low camber wing during ingress/egress phase of flight gives the MAV better rejection of wind gusts and reduces the time of ingress/egress. Reduced airspeeds, which are crucial during loiter, can be realized by an in-flight increase in wing camber. The MAV is a hand-launched vehicle without any special landing devices, and, therefore, a low approach speed is important for longer life of the vehicle. Decrease of the landing speed also can be achieved by increasing the camber to a higher value.

An onboard video camera and video transmitter are required for this project. With this equipment, images of the landscape, horizon, and target can be relayed to the control ground station in real time. Using the image, the pilot-operator will be able to control the vehicle via a radio transmitter/receiver, even when it is beyond line-of-sight operation. The video of the events at the target location will

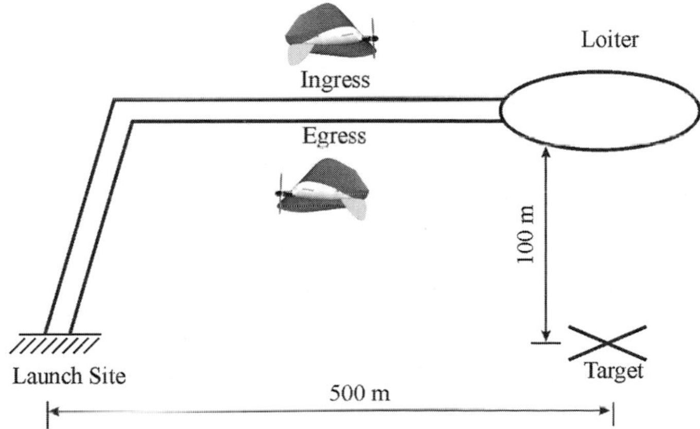

Fig. 6.2 Mission profile for an MAV.

be captured and stored in a computer. Thus, the standard payload for this MAV will be a color micro video camera with a video transmitter.

The mission requirements imposed on the design of MAVs typically consist of several distinct flight segments that generally conflict: the transit phases of flight require high speeds, but the loiter/surveillance and landing phases require lower flight velocities. Maximum efficiency must be sought in order to prolong battery life and aircraft endurance. To meet all of these requirements, the adaptive-wing MAV concept, specifically, a wing with a variable camber will be utilized in the present project.

6.2.2 Estimations of Wing Area and Aircraft Weight

Determination of the area of the wing and the mass of the vehicle is the next major decision. In 2000–2004, MAVs of 15–25.4 cm (6–10 in.) maximum dimension have been successfully flown in surveillance missions [9–12]. It was suggested that the adaptive-wing MAV's maximum size be constrained to 25.4 cm (10 in.), which was on the higher end of the best flyers at the time of the project. To maximize the wing area for a given maximum dimension of the aircraft, the planform shape of the MAV should be close to a circular disk, whose aspect ratio is $A = 4/\pi$. With the constraint on maximum dimension of 25.4 cm, the area of the circular wing is $S = 506\ \text{cm}^2$. For the same reason, the adaptive-wing MAV will utilize a flying-wing configuration.

By maximizing wing area for a given dimension, wing loading is minimized (assuming that the weight remains constant). Wing loading is defined as the weight of the aircraft divided by the wing planform area, or W/S. Wing loading has an effect on the flight speed of the aircraft, as well as on the controllability of the craft. A higher value for wing loading corresponds to a higher flight speed and makes the aircraft react more abruptly to control inputs from the pilot. Because these tiny flight vehicles are inherently difficult to control as a result of their small size, having too high a flight speed can make the aircraft virtually impossible to control without the implementation of some sort of an autopilot and/or stability augmentation system. Therein lies the principle design problem with MAVs—the aircraft must be light enough to keep the wing loading below a reasonable value.

Successful micro air vehicles developed in 2000–2004 [9–12] for surveillance missions had a wing loading in the range 30–40 N/m^2. In the present project, a value of wing loading of 30 N/m^2 was chosen in the conceptual design. Then, with a wing area of 506 cm^2, the weight of a MAV should be approximately $30 \times 0.0506 = 1.52$ N. (The mass is 155 g.)

6.2.3 Selection of the Airfoil, Planform, and Wing Surface Geometry

For a 10-in. MAV flying at a speed of 4–10 m/s, the root-chord Reynolds number is in the range 5×10^4–2×10^5. The selection of an optimum wing configuration for these low-Reynolds-number vehicles has proved challenging, particularly in the area of airfoil optimization for the different phases of MAV flight (i.e., ingress/egress, loiter, and landing). The main performance characteristics influenced by this selection are maximum lift coefficient $C_{L\,\max}$; lift-to-drag ratios C_L/C_D and $C_L^{3/2}/C_D$; and quarter-chord pitching-moment coefficient C_M.

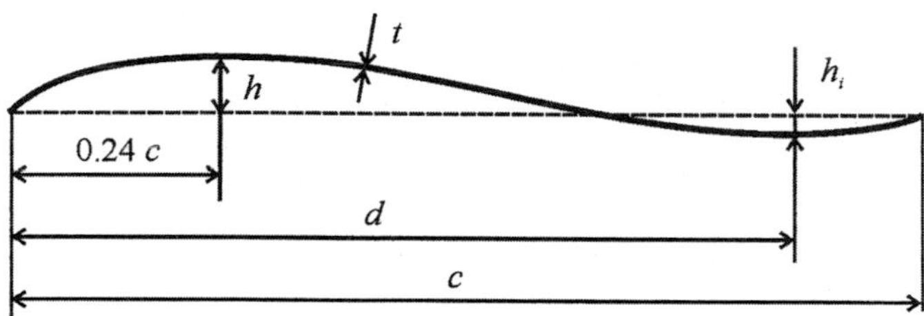

Fig. 6.3 S5010-TOP24C-REF airfoil.

For a flying-wing aircraft, particularly for an MAV, it is important that the wing does not have too much negative, or nose-down, pitching moment. A negative pitching moment, although necessary for pitch stability, tends to rotate the nose of the aircraft downward. Because the moment arm from the elevator control surface to the center of gravity of the MAV is relatively short, the elevator cannot adjust for too much of a nose-down tendency unless excessive, drag-producing elevator deflection is used. To avoid this situation, it is desirable to have a small amount of reflex in the trailing edge of the wing to help reduce the nose-down pitching moment.

The airfoil Selig 5010 (S5010) flying-wing airfoil [13] of 10% thickness has a successful history of being used on radio-controlled flying wings. Because the desired airfoil is a thin cambered plate, only the top surface of the S5010 was used as a basis for the design [7]. Utilizing the airfoil analysis program XFOIL [14], the aerodynamic characteristics of the top surface of the airfoil S5010 were analyzed. It was found that the lift and drag coefficients of the top surface were quite good, but the chordwise position of maximum camber still produced too strong of a negative (nose-down) pitching moment. To compensate for this, the airfoil was modified so that the chordwise position of maximum camber was moved to 24% of the chord instead of the 29% found on the standard S5010 top surface. XFOIL was again used to analyze the modified airfoil, and it was found that the lift and drag characteristics of the new airfoil were virtually unchanged, but the negative pitching moment, although still present, was substantially reduced. More reflex was added to the airfoil after numerous test flights showed that MAVs utilizing the previous airfoil needed excessive elevator deflections to maintain steady, level flight, resulting in decreased flight times. This new airfoil was designated as S5010-TOP24C-REF [8].

Finally, based upon the S5010-TOP24C-REF airfoil, four airfoils have been designed [6] with the maximum camber height-to-chord ratio $h/c = 3, 6, 9$, and 12%. The intermediate 6% camber is shown in Fig. 6.3. The relative thickness of the airfoil is $t/c \approx 0.25\%$. Its edges are rounded. A very significant feature of these airfoils is that the ratio of inverse camber to a maximum camber remains constant, $h_i/h \approx \frac{1}{3}$.

Wings designed with the same airfoil section can come in many shapes and sizes. Circular and Zimmerman wings were designed in the present study, both based upon the S5010-TOP24C-REF airfoil.

As mentioned earlier, the circular planform provides a maximum wing area and, therefore, a minimum wing loading, for a given size of the wing. The circular planform wing with diameter (wing span) b and root chord c_0 can be seen in

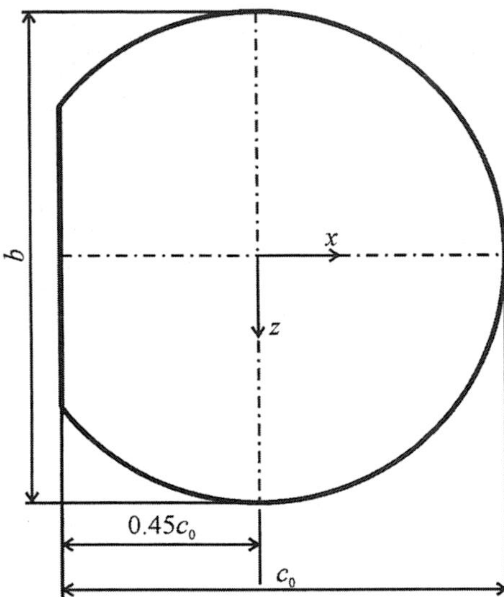

Fig. 6.4 Circular wing planform shape.

Fig. 6.4. A squared-off leading edge was introduced in order to provide a space needed for the propeller.

Based on the selected airfoil and planform, the wing top surface was designed as illustrated in Fig. 6.5. First, the S5010-TOP24C-REF airfoil (Fig. 6.3) with the root chord c_0 and maximum camber h_0 was placed in the vertical plane Oxy with the chord line aligned with the x axis. The airfoil was swept spanwise in the z direction to form a constant-chord rectangular wing. Then, the circular planform was projected onto the wing to form the wing surface shown in Fig. 6.5a. Because the wing has a constant thickness, the lower wing surface can be created by copying the upper surface in parallel.

The presented method for the wing surface design is easy to manufacture. The resulting wing features a variable chord of length c_z, a relative camber h_z/c_z decreasing to the wing tip, and a twist angle φ_z (see Fig. 6.5). The twist angle is increasing from $\varphi_z = 0$ deg at $2z/b = 0$ to $\varphi_z = 1$ deg at $2z/b = 0.6$ and then to $\varphi_z = 5$ deg at $2z/b = 0.8$. Aforementioned spanwise variations in the wing airfoil sections increase the effective angle of attack to the wing tip, and it is called a wash-in. On the other hand, the effect of wash-in is decreased towards the wing tip because the wing's area decreases. Although the presented method of construction of the wing is easy to implement, having a wash-in can cause tip sections to stall before inboard sections, causing the plane to roll fast in a stall. This drawback of the wing construction method was corrected in the later MAVs, and the modified method is described next for Zimmerman planform MAV.

The second planfrom selected for the present project was the Zimmerman planform shape. The Zimmerman planform (Fig. 6.6) is formed by two half-ellipses joined together along the maximum wing-span line passing through the $0.24c_0$ point (maximum camber point of the S5010-TOP24C-REF airfoil).

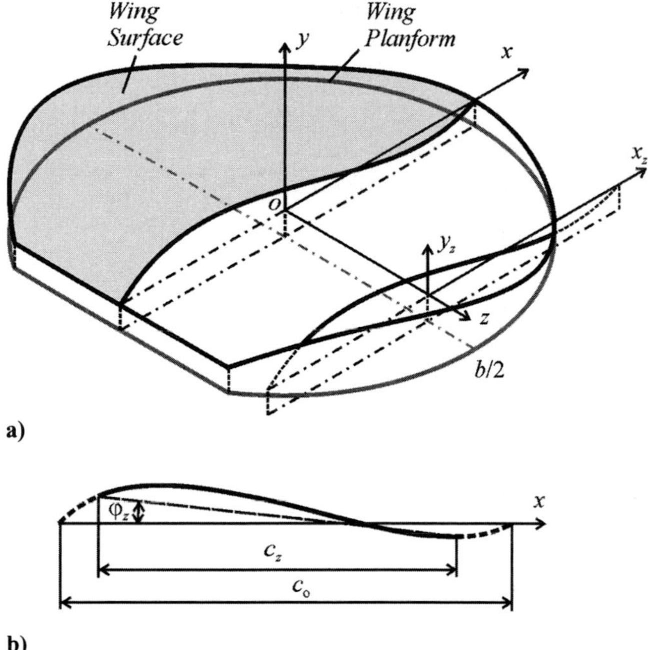

Fig. 6.5 a) Cambered circular wing; b) Representative cross section.

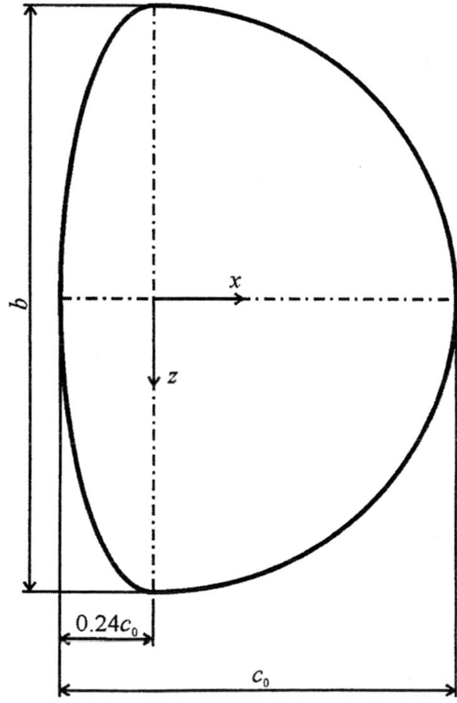

Fig. 6.6 Zimmerman wing planform shape.

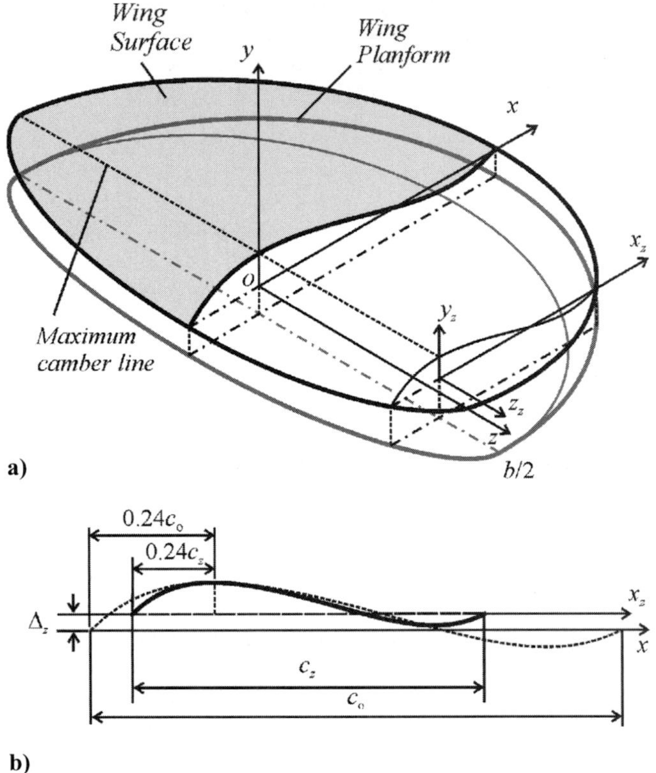

Fig. 6.7 a) Cambered Zimmerman wing; b) Representative cross sections.

The S5010-TOP24C-REF airfoil was utilized in designing the Zimmerman planform wing shown in Fig. 6.7a. There were two constraints imposed on the wing geometry. First, the wing surface was formed with a constant relative camber, $h_z/c_z = h_0/c_0 = \text{const}$. The second constraint was to have maximum camber points on the line spanwise, as shown in Fig. 6.7. For the arbitrary cross section z, the chord length is $c_z = c_0\sqrt{1-(2z/b)^2}$. To keep the relative camber constant, the chord line at the z cross section is shifted with respect to the root-chord line by $\Delta_z = h_0 - h_z = h_0[1-\sqrt{1-(2z/b)^2}]$, as depicted in Fig. 6.7b.

SolidWorks™ software was utilized for the construction of the wing geometry. First, 24 cross sections were drawn as just described and were placed uniformly over the half-span of the wing. The leading edge, trailing edge, maximum, and minimum points for each cross section were utilized in forming guide curves. To ensure the wing's geometry, guide curves along with 24 cross sections were used to develop the half-wing surface with the help of the Loft feature of SolidWorks™.

6.2.4 MAV's Propulsion

For the period 1994–2001, electric propulsion was not used for MAV applications because of energy storage problems. Since 2001, lithium-polymer batteries,

Table 6.1 Propulsion characteristics

Motor	RE-10[15]	Feigao[16]
Model	118392	1208436
Type	Metal brushes	Brushless
Size, cm	1×3	1.2×3
Mass, g	10	17
Speed, rpm/V	4410	4100
Current, A	0.7	4
Battery	145 mAh (3 cells)	640 mAh (2 cells)
Receiver	RX72 HYBRID	PENTA
Speed controller	RX72 HYBRID	ADVANCE 04-3P

especially those mass produced for cellular phones, have become an enabling technology for MAVs. [10–12] Lithium-polymer batteries are rechargeable and feature high current and high capacity. Their excellent power density (the capacity per unit weight) provides a maximum benefit in designing MAVs. In addition to powering the motor to drive a propeller, the batteries can also power all onboard electronics. Another advantage of the selection of electric propulsion is that electric motors are relatively quiet and easy to operate. When the pilot adjusts the throttle stick, the appropriate control signals are generated by a radio-control (RC) transmitter and sent to a speed controller via a receiver, both installed onboard the MAV. This allows controlling the speed of the motor in the throttle stick setting range of $T_S = 0$–100%.

Two electric motors were evaluated, tested, and used: Maxon RE-10 [15] and Feigao [16]. The Maxon RE-10 is a coreless dc motor and it has a small size and weight. The direct-drive propeller used with this motor is Union 80 with an 80-mm diam and 22-mm pitch. The Feigao has about the same dimensions as the RE-10 motor, yet it is 7 g heavier. The Feigao is a high-powered brushless motor used with a direct-drive propeller, EP-3020 with 76-mm diam and 50-mm pitch. Higher power and greater power efficiency are the advantages of a brushless motor. On the other hand, the motor has high current consumption and requires a certain type of speed controller unlike brushed motors.

The RE-10 motor was powered by three Kokam 145-mAh batteries wired in series. The batteries are connected to the motor via an FM receiver, RX72 HYBRID, with a built-in speed controller [17]. The Feigao motor was driven by two Kokam 640-mAh batteries wired in series and connected via an ADVANCE 04-3P speed controller [18] and PENTA receiver [19]. Characteristics of the two propulsion systems are summarized in Table 6.1.

6.3 Wind-Tunnel Testing

6.3.1 Wind-Tunnel Facility

The facility used to perform this series of tests is known as the low-speed wind tunnel and is located in the Department of Aerospace and Mechanical Engineering

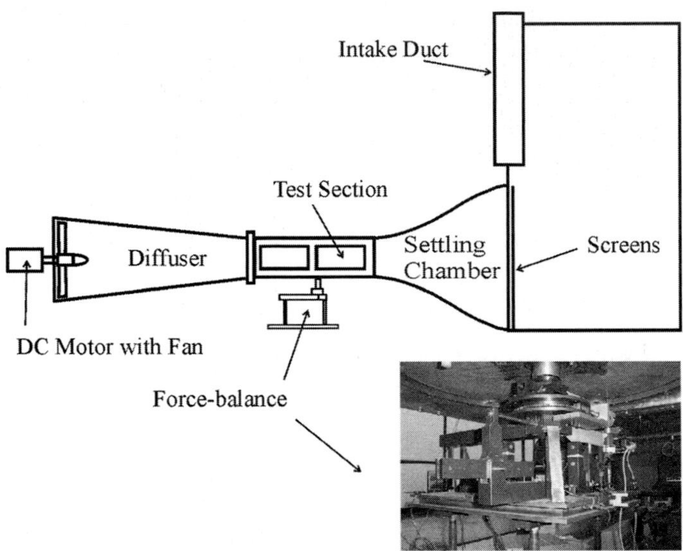

Fig. 6.8 Schematics of the low-speed wind tunnel and force-balance system.

at the University of Arizona. The tunnel is a suction-based, nonreturn tunnel with a test section of 4 × 3 ft (1.2 × 0.9 m) and is capable of speeds from 2 to 50 m/s. The wind tunnel is rated at a 0.3% turbulence level in the axial direction for the freestream velocities used in the present study. A schematic of the tunnel is seen in Fig. 6.8.

The force-balance system contains six precision strain gauges for measuring lift, drag, and side forces, and pitching, rolling, and yawing moments. For this sequence of tests, no side forces were analyzed because the models were symmetrical and rolling and/or yawing moments were not sought.

The data-acquisition system (DAQ) utilizes a National Instruments low-noise SCXI-1000 chassis that is capable of sampling rates up to 333,000 samples per second for each DAQ device. The acquisition devices themselves are two National Instruments SCXI-1321 terminal blocks. National Instruments LabView 6.1 software provides the user interface and is used for sampling the data from the DAQ devices and writing the sampled data to a Microsoft Excel spreadsheet for later aerodynamic analysis. The resolution of the DAQ is 16 bits.

To determine the error in the measured aerodynamic coefficients, a calibration of the wind tunnel and the balance was performed before each test series. Following procedures [20], standard deviations of lift, drag, pitching moment, and dynamic pressure were estimated. Utilizing calibration measurements [20] and the small-sample method [21], the uncertainty intervals in aerodynamic coefficients corresponding to a confidence level of 99% were determined. Solid blockage, wake blockage, and streamlined curvature corrections were estimated based on the methods [22] and found negligible.

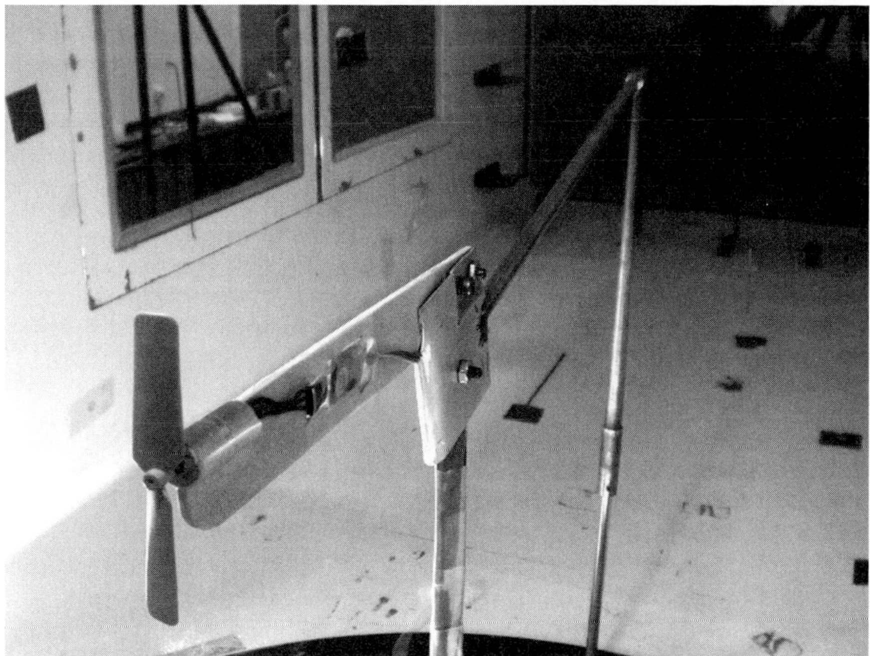

Fig. 6.9 Motor and propeller mounted to the top of the pylon in the test section.

6.3.2 Motor-Propeller Testing

Following the discussion in Sec. 6.2.4, two motors were selected for MAV propulsion: RE-10 and Feigao. To evaluate the power available in flight, prior knowledge of the static thrust on the propeller motor from the lithium-polymer batteries was needed. Preliminary static thrust testing was conducted by running the motor propeller installed on the electronic scale. Tests showed approximately 38 g of static thrust for the RE-10 and 74 g for the Feigao were available at a throttle setting of $T_S = 100\%$.

Both motors were tested in the wind tunnel for determination of thrust variation under flight speeds. First, dynamic tares with no motor were taken to isolate the drag of the pylon fixture in the tunnel. Then, the motor and propeller were mounted to the top of the pylon in the test section (Fig. 6.9), and an electrical connection was established between the motor and a power supply located outside the wind tunnel. The test power was increased to the motor so as to attain a static thrust that corresponds to $T_S = 100\%$ under battery power. With the power set and the motor running, the tunnel flow was started. Net load measurements were taken at tunnel speeds $V_T = 0\text{–}15$ m/s. The test was then repeated two more times, at throttle settings $T_S = 75$ and 50%, so that an accurate model of the thrust behavior of the motor-propeller combination could be obtained.

In general, the measured net force contains both thrust force from the propeller and drag forces caused by the flow over the motor. However, it will be referred

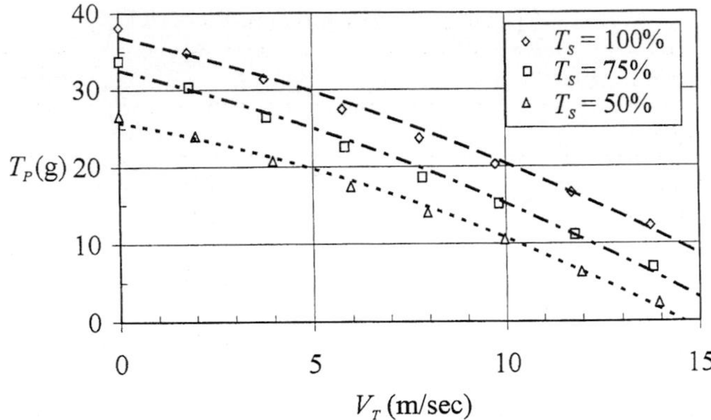

Fig. 6.10 Thrust vs wind-tunnel airflow velocity for the RE-10 motor with U-80 propeller.

to as just thrust force T_P. The experimental data points for thrust force T_P with wind-tunnel speed V_T are presented in Figs. 6.10 and 6.11 for RE-10 and Feigao, respectively. The trend lines in these figures show that thrust varies quadratically with speed. Note that these results are for off-the-shelf propellers that have not been optimized.

The propellers used have a relatively low pitch and are not well suited for high-speed use. To make use of the higher lift-to-drag ratio of the designed wing, a custom-designed propeller with a higher pitch would have to be implemented. Doing so though would result in an MAV with a higher overall flight speed, and low-speed flight would suffer accordingly.

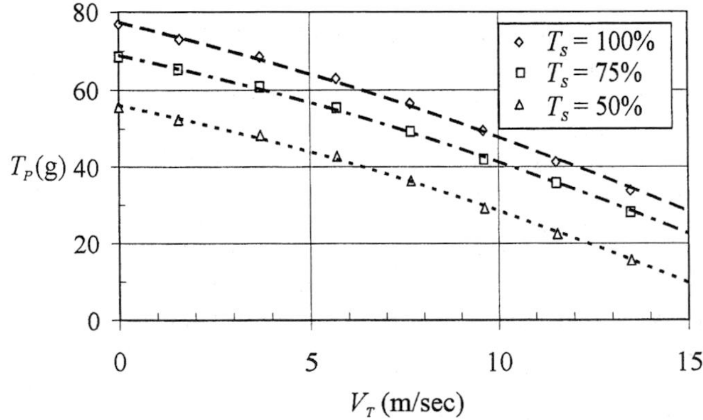

Fig. 6.11 Thrust vs wind-tunnel airflow velocity for the Feigao motor with EP-3020 propeller.

Table 6.2 Wind-tunnel models

Planform shape	Circular		Zimmerman	
Camber, %	3	9	3	9
Wing span b, cm	22.9	22.9	25.4	25.4
Root chord c_0, cm	20.6	20.6	16.9	16.3
Wing area S, cm^2	387	387	335	322
Camber height h_0, cm	0.62	1.86	0.511	1.47
Position of maximum reflex d, cm	18.6	17.8	14.5	13.9
Inverse camber h_i, cm	0.239	0.635	0.178	0.518
Wing aspect ratio A	1.35	1.35	1.91	1.98
Mean aerodynamic chord $\bar{c}$, cm	18.2	18.2	14.4	13.9

6.3.3 Aerodynamic Coefficients for Circular and Zimmerman Wings

Low aspect ratios and low Reynolds numbers are two major factors negatively influencing the aerodynamics of MAVs. Therefore, designing these flight vehicles requires significant research efforts in order to provide effective lift capabilities, performance, stability, controls, and handling qualities for piloting.

A series of wind-tunnel models of wings have been built with circular and Zimmerman planform shapes. A complete set of results of tests for circular wings with 3, 6, 9, and 12% camber is given [6]. In this section, only the data for circular wings with 3 and 9% camber are presented in relation to the present case study.

Aerodynamic data for Zimmerman wings with 3 and 9% camber were obtained in the present study. All wings were tested in the low-speed wind tunnel at angles of attack ranging from 0 to 35 deg and velocities corresponding to mean aerodynamic chord Reynolds numbers of 5×10^4, 7.5×10^4, and 1×10^5. Data for the aerodynamic coefficients C_L, C_D, and C_M and lift-to-drag ratios obtained are discussed here in detail.

6.3.3.1 Construction of wind-tunnel wing models. Wind-tunnel models for circular and Zimmerman planforms with 3 and 9% camber are all based upon the S5010-TOP24C-REF airfoil mentioned earlier. Geometrical parameter data are presented in Table 6.2.

Construction of the wind-tunnel models started with the design and construction of the molds for different cambered wings. Drawings were generated using SolidWorks™ as described in Sec. 6.2.3 in the file formats required by the equipment used in mold fabrication. Molds are constructed using the female airfoil mold. Either the female or male mold could have been used, but the female mold was selected because it results in a smoother surface on the top of the finished wing.

Because of its simplicity, the mold for the circular wing was made from foam using a hot wire. For the Zimmerman wing, the rapid prototyping technology and powder-based machine ZPrinter 310 System were employed in the present project. Because of the required high-quality surface, the product using the rapid prototyping technology required additional finishing of the mold surface. The mold

Fig. 6.12 Wing mold.

was coated with epoxy resin and polished to ensure smoothness of the surface. The final mold of the 9% Zimmerman wing is shown in Fig. 6.12.

The MAV wind-tunnel models were made of one layer of 0.02-g/cm^2 fiberglass cloth and three sheets of 0.018 g/cm^2 bidirectional carbon-fiber cloth laminated with epoxy resin. The use of fiberglass ensures that the top surface of the wing comes out smoother. The wing construction process begins with the application of a thin layer of wax on the mold. This process ensures the epoxy will not stick to the mold, making the wing easy to remove from the mold when cured. Then, the cloths are laid one by one onto the mold and laminated with the epoxy using a paint brush to spread the epoxy resin equally. Care is taken at this time to remove any excess resin from the carbon cloth so that the wing is as thin, smooth, and clean as possible when it has finished curing.

Then, the entire stack is placed in a vacuum bag to cure the epoxy. At this point, a piece of porous "peel ply" and paper towels are placed on top of the wing. With this system, any excess epoxy in the wing is squeezed out through the peel ply during the vacuum-bagging process and is absorbed by the paper towels, resulting in a thin, clean, and lightweight wing (extremely important for an actual flying model). Then, the mold and laminated wing are wrapped in a layer of breather cloth and placed into a vacuum bag. The breather cloth allows air to circulate over and around the mold evenly so that an even pressure is applied to the wing in all places during the curing process. With the wing and the mold in the bag, the vacuum pump is turned on and the ends of the bag are sealed. The vacuum in the bag is set to approximately 50 KPa, and the wing is allowed to cure at least 6 h. A vacuum setting of 50 KPa is sufficient to evenly compress the wing into the mold. Figure 6.13 shows the mold and the wing after curing.

When the wing has finished curing, the mold is removed from the vacuum bag, and all of the excess layers of material (peel ply and paper towels) are peeled off the wing. With the wing blank still affixed to the mold, a paper planform template is placed over the carbon material and is traced onto it. Then the carbon is removed from the mold, and the wing planform is cut from the blank using a pair of scissors. At this point, the basic wing is complete.

Because the main goal of this testing series was to obtain wing-only aerodynamic data, it was decided that the wing would be mounted in the wind tunnel as close to the trailing edge as possible and with an "aerodynamically clean" mount system. An aluminum mount was constructed with minimum frontal area and with a smooth, aerodynamic leading edge. Because the wing would be mounted near

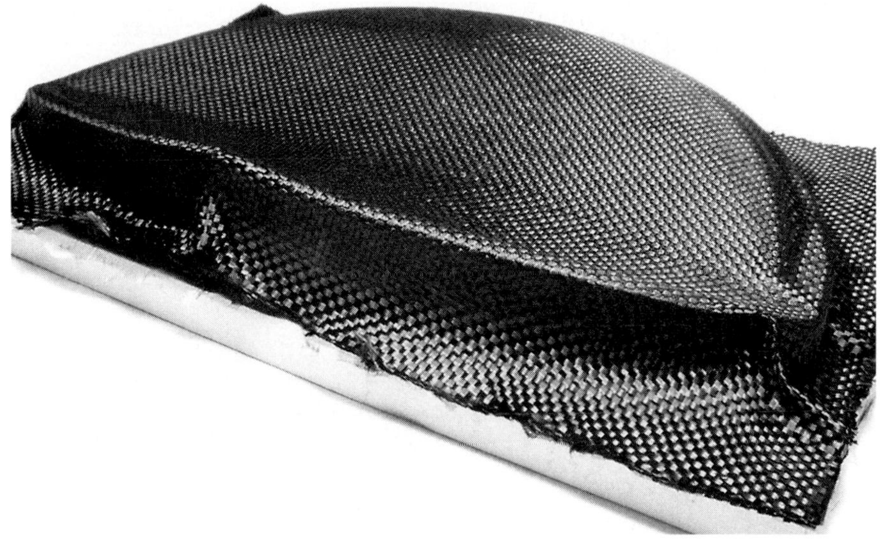

Fig. 6.13 Mold and the wing after curing.

the trailing edge, the wing was made stiffer by bonding a 1/16-in. plywood "rib" from the leading to the trailing edge. This rib also functioned as the mounting point for the wind-tunnel mount. A picture of a representative wing model in the wind tunnel is seen in Fig. 6.14.

6.3.3.2 Lift coefficients. The slope of the lift curve, stall angle of attack, and maximum lift coefficients are the parameters to look at when analyzing and comparing the lift performance of wings. The lift curves can be seen in Figs. 6.15 and 6.16 for the circular wing and for the Zimmerman wing, respectively. Numerical values for the lift-curve slopes are presented in Table 6.3.

The slopes of the lift curves were determined with the help of a least-squares linear regression analysis; data corresponding to the angle of attack from 1.3 to 10.4 deg are presented in Table 6.3. For circular wings and for Zimmerman wings at $Re = 5 \times 10^4$, the lift-curve slopes increase with a camber increase from 3 to 9%. Note that lift-curve slopes decreased for a circular wing with 12% camber in the previous study [6]. For the 3 and 9% circular wing and for the 3% Zimmerman wing, lift-curve slopes increase slightly with increased Reynolds number. Similar trends were observed by Torres and Mueller [5] utilizing, among others, flat-plate circular and Zimmerman planform models. Note that Torres and Mueller's lift-curve slope for these models at $Re = 1 \times 10^5$ coincides with data from the present study for the 3% camber Zimmerman. As for the effect of the aspect ratio, lift-curve slopes decrease with aspect ratio increases from $A = 1.35$ for the circular wing and $A = 1.9$ for the Zimmerman except for 3% camber at $Re = 7.5 \times 10^4$.

For the circular wing, it appears that as the camber increases the stall angle of attack decreases: stall angles of attack are approximately 33 and 27 deg for the 3 and 9% camber, respectively. Of interest is that the 3% camber circular wing does

Fig. 6.14 Wing model in the wind tunnel.

not show a true stall at the range of angles of attack studied. The maximum lift coefficients decrease as the Reynolds number decreases. For the same Reynolds number, 3% camber circular wings have a higher maximum lift coefficient than 9% camber wings. At $Re = 5 \times 10^4$, the maximum lift coefficients are 1.34 and 1.23 for 3 and 9%, respectively.

For the Zimmerman wing, some trends are the opposite of those observed with the circular wing. The stall angle of attack increases as the camber in the Zimmerman wing increases: at $Re = 5 \times 10^4$, stall angles of attack are 22.5 and 25 deg for the 3 and 9% cambers, respectively. The 9% camber shows a lift spike followed by stall, similar to the behavior that was observed in the 9% camber circular wing. It is unclear what is causing the erratic lift-curve behavior in the 9% camber. The 3% camber wings have smaller maximum lift coefficients than the 9% wings. At $Re = 5 \times 10^4$, the maximum lift coefficients are 0.81 and 1.08 for 3 and 9%, respectively, and they are less than those for the circular wing with the same camber values.

6.3.3.3 Lift-to-drag ratios. The lift-to-drag ratios C_L/C_D and $C_L^{3/2}/C_D$ are typically the best performance measures when it comes to choosing or designing a wing for virtually any flight vehicle. Micro air vehicles are no exception, and, in fact, they are likely more adversely affected by low lift-to-drag parameters. With their inherently low aspect ratios and low-Reynolds-number flight environment, MAVs with inefficient wings (poor lift-to-drag ratios) are doomed to be poor performers with reduced endurance, low climb rates, and slow flight speeds.

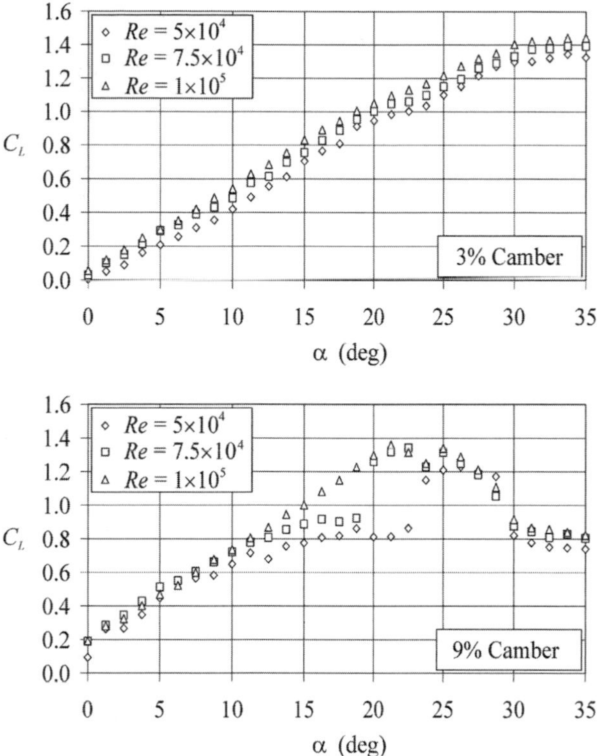

Fig. 6.15 Variation of lift coefficient with Reynolds number for the circular wing.

The C_L/C_D curves can be seen in Figs. 6.17 and 6.18 for the circular wing and for the Zimmerman wing, respectively, while $C_L^{3/2}/C_D$ plots are presented in Fig. 6.19 for the Zimmerman wing with 3 and 9% camber at $Re = 5 \times 10^4$.

One item of particular note is that for circular wings at a speed corresponding to $Re = 7.5 \times 10^4$–1×10^5, the maximum C_L/C_D occurs at an angle of attack of 5 deg. For the Zimmerman wing at the same Reynolds number, the maximum C_L/C_D corresponds to a higher angle of attack of approximately 8 deg. It can be seen from Figs. 6.17 and 6.18 that the 3% camber Zimmerman wing gives the best lift-to-drag ratio ($C_L/C_D \approx 10$) and theoretically would be the optimal choice for high-speed, efficient flight. For a MAV to take advantage of these high values, it needs to fly quite fast. To fly fast, the motor and propeller combination must be capable of providing enough thrust to overcome drag at high speeds, and in general high flight speeds correspond to an aircraft that reacts very quickly to pilot control inputs. In the interest of making an MAV that is somewhat easy to control, it might be beneficial to sacrifice some lift-to-drag performance for a reduced flight speed. The 9% camber C_L/C_D plots for both wings (Figs. 6.17 and 6.18) feature a second maxima at higher angles of attack.

The minimum power required for steady level flight occurs at maximum of $C_L^{3/2}/C_D$. For slow flight speeds, as in a loiter situation, it appears that the 9%

Table 6.3 Lift-curve slopes

Planform shape	Circular		Zimmerman	
Camber, %	3	9	3	9
$Re = 5 \times 10^4$	0.041	0.049	0.037	0.047
$Re = 7.5 \times 10^4$	0.045	0.050	0.047	0.045
$Re = 1 \times 10^5$	0.048	0.053	0.047	0.045

camber Zimmerman will perform the best. This wing gives the highest ratio ($C_L^{3/2}/C_D = 3.5$) at its stall angle of attack of 25 deg (Fig. 6.19). For comparison, the 3% camber wing has $C_L^{3/2}/C_D = 2.5$ at its stall angle of attack of 22.5 deg.

6.3.3.4 Moment coefficients. Plots of the pitching-moment coefficients about the quarter point of the root chord C_M for the circular and Zimmerman wings can be seen in Figs. 6.20 and 6.21, respectively.

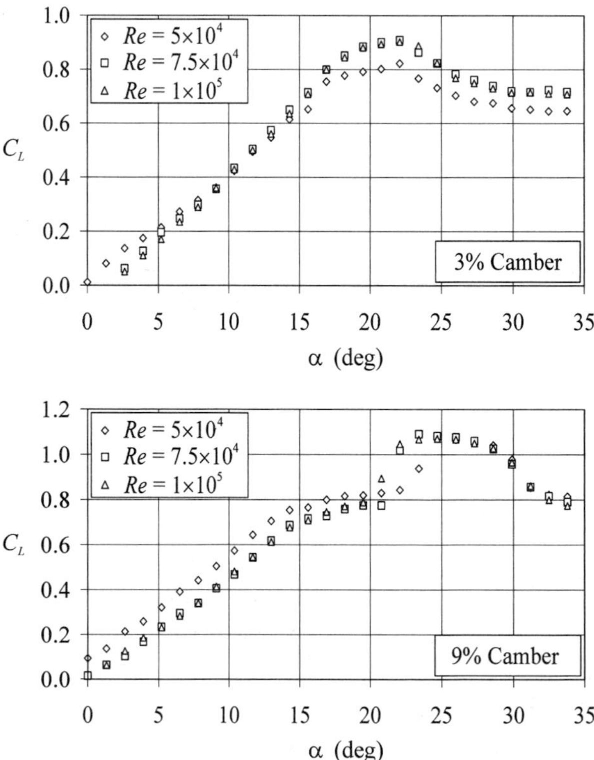

Fig. 6.16 Variation of lift coefficient with Reynolds number for the Zimmerman wing.

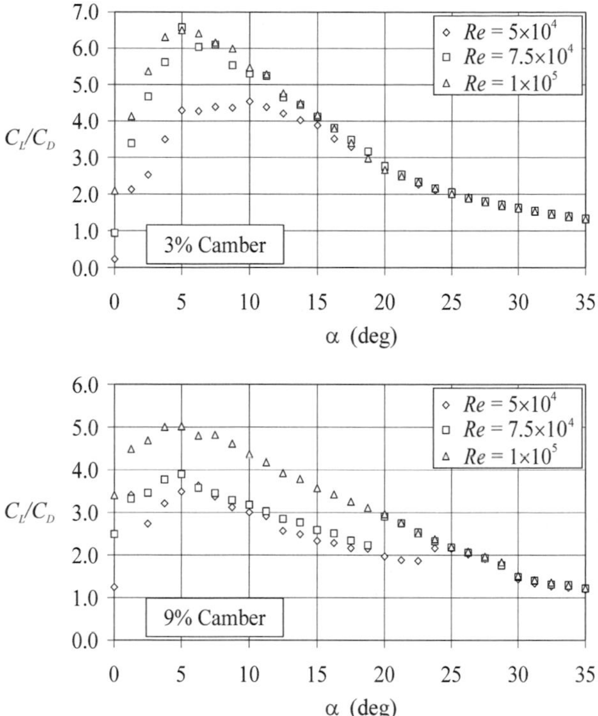

Fig. 6.17 C_L/C_D for the circular wing.

In the plots for the 3 and 9% circular wings (Fig. 6.20), it is noticeable that the pitching-moment coefficients C_M are relatively high (in the positive, nose-up direction) in the angle of attack range $\alpha = 0$–20 deg at the lowest of the Reynolds-number values tested. This phenomenon has been seen during flight tests of MAVs implementing these wings; during slow-flight maneuvers, the nose of the aircraft tends to rotate upward toward the stall point and requires downelevator control surface movements from the pilot to keep the aircraft from stalling. At higher flight speeds, this phenomenon has not been seen.

Pitching-moment coefficients for the Zimmerman wings (Fig. 6.21) are lower toward the negative nose-down direction, especially for 9% camber. For the 3% Zimmerman wing, the C_M coefficients are positive in the angle-of-attack range $\alpha = 0$–15 deg. For the 9% Zimmerman wing, the C_M coefficients are close to zero or negative for all Reynolds numbers and angles of attack. It might require a further increase of the reflex and, therefore, a sacrifice of some lift-to-drag performance to make the 9% camber Zimmerman wing MAV stable in pitch and easy to control.

6.4 Design of Adaptive-Wing MAVs

In the course of the project at the University of Arizona, two adaptive-wing micro air vehicles were designed, built, and tested: circular UA-AW during the first phase

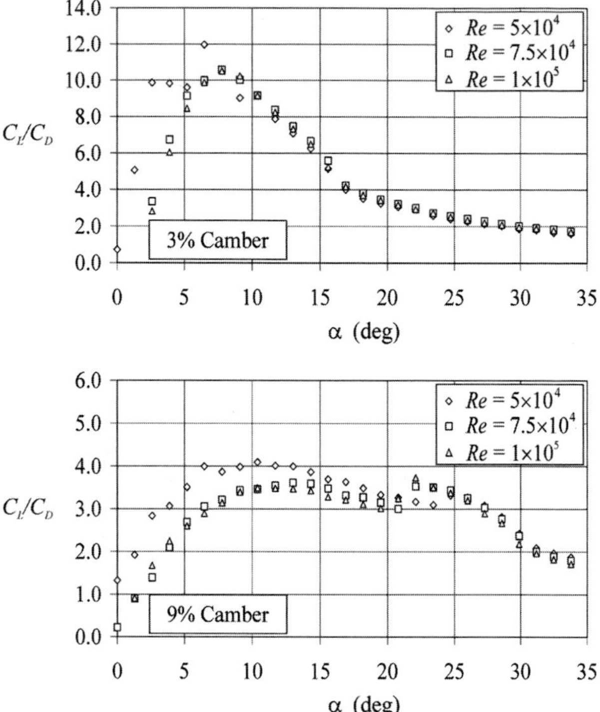

Fig. 6.18 C_L/C_D for the Zimmerman wing.

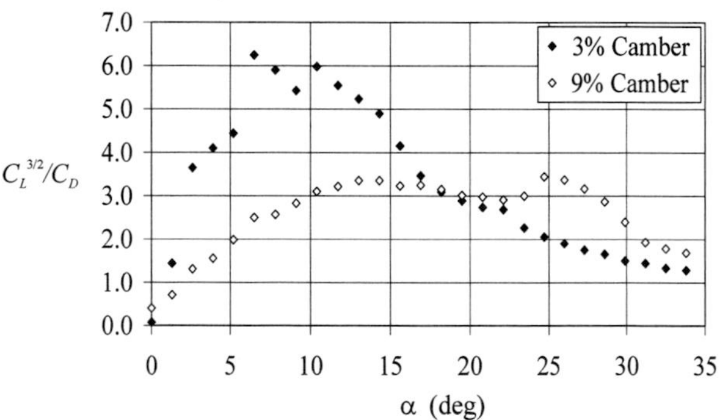

Fig. 6.19 $C_L^{3/2}/C_D$ for the Zimmerman wing at $Re = 5\times10^4$.

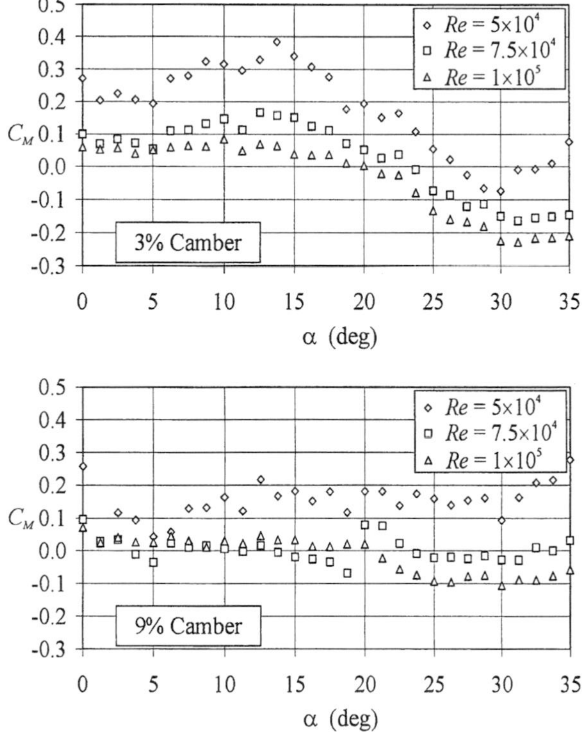

Fig. 6.20 C_M for the circular wing.

of the project in 2002–2003 and Zimmerman UA-AW during the second phase in 2003–2004. This section will outline both designs. It will describe selection and arrangements of components, camber change mechanism functioning, and details of construction. Throughout this discussion, experience and achievements gained in MAV technology will be illustrated as well.

6.4.1 Circular UA-AW MAV

The first-generation circular UA-AW MAV (Figs. 6.22 and 6.23) was designed, built, and flight tested during the first phase of the project [23] and [24]. This phase pursued two objectives: to prove the feasibility of the concept for the speed change by using a variable camber wing and to test the mechanism for the camber change. Geometric parameters of the wing, elevator, fin, and rudder are given in Table 6.4. The wing's planform parameters were selected from Table 6.2 corresponding to the 3% camber: wing span of 22.9 cm and wing aspect ratio of 1.35.

All components of the airplane and their corresponding masses are presented in Table 6.5. The total mass of the vehicle is 72.8 g, providing the wing loading $W/S = 18.2\,\mathrm{N/m^2}$. The RE-10 motor with Union 80 propeller was installed at the nose of the vehicle. Conventional control surfaces elevator and rudder provide pitch and yaw control for the vehicle. The control surfaces are taped to the trailing

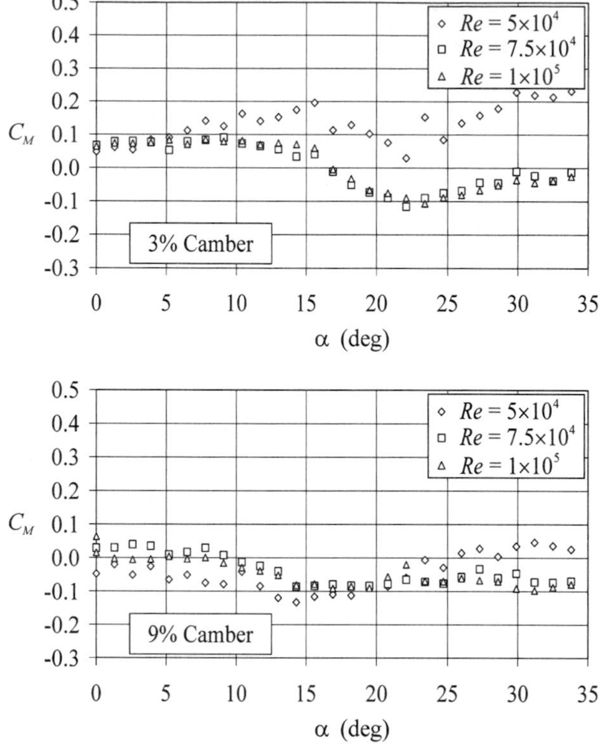

Fig. 6.21 C_M for the Zimmerman wing.

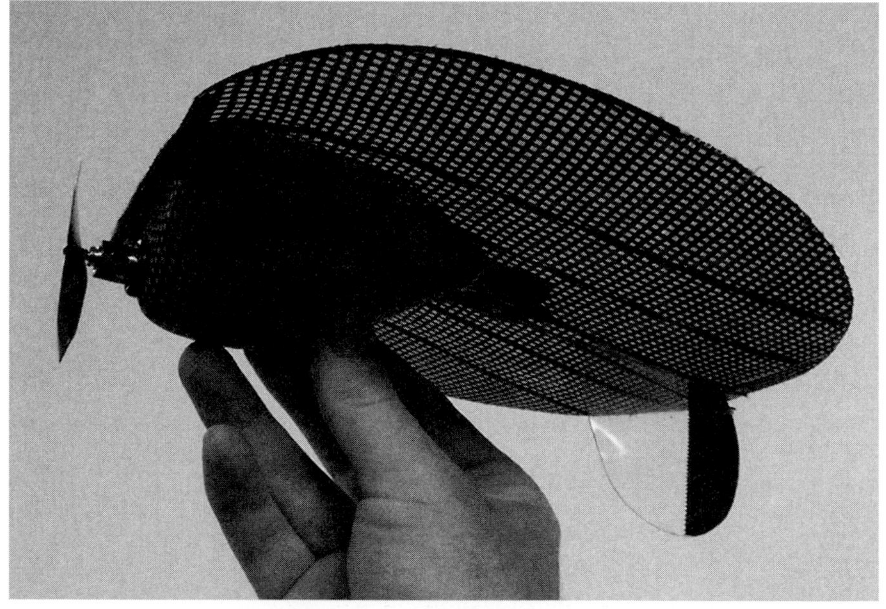

Fig. 6.22 UA-AW with a circular wing.

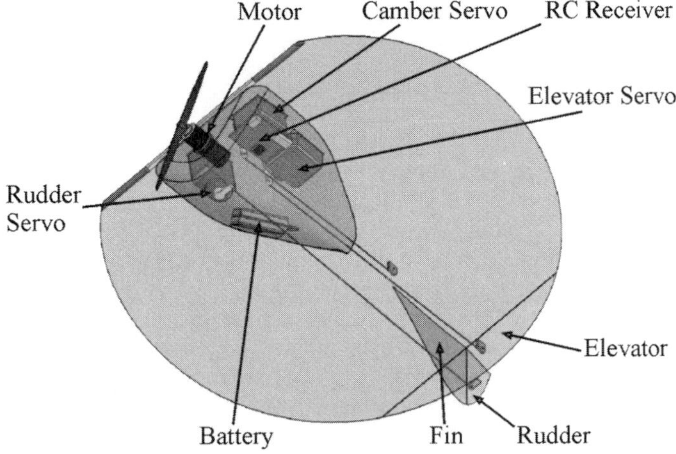

Fig. 6.23 Variable-camber MAV internal arrangement drawing.

edge of the wing and the fin, and they are actuated by servomotors providing maximum deflections of ±25 deg. No video camera was installed onboard the MAV at that time.

The basic premise of the design was to create an MAV capable of changing its camber in flight from 3 to 9%. Much of the MAV's look and characteristics would be identical to the conventional MAV after the switch to a camber-changing wing; however, several issues have to be addressed to integrate the changes into a successful flying machine.

The internal arrangement drawing of the MAV is shown in Fig. 6.23. The vehicle was built by employing composite materials construction methods outlined in Sec. 6.3.3.1. The MAV has a small fuselage serving as a compartment for the motor, battery, RC receiver, and servos. When the camber of the wing had been changed during flight, a drag-inducing gap had formed between the wing and the fuselage. This was compensated by employing a two-layer fuselage. The outer

Table 6.4 Circular UA-AW MAV specifications

Parameter	Value
Wingspan, cm	22.9
Length, cm	20.7
Height, cm	8.26
Wing area, cm^2	387
Elevator area, cm^2	14.7
Fin area, cm^2	19.8
Rudder area, cm^2	4.85

Table 6.5 Components of the circular UA-AW MAV

Component	Description	Mass, (g)
Airframe	——	32.4
Motor/propeller	Maxon Re-10/ Union-80	11.1
3-cell lithium-polymer battery	Kokam, 145 mAh	10.5
3 micro servos	Hitec, HS-50	16.2
Receiver/Speed Controller	Skyhooks, RX-72	2.6
Total	——	72.8

layer, a complete fuselage, was molded at the maximum camber to which the wing would be deflected, 9%. The inner fuselage served as a flap that closed the gap that opened when the wing was contracted. The inner layer was the bottom portion of a fuselage of 2-cm height, also molded at 9%. The inner layer had two slits cut in each side of it at regular intervals and was then spread out and attached to a 3% wing that was still attached to the mold. This was done so that when the 3% wing was deflected to 9% camber the slits in the inner fuselage would close up, reducing drag. This is a simple, yet effective method of solving the geometry problem of a wing that changes shape that is attached to a fixed-geometry fuselage.

Addition of a third servo is the means by which the MAV's camber is changed during flight. It is mounted on the underside of the MAV's wing (Fig. 6.23). The camber change control horn is placed near the inflection point of the wing's airfoil. This causes the camber change to be tailored so that the wing's maximum camber point remains close to 24% of chord rather than simply deflecting the trailing edge of the wing down. Strictly pulling down the trailing edge results in a large nose-down pitching moment, which is undesirable. Carbon braces prevent bending over the middle of the wing and encourage the maximum camber to occur at the desired point. Reflex in the airfoil must also be increased to avoid large increases in pitching moment.

To combat increasing pitching moment during camber increase, the elevator compensates adding reflex to the wing. The elevator's control rod is fixed in length. Because the elevator servo is attached to the wing as well, when the camber servo contracts the wing the control rod pushes the elevator upwards. The aircraft has just the right amount of reflex at its 3 and 9% camber modes by virtue of mixing accomplished in the digital Futaba T6XA transmitter.

Although the elevator depends on the camber change for adjusting its reflex, it is essential that the rudder does not deflect when the camber of the wing is adjusted in flight. Therefore, the rudder servo and the vertical tail were attached to the fuselage. This ensures that their functions are independent of any camber increase in the wing. The geometry of the vertical tail is such that it can accommodate an increase in camber of up to 9%.

A schematic of the electrical components is shown in Fig. 6.24. Electric power is supplied from the lithium-polymer batteries connected in series. The batteries supply the power into the speed controller, and then the speed controller regulates correct polarity to the motor. Also the speed controller distributes the power to the receiver and three servos.

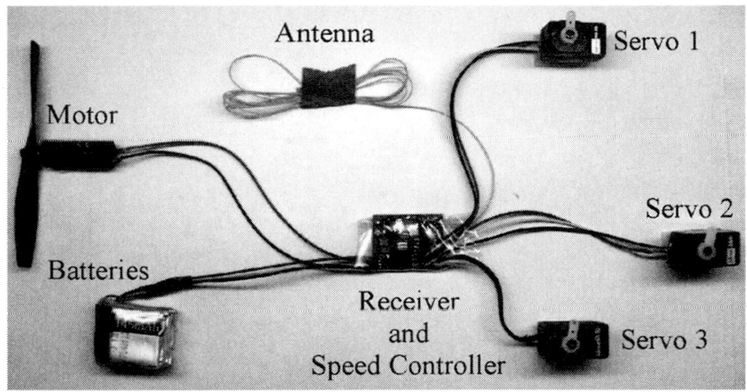

Fig. 6.24 MAV electrical components.

6.4.2 Zimmerman UA-AW MAV

The second-generation UA-AW can be seen in Figs. 6.25–6.27. The Zimmerman planform for the wing was incorporated into a flying-wing configuration. Camber change and corresponding reflex change are illustrated in Fig. 6.25. Other features of the design, in comparison with the circular UA-AW, include installation of the more powerful and efficient Feigao brushless motor and use of elevons for the control. The fin provides the stabilization in roll and yaw, while elevons control the vehicle in pitch and yaw.

At 3% camber, the wing span is 25.4 cm, wing area is 335 cm^2, and the aspect ratio is 1.91. Other geometric parameters of the wing, fin, and elevons are presented in Table 6.6. All components of the airplane and their masses are presented in Table 6.7. The total mass of the MAV is 120 g. The wing loading of $W/S = 35$ N/m^2 is almost two times greater than that of the circular UA-AW, which with a more powerful motor gives the Zimmerman UA-AW higher speed. A video system of mass 5.2 g was installed onboard as a payload.

Figures 6.26 and 6.27 illustrate the internal arrangement of the Zimmerman UA-AW. The wing in this design consists of two halves attached to opposite sides of the fuselage. The attachments allow deforming the wing from 3 to 9% camber. A new fuselage was designed to help eliminate some of the shortcomings of the previous design, and a novel rod-and-driving tube hinge was designed to securely join the wing and fuselage together, but allow easy camber change.

The fuselage is composed of two sections: nose and main section. The nose section is made from several layers of fiberglass to achieve a streamlined shape to decrease drag as much as possible. The main section is constructed from thin balsa wood sandwiched between two layers of fiberglass. The vertical fin is finished with the same material as the main section of the fuselage.

The wing construction is a sandwich of Kevlar. At the leading edge and in front of the elevons, the core material is composed from carbon tissue and hard foam. The wing is connected to the fuselage with four rods: one rod at the leading edge, one rod at the maximum camber of the wing, and two rods at the tail. The maximum camber rod is glued to the driving tube. Two rods at the tail are glued to the fuselage.

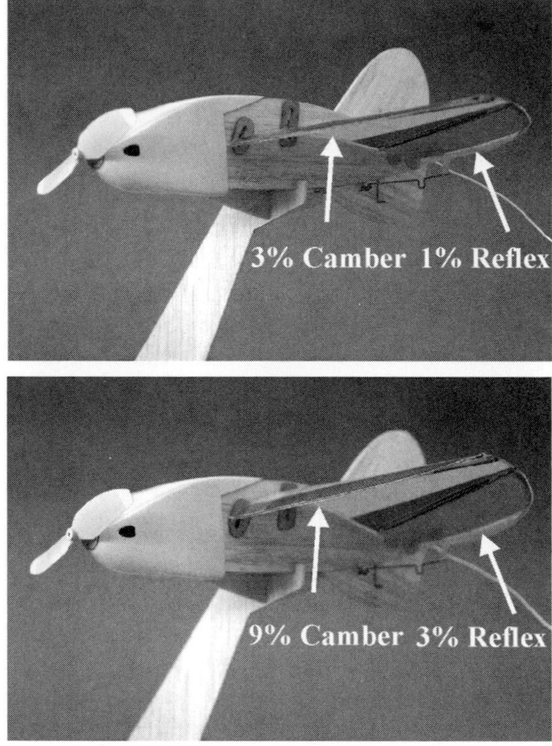

Fig. 6.25 UA-AW with Zimmerman wing (photo courtesy of E. Stiles).

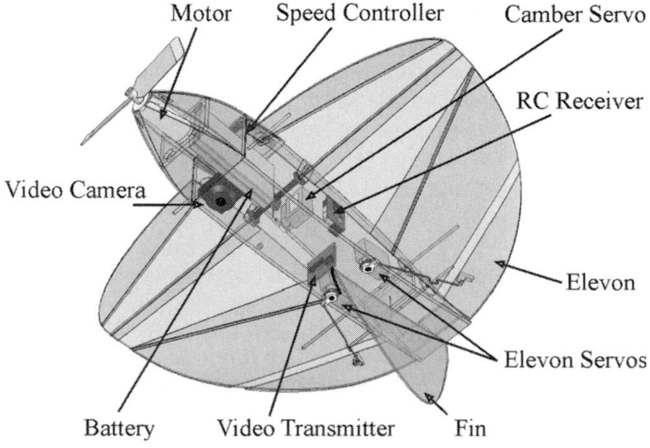

Fig. 6.26 Zimmerman wing MAV internal arrangement drawing.

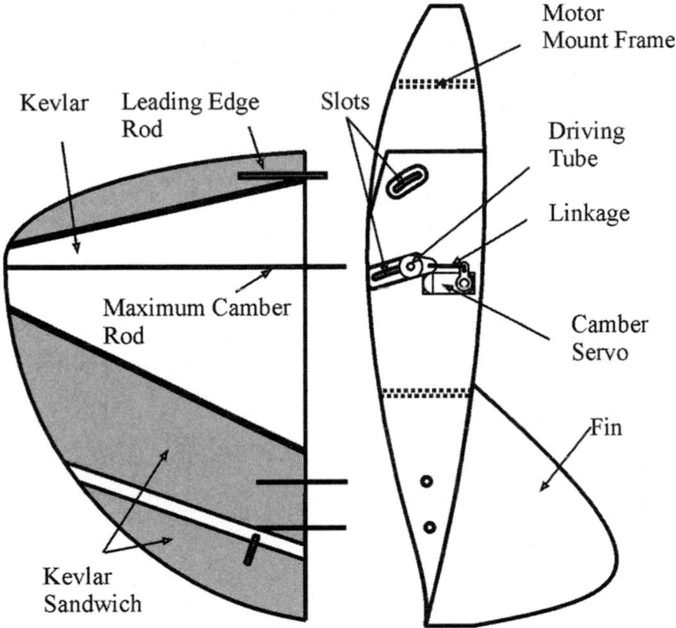

Fig. 6.27 Structure drawing of the Zimmerman UA-AW.

The mechanism of changing the camber is achieved by a camber servo. There is a short linkage to connect the servo arm to the driving tube as shown in Fig. 6.27. The arm is aligned with the horizontal at 3% camber, and as the servo rotates in the clockwise direction, it increases the camber height up to the 9%. There are two slots in the fuselage: at the maximum camber position of the wing in the main section and at the aft of the nose section. The maximum camber rod is driven by a driving tube that follows the first slot and allows the camber height change. The leading-edge rod freely slides in the second slot in the fuselage providing airfoil shape. Then the reflex of the wing is achieved by the changing the elevon deflection. This is done by the mixing feature of the radio transmitter between elevon servos.

Table 6.6 UA-AW with Zimmerman wing specifications

Parameter	Value
Wing span, cm	25.4
Length, cm	23.2
Height, cm	9.06
Wing area, cm^2	335
Fin area, cm^2	34.9
Elevon area, cm^2	29.6

Table 6.7 Components of the Zimmerman UA-AW MAV

Component	Description	Mass, g
Airframe	——	55.4
Motor/propeller	Feigao/EP-3020	17.4
2-cell lithium-polymer battery	Kokam, 640 mAh	28.1
3 micro servos	Cirrus CS-4.4	10.5
Receiver/speed controller	PENTA/JETI	8.6
Total	——	120

The motor is aligned 5 deg downthrust to ensure trimmed flight at all throttle settings. With the more powerful brushless motor and Pitot system or video camera installed, the vehicle requires more electric power from the battery (see Tables 6.5 and 6.7). The motor and battery are placed in the nose of the fuselage, providing a forward location for the center of gravity of the airplane.

6.5 Flight Testing

Flight testing is an important step in bringing together the results of wind-tunnel tests, component selection, and design. A pilot provides a designer with valuable feedback about the performance characteristics of the designed aircraft, however, these observations are subjective in nature, and therefore the possibilities of utilizing data recording devices are also considered in planning flight tests.

A series of flight tests were conducted in the course of the project. These tests can be grouped as follows: 1) trimming the MAV at 3, 6, and 9% camber and making final improvements in the design; 2) endurance measurements; 3) airspeed measurements; and 4) in-flight video capturing.

6.5.1 *Circular UA-AW MAV*

Flight tests of the circular wing MAV were conducted in an outdoor park. After the MAV was properly balanced and trimmed on the ground, it was hand launched at the throttle setting $T_S = 60\%$ and at the 3% camber configuration. While in the air, the pilot adjusted control surface throws and trims. Fine tuning and neutrals adjustments were made between flights. For the second flight, the airplane took off and flew at 6% camber position, which seemed easier to handle. During the flight, the camber was changed from 6 to 3% and back, and then from 6 to 9% and back. Per pilot's feedback, the overall control response worsened with camber increase, and more thrust was needed. When the throttle was increased, the airplane started climbing instead of accelerating. About 10 deg of downthrust was introduced into the motor installation, which made the MAV fly straight and level flight at all speeds and cambers. After these adjustments, the MAV demonstrated a flight endurance of 5 min at 6% camber and at $T_S = 70\%$.

The MAV was also tested for flight velocity changes with camber varying during the flight. The aircraft was flown at minimum and maximum cambers between two flags, 60 ft apart. Velocity was determined by noting the time it took to fly between the flags on a digital video camera. Readings were backed up by use of a

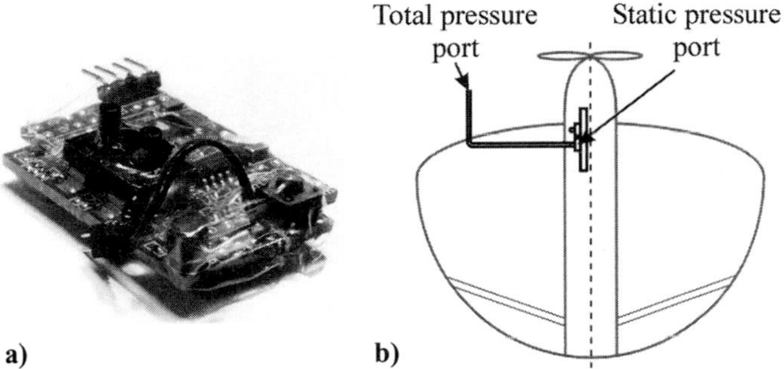

Fig. 6.28 Pictures of a) flight data recorder and b) its mounting.

K-band radar gun. However, it is hard to aim the aircraft, and wind gusts in open air adversely affected the readings and led to inconclusive results. No airspeed recording sensors were available to the project team at that time.

6.5.2 Zimmerman UA-AW MAV

After trim and balance were achieved on the Zimmerman wing MAV, it was flown with a constant 6% camber while testing two propellers: EP-3030 and EP-3020. There were no noticeable flight speed changes between the two propellers and flight throttle settings. The EP-3030 propeller has a higher pitch compared to EP-3020 and produces excessive torque depending on the throttle changes. Because of the torque issue of EP-3030, the pilot had a hard time flying the aircraft. For this reason, EP-3020 was selected as the propeller for this airplane.

Initially, the airplane had two vertical surfaces, on the top and bottom of the fuselage. A large top vertical fin seemed to cause a side slip in turns. After the fin was moved from the top (as in Fig. 6.25) to the bottom of the fuselage (as in Figs. 6.26 and 6.27), the aforementioned problem disappeared. Camber changes in flight from 3 to 9% were attempted, and it was found that more mixing between elevon and camber servos was needed on the transmitter to get the reflex right at all positions.

Airspeed measurements of the fixed-wing MAV were conducted with the help of a lightweight flight data recorder [25] with dimensions 4.3 × 3.0 × 1.4 cm and mass of 8.4 g (Fig. 6.28). The Pitot-tube-type recorder is capable of airspeed measurements from 4 to 130 m/s [25]. The tube itself was made from 1.5-mm outer diameter (0.8-mm inner diameter) brass tubing and mounted below the wing as can be seen in Fig. 6.28. The recording device, also seen in Fig. 6.28, was placed in the fuselage. Preliminary flights showed some sideslip again that could be a product of the center of gravity moving or increased drag on the right wing because of the Pitot-tube mount. The 3% camber showed an instability in roll compared to the 6 and 9% configurations. The 9% camber demonstrated a good speed range, but control was still lacking, therefore, rate mixing was added for the 9% configuration. In future applications, a smaller tube construction should be considered for decreasing the drag caused by the Pitot tube. After the flight, the data were downloaded to the computer via a parallel port for further analysis.

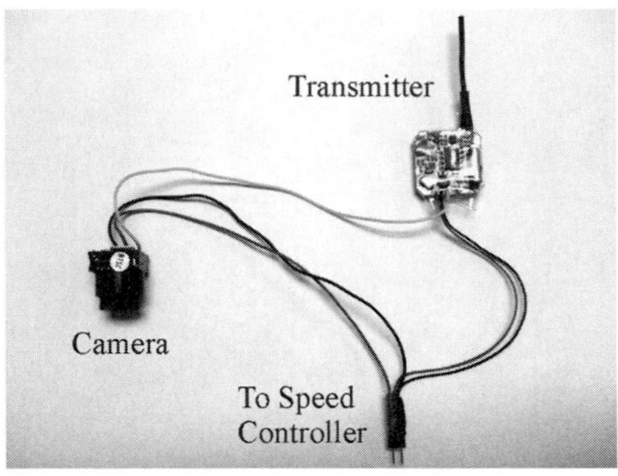

Fig. 6.29 Video system.

The series of flight tests involved Pitot-tube measurements of airspeed were conducted at full throttle setting. Based on in-flight measurements, the following average speeds were determined: 8.2, 6.1, and 4.6 m/s, for 3, 6, and 9% camber, respectively. These results proved that the flight speed can be effectively controlled and changed through the camber adjustments.

6.5.3 Video-Operated Flight

A video system is a standard MAV's payload serving a twofold purpose: 1) MAVs need to be "flown by camera" into the mission operating environment and 2) transmission of images to the ground station. The MAV video system consists of a MAV-mounted camera and video downlink and a head-mounted personal display or a laptop computer. Figure 6.29 shows the configuration of the video system mounted on the Zimmerman wing MAV that includes a microcolor CMOS camera with pinhole lens with dimensions $12 \times 12 \times 13$ mm. The camera is connected to a 2.4-GHz transmitter with an antenna [26]. The system can operate on less than 5 V, and it is fed from the speed controller. The video camera was attached to the floor of the fuselage behind its nose section, and the video transmitter was mounted in the middle of the fuselage next to the fin as illustrated in Fig. 6.26. The total mass of the video system onboard the MAV is 5.2 g. In the short-range flight mission, the camera was mounted directly downward (Fig. 6.26), while for the long-range flights it was pointed forward and used by the pilot to identify the horizon and to keep the plane in level flight.

The video system onboard the MAV transmits video images to the ground station (Fig. 6.30). The ground station antenna receives the signal from an MAV and relays it via a video receiver to a laptop computer or personal video display. A 2.4-GHz circular polarized antenna [27] and 2.4-GHz video receiver [28] were utilized in the project. With the present video system, video receiving range was approximately 1000 m along the straight line with no obstacles.

Video images were digitized and recorded on a laptop computer for postflight viewing and analysis. The live video signal was also fed into the head-mounted personal video display DH-4400VP [29] shown in Fig. 6.30. When the target is located at a distance far from the aircraft launch point, it requires the pilot to fly

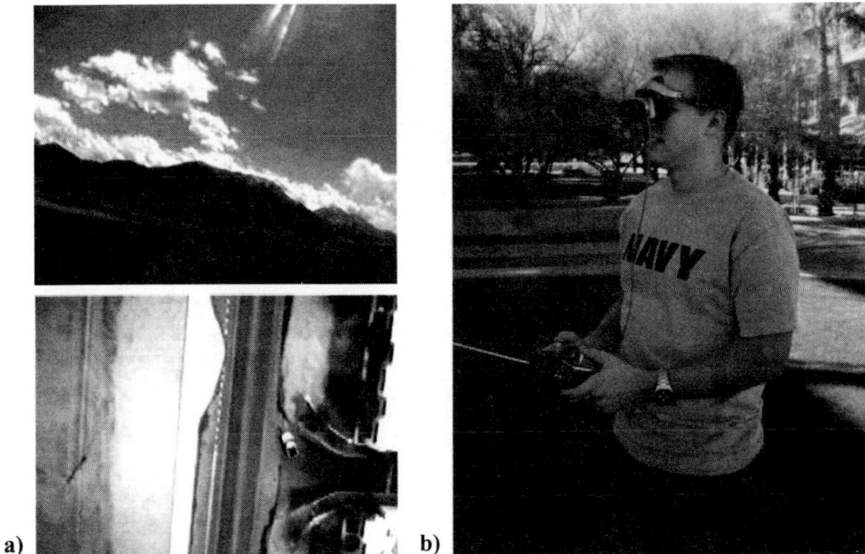

Fig. 6.30 Teleoperation of MAV: a) video captures and b) a pilot with the head-mounted personal video display.

the craft via "camera," that is, to fly the MAV by reference to the onboard video image transmitted to the personal video display.

For a typical mission, the target was located about 500 m from the aircraft launch point. The MAV pilot flew the vehicle out to an area of interest at high speed at 3% camber setup and then went into loiter mode by changing the camber from 3 to 9%, capturing live video images of the area of interest. The typical flight time of the Zimmerman wing micro air vehicle was approximately 10 min. A slow flight speed during the loiter phase of the mission ensured that quality video images were relayed to the ground station.

6.6 Conclusion

To achieve the desirable flight performance, a MAV can be coordinated and controlled by deformation of the wing structure. This chapter has presented a case study of the innovative design of the University of Arizona MAV with variable camber wing. The wing design features a change in its cross-sectional geometry through a synchronous increase/decrease of camber and reflex.

The design constraint on the maximum dimension dictates the selection of the circular or the Zimmerman wing planform; for the same reason, the MAV utilizes a flying-wing configuration. The biggest factor driving the design of the flying-wing aircraft is pitch stability and control. A negative, nose-down pitching moment for cambered, low-aspect-ratio wings at a low Reynolds number is compensated for by introducing a wing with inverse camber. The airfoil was designed featuring the ratio of inverse camber to maximum camber of 1:3.

The effects of camber and Reynolds number on the aerodynamic coefficients for the circular and the Zimmerman planform shape wings were studied with the help of wind-tunnel testing. From an analysis of the wind-tunnel data, it appears that for high-speed flight (i.e., ingress/egress for a typical MAV mission) the 3% camber wing will give the best performance as a result of its high lift-to-drag ratio. For slow-speed flight, as in loitering or landing, it appears that the 6 and 9% cambers will perform the best.

Two micro air vehicles were designed and built with the circular and the Zimmerman planform shapes. A series of flight tests with in-flight changing of the camber were conducted with measurement of the endurance and airspeed and in-flight video capturing. These tests determined that the proposed adaptive-wing concept meets the specifications and that it is a promising configuration for MAV applications.

Acknowledgments

This research and development work was sponsored by NASA Langley Research Center under Grant NAG-1-03045. The authors also would like to thank to the rest of the Micro Air Vehicle Project at the University of Arizona for their contributions to this project: Wayne Jouse, Bret Becker, Anton Kochevar, William Null, Bill Silin, Jeremy Tyler, and Matt Wagner.

References

[1]Schmitz, F. W., "Aerodynamics of the Model Airplane. Part I. Airfoil Measurements, 1941," Redstone Arsenal, AL. Translation, RSIC-721, 22 Nov. 1967.

[2]Laitone, E. V., "Wind Tunnel Tests of Wings at Reynolds Numbers Below 70000," *Experiments in Fluids*, Vol. 23, No. 5, 1997, pp. 405–409.

[3]Selig, M. S., Gopalarathnam, A., Giguère, P., and Lyon, C. A., "Systematic Airfoil Design Studies at Low Reynolds Numbers," *Fixed and Flapping Wing Aerodynamics for Micro Air Vehicle Applications*, edited by T. J. Mueller, Progress in Astronautics and Aeronautics, Vol. 195, AIAA, Reston, VA, 2001, pp. 143–167.

[4]Pelletier, A., and Mueller, T., "Low Reynolds Number Aerodynamics of Low-Aspect-Ratio, Thin/Flat/Cambered-Plate Wings," *Journal of Aircraft*, Vol. 37, No. 5, 2000, pp. 825–832.

[5]Torres, G. E., and Mueller, T. J., "Aerodynamic Characteristics of Low Aspect Ratio Wings at Low Reynolds Numbers," *Fixed and Flapping Wing Aerodynamics for Micro Air Vehicle Applications*, edited by T. J. Mueller, Progress in Astronautics and Aeronautics, Vol. 195, AIAA, Reston, VA, 2001, pp. 115–141.

[6]Null, W., and Shkarayev, S., "Effect of Camber on the Aerodynamics of Adaptive-Wing Micro Air Vehicles," *Journal of Aircraft*, Vol. 42, No. 6, 2005, pp. 1537–1542.

[7]Null, W., Wagner, M., Shkarayev, S., Jouse, W., and Brock, K. "Utilizing Adaptive Wing Technology in the Control of a Micro Air Vehicle," *Smart Structures and Materials 2002: Industrial and Commercial Applications of Smart Structures Technologies*, edited by Anna-Maria R. McGowan, Proceedings of SPIE, Bellingham, WA., Vol. 4698, 2002, pp. 112–120.

[8]Shkarayev, S., Jouse, W. C., Null, W., and Wagner, M., "Measurements and Performance Prediction of an Adaptive Wing Micro Air Vehicle," *Smart Structures and Materials 2003: Industrial and Commercial Applications of Smart Structures Technologies*, edited by Eric H. Anderson, Proceedings of SPIE, Vol. 5054, 2003, pp. 53–65.

[9]Grasmeyer, J. M., and Keennon, M. T., "Development of the Black Widow Micro Air Vehicle," *Fixed and Flapping Wing Aerodynamics for Micro Air Vehicle Applications*, edited by T. J. Mueller, Progress in Astronautics and Aeronautics, Vol. 195, AIAA, Reston, VA, 2001, pp. 519–535.

[10]*Micro Air Vehicle Design Papers*, 6th International MAV Competition, Brigham Young University, Provo, UT, April 2002.

[11]*Micro Air Vehicle Design Papers*, 7th International MAV Competition, University of Florida, Gainesville, FL, April 2003.

[12]*Micro Air Vehicle Design Papers*, 8th international MAV Competition, University of Arizona, Tucson, AZ, April 2004.

[13]Selig, M. S., Lyon, C. A., Giguère, P., Ninham, C. N., and Guglielmo, J. J., *Summary of Low-Speed Airfoil Data*, Vol. 2, SoarTech Publications, Virginia Beach, VA, 1996, 292 p.

[14]Drela, M., and Youngren, H., "XFOIL: Subsonic Airfoil Development System," ver. 6.94, Dept. of Aeronautics and Astronautics, MIT, Cambridge, MA, 2000.

[15]Maxon Motors Products, Maxon Precision Motors, Inc., Burlingame, CA, 2006.

[16]Feigao Electric Motors, Feigao Electric Motors Co., Ltd., Changsha City, HuNan Province, People's Republic of China, 2006.

[17]Sky Hooks and Rigging, "CATALOG 2003," Oakville, Ontario, Canada, 2003.

[18]Jeti Programmable Brushless Controls, Hobby Lobby International, Inc., Brentwood, TN, 2006.

[19]Gasparin Micro Solutions, Products, Ricany, Czech Republic, 2006.

[20]Aki, M., and Noseck, A., "Calibration of Low Speed Wind Tunnel," Aerospace and Mechanical Dept., Rept. Univ. of Arizona, Tucson, Aug. 2004.

[21]Kline, S. J., and McClintock, F. A., "Describing Uncertainties in Single-Sample Experiments," *Mechanical Engineering*, Vol. 75, No. 1, 1953, pp. 3–8.

[22]Barlow, J. B., Rae, W. H., Jr., and Pope, A., *Low-Speed Wind Tunnel Testing*, 3rd ed., Wiley, New York, 1999, 713 p.

[23]Shkarayev, S., Null, W., and Wagner, M., "Development of Micro Air Vehicle Technology with in-Flight Adaptive-Wing Structure," NASA/CR-2004-213271, Oct. 2004.

[24]Null, W., "The Design and Development of an Adaptive Wing Micro Air Vehicle," M.S. Thesis, Dept. of Aerospace and Mechanical Engineering, Univ. of Arizona, Tucson, December 2003.

[25]Data Sheet for the USB Flight Data Recorder, Eagle Tree Systems, LLC, Bellevue, WA, 2003.

[26]Micro Video Transmitter with Receiver, MicroTek, Knoxville, TN, 2006.

[27]Wireless Circular Polarized Antenna, HyperLink Technologies, Boca Raton, FL, 2006.

[28]Video Receiver, Klarich Electronics, LLC, Okawville, IL, 2006.

[29]Personal Video Display, Cybermind Interactive Nederland, Maastricht, The Netherlands, 2006.

7
Summary

Interest in unmanned air vehicles was stimulated by World War I. In the beginning full sized airplanes were used with primitive controls capable of stabilizing and navigating without a pilot on board. The conversion of manned airplanes to target drones continued during the 1920s and 1930s. A number of radio-controlled, radar-controlled, and television-controlled glide bombs were used in World War II.

Pioneers in aviation used model airplanes to help them understand aerodynamic forces and how to control them. The availability of small internal combustion engines and small radio receivers and transmitters and the invention of control surface actuators in the 1930s led to the era of radio-controlled model airplanes in 1936. Continuous improvements in RC model equipment, including the introduction of electric motors, plus advances in micro-mechanical systems, micro-electronic components, and sensors led to the feasibility of small unmanned air vehicles and then micro air vehicles in the 1990s. Studies by the RAND Corporation in 1992 and 1993 of the feasibility of very small controlled or autonomous vehicles were followed by a more detailed study by the MIT Lincoln Laboratory in 1995. These studies led to a DARPA workshop in the fall of 1995 and the proposal by R. J. Foch of the Naval Research Laboratory and M. S. Francis of DARPA to develop a 15.24 cm (6 in.) flying vehicle with a useful payload. One of the fundamental disciplines required to design a micro air vehicle was the low aspect ratio wing operating at low Reynolds numbers.

Except for the pioneering research of F. W. Schmitz in the 1930s, the most significant research to understand the performance deterioration of low Reynolds number airfoils and wings took place in the 1980s. Initially, existing airfoils designed for Reynolds numbers of 500,000 or greater were studied at lower Reynolds numbers where lift decreased, the drag increased, and hysteresis in these forces often occurred. These problems were better understood and solved by using the design methods of Eppler and Drela to design profiles for a specific low Reynolds number range. A large number of careful wind-tunnel experiments were used to determine and verify the lift and drag performance, and the effects of laminar separation bubbles and tip vortices for low aspect ratio wings. Experimental results were also useful in the development of numerical methods that can now be used in the MAVdesign process.

The choice of a propulsion system was an important early decision. Electric motors have significant advantages over internal combustion engines for MAV propulsion. Recent advances in light-weight electric motors, and batteries with increased energy densities have made electric propulsion the predominant choice for MAV propulsion systems.

Due to their small size, MAVs are prone to unsteady behavior with high frequency oscillations, limiting the usefulness of their application. Thus, an automatic flight-control system is needed for the progress of MAV technology. The University of Arizona MAV project has been focusing on the development of a fully autonomous MAV since 2003. A systematic approach for the integration of the Paparazzi autopilot into an MAV was developed based on a discrete control design and linearized equations of motion of the aircraft. The open-loop stability and control laws for closed-loop stability were investigated and control gains were determined to satisfy the stability criteria in discrete time. These gains were then used as initial values by the autopilot.

Two fully autonomous MAVs were designed: 59-cm (23.2 in.) Zagi and 30-cm (11.8 in.) Dragonfly. In a series of flight tests of the MAVs, telemetry data on control actuation, altitude, attitude, and GPS location of the airplane were collected. The flight-control system that was developed was tested in all flight phases, including straight level dashes, correct banked turns, control pulses simulating wind gusts, waypoint navigation, fast climbing, and autonomous landing. Based on these data, Proportional-Integral-Derivative (PID) controllers were adjusted and flight performance characteristics were determined for autonomous MAVs. Future research in this area will focus on increasing the robustness of the automatic flight-control system. Enhanced control laws will be developed to satisfy the needs for flying very aggressively: the autonomous MAV will be capable of sharp turns and pull-outs, steep climbs and descents, and spiraling up and down in narrow spaces.

The first MAV case study describes the rigid wing MITE series of vehicles developed by the Naval Research Laboratory from 1996–2002. The purpose of the MITE program was to develop enabling technologies to create the smallest possible air vehicle capable of carrying out a Navy/USMC mission. The main driving factors for the design were: compact size, requiring a low Reynolds number configuration; expendability, requiring low cost; unobtrusive operation, requiring near-silent flight; inherent stability, permitting the use of simple automatic guidance and flight control systems; and adaptability to a variety of micro payloads. The MITE was a low aspect-ratio rigid flying wing, with twin, counter-rotating propellers driven by coreless electric motors and powered by batteries. Design optimization and determination of the aerodynamic characteristics were accomplished by computational fluid dynamics. MITEs were flown by radio control, with the pilot observing the aircraft from a vantage point on the ground or watching a video image transmitted from the aircraft. Autonomous flight was also achieved via a low-cost heading and altitude hold system with infrared stabilization for pitch and roll and a gyro damper for yaw. Construction was similar to that of small flying model airplanes, with a balsa structure of ribs and spars covered by doped polyester tissue. Many off-the-shelf model airplane components were used in the MITEs, and the flying model experience of most of the MITE program engineers proved extremely valuable.

The enabling technologies that were developed, refined, and/or demonstrated on the MITEs were: the MITE vehicle configuration, reliable battery-powered electric propulsion, a wing blown by the propeller wash, simple navigation and guidance systems, and micro video and electronic warfare payloads. MITE 4, the optimal MITE tested during the program, had a 47 cm (18.5 in.) span, 25 cm (9.8 in.) chord, and a gross weight of 300 g (10.6 oz) including a payload of about 70 g (2.5 oz). Endurance on lithium-ion polymer batteries was 20 minutes. The MAV enabling technologies developed for the MITE led directly to the world's first operational electric-powered aircraft, the USMC's Dragon Eye.

In 1999 the University of Florida began the development of a series of flexible wing MAVs. By modifying the compliance of a MAV wing described in the second case study, one can achieve vastly different aerodynamic characteristics, as can be seen in the wind-tunnel data. Pilots also report significant differences as a function of wing compliance. Many of the flexible structures have resulted in vehicles that have improved characteristics over rigid wing versions. A perimeter-reinforced flexible wing with a compliant elastic membrane provided improved payload capacity, static longitudinal stability, while a batten-reinforced flexible wing provided better gust rejection over a nominally rigid version of the same geometry. These flight characteristics, although difficult to quantify in the traditional manner (wind-tunnel sting balance measurements), are documented in terms of deformation measurements and flow visualization as well as pilot feedback. The physics of how these wings perform has been documented and modeling efforts are underway to predict how future versions of the compliant wing behave. The long-term goal is to be able to optimize the compliance for a specified benefit, whether for improvement in control or for enhanced efficiency in specific flight regimes. Improvements in analytical models and validation through advanced experimental methods are in progress and should result in optimizing the compliance for each particular application. As a validation of the benefits of the flexible wing concept, the University of Florida MAV team has won the overall title of the International MAV Competition, using the flexible wing concepts, from 1999–2006. They also won the surveillance aspect of the competition six out of the past eight years. Vehicles as small as 11.4 cm (4.5 in.) maximum dimension are capable of carrying a color video camera and have a 1 km (0.62 mi.) line-of-sight range and can fly in somewhat blustery conditions.

In order to achieve the desirable flight performance, a MAV can change its geometry by deformation of the wing structure. The University of Arizona, in the third case study, presents a description of the innovative MAV design with a variable camber wing. The wing design features a change in its cross-sectional geometry through a synchronous increase/decrease of camber and reflex.

The design constraint on the maximum dimension dictates the selection of the circular or the Zimmerman wing planform; for the same reason, the MAV utilizes a flying wing configuration. The biggest factor driving the design of the flying wing aircraft is pitch stability and control. A negative, nose-down pitching moment for cambered, low aspect-ratio wings at a low Reynolds number is compensated for by introducing a wing with inverse camber. The airfoil was designed featuring the ratio of inverse camber to maximum camber of 1:3. Two MAVs (i.e., with circular and Zimmerman planforms) were designed and built. Flight tests indicated that the in-flight changes in camber were successful. The 3% camber was found to be the best for high speed flight while 6% and 9% camber were best for slow flight such as in loitering or landing.

In the decade since the MAV was first proposed, a large number of fixed-wing configurations have been designed and successfully flown. The variety of these configurations clearly demonstrates the innovative ability of their designers. Almost all of these vehicles were able to carry a payload consisting of a video camera and transmitter. A large number of these vehicles were designed by student groups in order to compete in the annual competitions in the U.S. and Europe. It is also important to note that the research directed at MAV design was useful in developing somewhat larger vehicles such as the Dragon Eye and Wasp.

The goal of aircraft design at any scale is to manage a multitude of tradeoffs (lift for drag, aerodynamic optimization for weight, agility for stability, payload for size, performance for cost, etc.) to achieve the best balance for a given set of requirements. There is no ideal MAV design, just as there is no ideal human-carrying aircraft design; each design is optimized for its intended task and desired cost. Form is driven by function, within the constraints of the laws of physics. For MAVs, this form may be rigid wing, aeroelastic flexible wing, or adaptive wing, as well as flapping wing, rotary wing, or a number of alternatives. The designer must therefore be familiar with the range of what is possible in order to decide what is most practical for a specific application.

Index

Supporting Materials

Many of the topics introduced in this book are discussed in more detail in other AIAA publications. For a complete listing of titles in the AIAA Education Series, as well as other AIAA publications, please visit http://www.aiaa.org.